Fundamentals of Structural Dynamics

Fundamentals of Structural Dynamics

ZHIHUI ZHOU
Central South University, Changsha, Hunan, China

YING WEN
Central South University, Changsha, Hunan, China

CHENZHI CAI
Central South University, Changsha, Hunan, China

QINGYUAN ZENG
Central South University, Changsha, Hunan, China

ELSEVIER

Elsevier
Radarweg 29, PO Box 211, 1000 AE Amsterdam, Netherlands
The Boulevard, Langford Lane, Kidlington, Oxford OX5 1GB, United Kingdom
50 Hampshire Street, 5th Floor, Cambridge, MA 02139, United States

British Library Cataloguing-in-Publication Data
A catalogue record for this book is available from the British Library

Library of Congress Cataloging-in-Publication Data
A catalog record for this book is available from the Library of Congress

ISBN: 978-0-12-823704-5

For Information on all Elsevier publications
visit our website at https://www.elsevier.com/books-and-journals

Publisher: Glyn Jones
Editorial Project Manager: Naomi Robertson
Production Project Manager: Vijayaraj Purushothaman
Cover Designer: Mark Rogers

Typeset by MPS Limited, Chennai, India

Contents

About the authors

Dr. Zhihui Zhou is currently an associate professor at the School of Civil Engineering, Central South University (CSU), in China. He received a PhD in Civil Engineering from CSU in 2007 under the supervision of Prof. Qingyuan Zeng. He was invited to study at the University of Kentucky in 2014. Dr. Zhou's research interests include train derailment and dynamics of train–bridge (track) systems. He has been the principal investigator of several research grants, including the research project of National Natural Science Foundation of China (a study on the control theory of running safety and comfort for high-speed trains on bridges), a scientific research project of China's Ministry of Railways (a study on safety of running trains on large span cable-stayed bridges), special and general projects of the Chinese Postdoctoral Science Foundation, and some other scientific research projects. Dr. Zhou has published over 30 journal papers as the first author, and two monographs entitled "Lectures on dynamics of structures" and "Theory and application of train derailment." He won the first prize of the Science and Technology Progress of Hunan Province for his study "Theory and application of train derailment" in 2006.

Dr. Ying Wen was employed in the School of Civil Engineering, CSU, in China, after obtaining his PhD in 2010, and he was promoted to associate professor in 2012. He became a research associate in the Department of Civil and Structural Engineering, The Hong Kong Polytechnic University in 2011. In 2014 Dr. Wen was invited to visit the Department of Aerospace and Mechanical Engineering, University of Southern California, for a collaborative research on the problem of moving loads on structures. After he returned to CSU in 2015, Dr. Wen was appointed as the vice director of the Key Laboratory of Engineering Structures of Heavy-haul Railway, Ministry of Education. Dr. Wen has interests in fields of various structural dynamics and stability, especially nonlinear mechanics of long-span bridges and their dynamic stability under moving trains. Dr. Wen has published more than 20 journal papers, one of which is listed as the Top 25 Hottest articles published in "Finite Elements in Analysis & Design." He has also published three Chinese monographs about statics and dynamics of structures as a coauthor. Dr. Wen has received the awards of the Science and Technology Progress of Hunan Province (2006) and Zhejiang Province (2011).

Dr. Chenzhi Cai received his BS degree in civil engineering and MS degree in road and railway engineering from CSU, in China in 2011 and 2015, respectively. He graduated from The Hong Kong Polytechnic University with a PhD in civil engineering in 2018 and joined the Department of Bridge Engineering as well as the Wind Tunnel Laboratory of CSU as an associate professor later that year. Dr. Cai's main research interests are the fields of noise and vibration control, train-bridge interaction dynamics, and train-induced ground vibration isolation. He has participated in several research projects funded by the Hong Kong government and has also received research funding from the National Natural Science Foundation of China and Hunan Provincial Natural Science Foundation of China. Dr. Cai has published more than 20 papers in international journals, and some of his work is under consideration for acceptance by the UK CIBSE Guide.

Prof. Qingyuan Zeng is a distinguished scientist on bridge engineering at Central South University, in China. He obtained his BS and MS degrees from the Department of Civil Engineering, Nanchang University and Department of Engineering Mechanics, Tsinghua University, in 1950 and 1956, respectively. He was elected as a member of the Chinese Academy of Engineering in 1999 for his great contributions to local—global interactive buckling behavior of long-span bridge structures, train—bridge interaction dynamics and the basic theory of train derailment. He presented the principle of total potential energy with a stationary value in elastic system dynamics and the "set-in-right-position" rule for assembling system matrices, which is a significant improvement of the classical theory of structural dynamics and finite element method. Prof. Zeng has an international reputation for his originality in the transverse vibration mechanism and time-varying analysis method of the train—bridge system. He has authored and coauthored more than 100 journal papers, three monographs, and three textbooks. He received numerous awards, including the State Science and Technology Progress Award, Distinguished Achievement Award for Railway Science and Technology from Zhan Tianyou Development Foundation, and Honorary Member Award from the China Railway Society. He has supervised more than 16 MS students and 30 PhD students in the past three decades.

Preface

Nowadays, the design of engineering structures, for example, long-span bridges, high-rise buildings, stadiums, airport terminals, and offshore platforms, seeks a large ratio of their load carrying capacity to self-weight to achieve esthetic pleasure and economy. However, the type of these lightweight and flexible structures will lead to a large deformation and excessive vibrations under loading. In addition, these structures may suffer from some extreme excitations, for instance, strong winds, seismic actions, high-impact collisions, and impacts of water wave flow. Therefore, investigation of structural behaviors under dynamic loads is essential in order to achieve a good performance of the structure when satisfying the requirement of designed service. The basic concept of structural dynamics is of great help to engineers in understanding structural vibration and taking appropriate measures.

This book introduces the fundamental concepts and basic principles of the "dynamics of structures." Although the book focuses on the linear problem in structural dynamics, solutions for some nonlinear problems have also been briefly introduced. It should be noted that random vibration is beyond the scope of this book and is not included here. The main content of this book includes the overview of structural dynamics, the formulation of equations of motion of systems, the analysis of dynamic response of SDOF systems, the analysis of dynamic response of MDOF and continuous systems, the mode superposition method, the approximate evaluation of natural frequencies and mode shapes, and the step-by-step integration method.

Three original contributions have been proposed in this book, namely, the principle of total potential energy with a stationary value in elastic system dynamics, the "set-in-right-position" rule for assembling system matrices, and the method of computer implementation in Matlab. Moreover, this book introduces the fundamental concepts of structural dynamics in a concise way rather than with a detailed description, which is more efficient for abecedarians in understanding the basic concepts and methods of vibration analysis.

Participants in the writing of this book include Zhihui Zhou, Ying Wen, Chenzhi Cai, and Qingyuan Zeng from Central South University. The specific division of the organization and writing of this book is as follows: Zhihui Zhou is responsible for the writing of Chapters 1 to 4; Ying Wen has fulfilled Chapters 5 and 6; Chenzhi Cai has completed Chapter 7, and Qingyuan Zeng supplied the original manuscript of the book.

The authors wish to express their sincere thanks and appreciation to Prof. Xiaojun Wei from Central South University, Prof. Tong Qiu from The Pennsylvania State University, and PhD student Juanya Yu from University of Illinois at Urbana-Champaign for valuable advice in the process of writing. The authors are also grateful to Mr. Lican Xie, Ms. Manxuan Yang, Mr. Liang Zhang, Mr. Bao Zhang, Mr. Xuanyu Liao, Mr. Chenlong Tang, Mr. Zhenhua Jian, Mr. Xiaojie Zhu, and other graduate students from Central South University for their contributions in different ways to the content of this book.

This book can be used as a textbook for both postgraduates and undergraduates majoring in civil engineering, engineering mechanics, mechanical engineering, and other related fields in general colleges and universities. It can also be a reference for teachers, general students, and short-term trainees in institutions of higher vocational education.

The authors cordially invite the audience of this book to contact with us (Zhihui Zhou: zzhyy@csu.edu.cn) if you have any suggestions for improvements and clarifications in the content organization, and even to help identify errors. All the above efforts and comments are sincerely acknowledged.

Zhihui Zhou
Ying Wen
Chenzhi Cai
Qingyuan Zeng

CHAPTER 1

Overview of structural dynamics

1.1 Objective of structural dynamic analysis

Dynamic analysis of the train–bridge system originated from the collapse of the Chester Railway Bridge in the United Kingdom due to a train passing over the bridge. In November 1940 the engineering community was astonished by the dynamic instability of the Tacoma suspension bridge in the United States under strong wind with a speed of 17–20 m/s. A large crowd of people participated in the opening ceremony of Wuhan Yangtze River Bridge in 1957, resulting in continuous swaying of the newly opened bridge. The swaying came to an end when the crowd went away at night. In 2011 the administrator of the Shanghai Railway observed the excessive transverse vibration of the Nanjing Yangtze River Bridge under the condition of a cargo train passing over the bridge. The transverse amplitude of the oscillated bridge exceeded 9 mm, which led to concerns over the safety of running trains on the bridge. Therefore the assessment of the safety and comfort of running trains on this bridge was conducted [1,2].

Seismic activity has been relatively active in recent decades, for instance, the Chilean earthquake in 1960, the Tangshan earthquake in China in 1976, the Mexico earthquake in 1985, the Osaka–Kobe earthquake in Japan in 1995, the India earthquake in 2001, and the Sichuan earthquake in China in 2008. In addition to serious disruption to the local economy, these disasters threatened the safety of residents and their properties in the concerned areas. Thus the aseismic design of infrastructures in seismically active areas is necessary to reduce or avoid severe earthquake damage for major projects. In addition, many airplane accidents have been caused by the flutter of aircraft wings or the abnormal vibration of engines. In mechanical engineering, vibrations may bring about negative effects on the performance of some precision instruments, for instance, these vibrations may increase abrasion and fatigue, or reduce machining accuracy and surface finish. However, some manufacturing facilities, for example, transmission, screening, grinding, piling, and so on, as well as various generators and clocks, benefit from the positive aspects of vibrations [3].

Fundamentals of Structural Dynamics
DOI: https://doi.org/10.1016/B978-0-12-823704-5.00001-X

The investigation of structural dynamics focuses on understanding the basic mechanism of vibrations and presenting the corresponding processing methods. These methods can be adopted to eliminate of the negative vibration effects of machines, prevent dynamic instability of bridges and improve the tamping and compaction performances of the road construction machinery, and so on.

1.2 Characteristics of structural dynamics

The main differences between statics and dynamics can be addressed in the following aspects: (1) in dynamics, both the loads and responses of structures are time-varying, which implies that, unlike static problems, the solution of dynamics cannot be a single one. Therefore the dynamic analysis of structures presents a more complex and time-consuming process when compared with the static analysis of structures; (2) acceleration is significant in dynamics. The so-called inertial force produced by acceleration acts in the opposite direction of the acceleration. As illustrated in Fig. 1.1A, the internal moment and shear of the cantilever beam should equilibrate the applied dynamic load, $F(t)$, as well as the inertial force associated with the acceleration. In Fig. 1.1B, the internal moment, shear, and deflection of the cantilever beam under a static load F depend only on the applied load itself. In general, once the inertial force accounts for a relatively large proportion of the forces equilibrated by the elastic internal force, the dynamic characteristics should be taken into account in the structural analysis. When applied loads do not change significantly, the dynamic responses are minor and the inertial forces can be neglected. Thus the static analysis procedure could be applied at any desired instant of time in these cases. If the exciting frequency is less than one third of the first natural frequency of the structure, the analysis of the structure

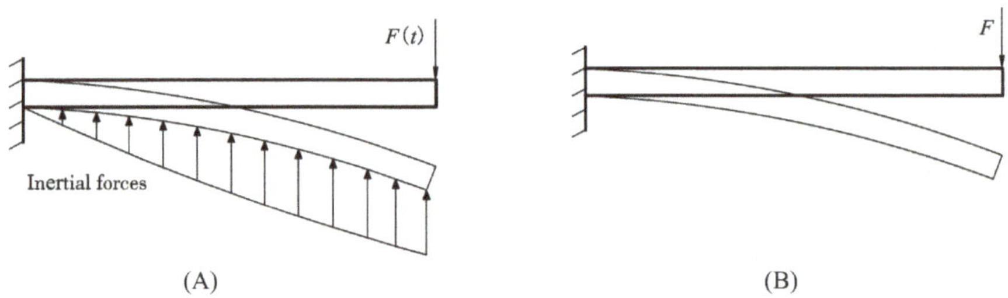

Figure 1.1 Cantilever beam subjected to (A) dynamic load and (B) static load.

could be treated as a static problem (a better understating of this concept can be achieved by means of Fig. 3.14); (3) damping is also an indispensable factor in dynamic problems. Energy will be dissipated in the vibration of structures. Structural damping is frequently ignored in the analysis of the natural dynamic properties and the dynamic response over a relatively short duration (such as the action of impulsive loads). However, structural damping must be taken into account when large damping exists or vibration lasts a long period, as well as in the analysis of the vibration in the resonance region.

1.3 Classification of vibrations

1. The vibrations could be classified as either deterministic or random vibrations according to the deterministic or random characteristics of the dynamic responses.
   a. Deterministic vibration: the structural responses are deterministic functions of time due to the determined load and system.
   b. Random vibration: the structural responses are random due to the uncertainty of load or system. However, the responses usually comply with certain statistical rules and can be analyzed with statistical probability methods. For instance, the vibrations of aircraft owing to aerodynamic noise, the vibrations of the train–track–bridge system caused by track irregularity, etc., are all regarded as random vibrations.
2. The vibrations could be classified as either free vibrations, forced vibrations, self-excited vibrations, or parametric vibrations.
   a. Free vibration: external perturbation makes the system deviate from the initial equilibrium position or have initial velocity. When the perturbation is rapidly removed, the system will vibrate due to initial displacement or velocity, which is called free vibration.
   b. Forced vibration: the vibration of the structure is caused by a continuously applied load, which is called forced vibration. The response of a forced vibration consists of two components. One is the transient response related to the initial conditions and the other is the steady-state response with the same frequency as the applied load. Since transient vibrations decay rapidly due to the damping effect, forced vibrations are often referred to as steady-state vibration.
   c. Self-excited vibration: the vibration is excited and controlled by the system motion itself, which is called self-excited vibration. In the analysis of self-excited vibrations, the components of the

system should be determined first. Then, the interaction among these components should be fully understood, as well as the process of the input and dissipation of system energy. In self-excited vibrations, energy is obtained from the periodic vibration of a part of the system. The excitation is a function of the displacement, velocity and acceleration of the system. It is common to encounter self-excited vibration phenomena in nature, engineering, and daily life, for example, the piston motion of engines, the working principle of clocks, the wind-induced motion of the Tacoma Bridge, and the vibration of leaves in the breeze.

Through the observation of the swing of leaves under the excitation of the wind, it can be noted that the wind angle of leaves standing against the wind will be changed due to bending of branches. Therefore part of the air flow along the leaves and the wind pressure on the leaves would be reduced. However, the elastic resistance of the branch forces the leaves to return to their initial positions. Such a process is repeated over and over. It can be concluded from the above description that the external wind itself does not vary periodically, while the wind excitation on the leaves is periodic. This is because the motion of the leaves controls the wind actions on the leaves. This type of vibration is referred to as self-excited vibration.

d. Parametric vibration: system parameters change with a certain rule due to the action of applied load, and the vibration is excited by the changing system parameters, which is called parametric vibration. The motion of a single pendulum with the time-varying length is a typical example, as illustrated in Fig. 1.2A. Considering a small-amplitude motion of a single pendulum, its equation of motion could be derived as $\ddot{\varphi} + 2(\dot{l}/l)\dot{\varphi} + (g/l)\varphi = 0$ (φ is the rotation of the pendulum; l is the time-varying length of the pendulum; g is the acceleration of gravity; the detailed derivation can be found in Example 2.6). It can be observed from the equation that the system parameters vary with the length of the pendulum l. The external force is not present in the load term of the equation of motion. Another typical example is the transverse vibration of a straight bar to a periodic axial force, as shown in Fig. 1.2B. The periodic axial force results in periodic variation of parameters in the equation of transverse bending (detailed information can be found in Chapter 17 of Ref. [4]), which leads to the vibration of the straight bar in the transverse direction. Once the frequency of the applied force $\overline{\omega}$, and the natural frequency associated with transverse bending of the bar, ω, satisfy the relation of $\overline{\omega} = 2\omega/K$, $K = 1, 2, \cdots$, the transverse amplitude of the bar would become larger and larger and instability would occur eventually. That is parametric resonance of the bar in the transverse direction due to the periodic excitation in the direction of the bar axis, as shown in Fig. 1.2C.

3. According to the linear or nonlinear differential equations of a system, the vibrations can be categorized into linear vibrations and nonlinear vibrations:

a. Linear vibration: the inertial force, damping force, and elastic resistance of the system are linearly related to the acceleration, velocity, and displacement, respectively. The vibration of a system is governed by a linear differential equation. Instead of second- and higher-order terms, only the first-order terms with respect to acceleration, velocity, and displacement are present in the differential equation. This book focuses on the investigation of the linear vibration.

b. Nonlinear vibration: in contrast to the linear vibration, the inertial force, damping force, or elastic resistance of the system are nonlinear with respect to acceleration, velocity, or displacement, respectively, and the corresponding vibration can only be governed by nonlinear differential equations. For instance, both the collapse of infrastructure due to earthquakes and large amplitude vibration of flexible structures due to strong winds are examples of the nonlinear vibration.

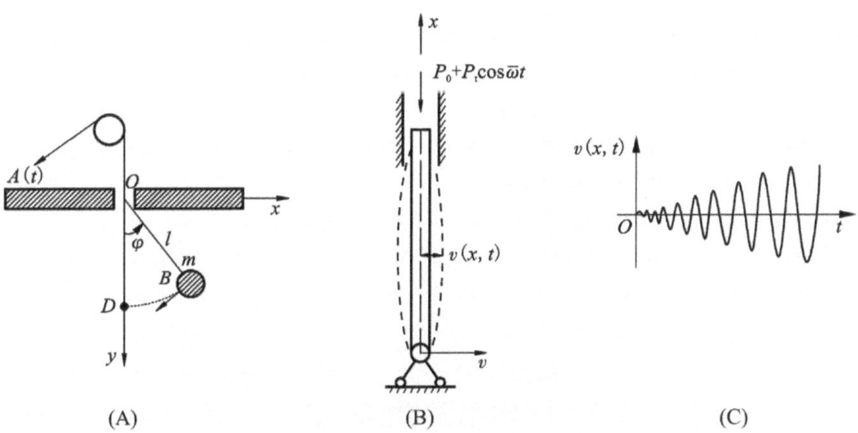

Figure 1.2 Examples and response characteristics of parametric vibration: (A) motion of a single pendulum with time-varying length; (B) transverse instability of a straight bar; (C) vibration response due to parametric vibration.

1.4 Vibration problems in engineering

In the analysis of vibration, the investigated object (the engineering structure) is generally referred to as the vibrating system, and can be described by the mass M, stiffness K, and damping C. The external loads that act on a system or the factors that lead to the vibration of a system are called the excitation or input. The dynamic responses of the system subjected to such an excitation or input, for instance, accelerations, velocities, and displacements, are regarded as the responses or output. The excitation (input) is connected with the responses (output) by means of the properties of the vibration system, as shown in Fig. 1.3.

The investigation of system vibration boils down to the analysis of the relationships among the system, input, and output. Theoretically, once two of these three factors are determined, the remaining one can be obtained. Therefore vibration problems in engineering can be classified into the following four types:

1. Response analysis: based on the given physical properties of the structural system and the applied loads, the responses, including the acceleration, velocity, and displacement, etc., are solved. Response analysis provides basic information for analyzing the strength, stiffness, and vibration state of a system. This book mainly focuses on the response analysis.
2. Environment prediction: based on the given properties and responses of the structural system, the input is to be determined, and the characteristics of the environment where the system is located may be identified.
3. System identification: the input and output are known, that is, the dynamic loads and responses of the system are known. Therefore the properties of the system can be obtained by using the system identification method. The identified parameters include both physical properties (mass, stiffness, damping, etc.) and modal parameters (natural frequencies and mode shapes).
4. System design: in many cases of engineering applications, the properties of the system can be designed based on the given input and required criteria of responses. In general, the system design depends on the response analysis. System design and response analysis are often conducted alternately in practical engineering.

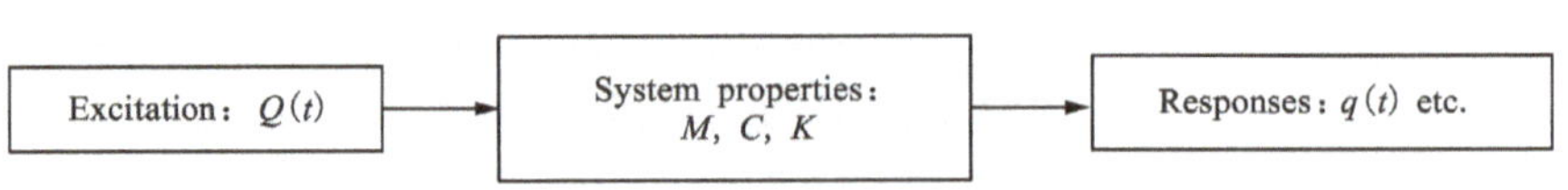

Figure 1.3 Three factors representing system vibration.

1.5 Procedures of dynamic response analysis of structures

1.5.1 Description of system configuration

Evaluation of the system responses of is a significant objective in structural dynamics. The prerequisite for finding the solutions of the structural responses is to formulate the dynamic equilibrium equation, that is, the equation of motion of the system, by considering the inertial, damping, elastic, and external forces. The inertial, damping, and elastic forces are directly related to the displacements, velocities, and accelerations of the system, as well as its physical properties. Therefore it is necessary to describe the configuration of the system at any instant of time. Generally, a vibration configuration is determined from the positions of all particles of the system. Practical structures are generally continuous systems, and infinite displacement variables are required to represent their vibration configuration theoretically. For example, the position coordinates v_k, $k = 1, 2, \cdots$, of all continuous particles distributing along the length of the beam should be obtained for the sake of accurate description of the vibration of the simply supported beam in the vertical plane, as shown in Fig. 1.4A. However, it is difficult and unnecessary to do so in vibration analysis of engineering structures. An approximate estimate of structural configuration can often satisfy the requirement of accuracy in practical engineering. It is both efficient and possible to discretize a simply supported beam into finite elements and use the displacements of nodes to describe the configuration of the beam, as shown in Fig. 1.4B. The selection of the appropriate coordinates that represent vibration configuration of structures is the preliminary and most important step for the modeling of practical structures, which is associated with computational effort and accuracy.

1.5.2 Analysis of excitation

Excitation is defined as the external actions which induce structural vibrations. The excitation of structural vibration is complex and affected by many random factors. For instance, the dynamic actions of a train running

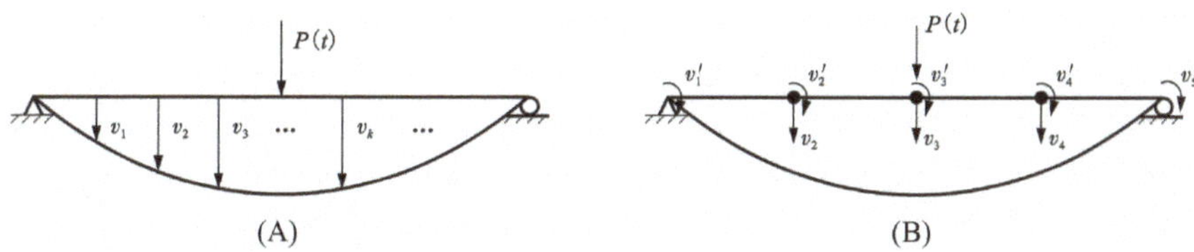

Figure 1.4 Configuration of a simply supported beam: (A) accurate description; (B) approximate description using the finite element method.

on a bridge include the wheel–rail contact forces caused by the hunting movement of wheelsets, eccentric loads of vehicles, and additional forces generated by track irregularities. It is difficult to identify these excitations with specific expressions quantitatively; however, these excitations satisfy certain statistical rules. Although seismic acceleration waves can be adopted for the input of earthquake actions on structures, there are no uniform mathematical models for seismic acceleration waves for different regions, even for earthquakes of the same magnitude. The seismic actions on the structure are random, as well as the wind actions. These dynamic loads are called random loads.

Some special excitations are present in engineering, which can be described with sufficient precision by a specific time-domain function. Harmonic excitation caused by the eccentric rotor with a constant angular speed is a typical example of this.

According to whether excitations can be described by a deterministic mathematical model or not, excitations can be classified into two types, namely, random dynamic load and prescribed dynamic load.

1. Random dynamic load: a time-varying random dynamic load cannot be represented deterministically. The differences of loads in each experiment are obvious. However, probability theory can be adopted to describe the statistical characteristics of these loads.
2. Prescribed dynamic load: the time variation of a deterministic dynamic load is specified. The obtained results of these kinds of loads in different experiments are nearly identical when considering the experimental error. Fig. 1.5 shows some typical prescribed dynamic loads.

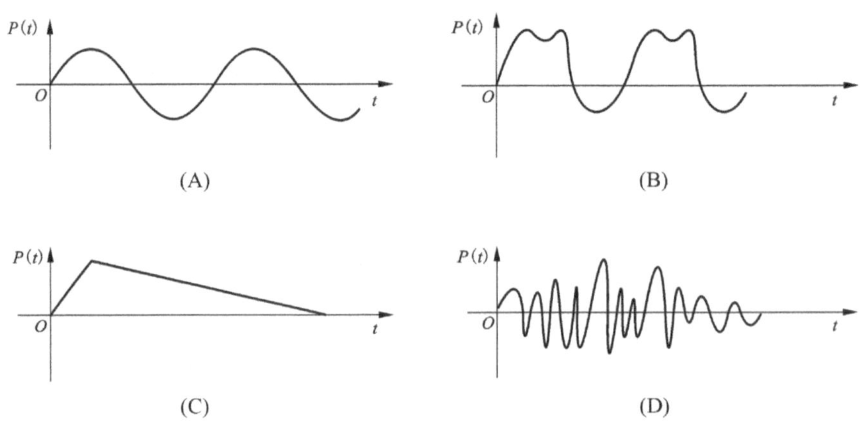

Figure 1.5 Typical prescribed dynamic loads: (A) harmonic load; (B) arbitrary periodic load; (C) impulsive load; (D) arbitrary nonperiodic load.

Prescribed dynamic loads include both periodic and nonperiodic loads. Periodic loads can be categorized into simple harmonic loads (Fig. 1.5A) and arbitrary periodic loads (Fig. 1.5B). Nonperiodic loads can be categorized as impulsive loads with an extremely short duration (such as a shock wave and explosion wave, as shown in Fig. 1.5C), and arbitrary nonperiodic loads with a specified duration (such as measured seismic excitations, as illustrated in Fig. 1.5D, which is regarded as a prescribed dynamic load in the analysis of deterministic vibrations).

1.5.3 Mechanism of vibration energy dissipation

The mechanism of energy dissipation is complex and not fully understood. Energy dissipation in structural vibration is related to the damping force. The damping force is mainly caused by the internal friction due to the deforming of solid material, the friction at connection points of structures (such as the friction at bolt joints of steel structures), the opening and closing of microcracks in concrete, and the friction due to external media around structures (such as the effects of air and fluids), etc. In reality, it is difficult to simulate damping accurately due to the combined effects of several factors. If only one kind of damping dominates the effects, it would be possible to find a reasonable model for the damping force. For instance, viscous damping force is proportional to the magnitude of velocity, that is, $F_{vd} = c\dot{v}$, and it opposes the velocity. Detailed information about the damping will be given in Section 3.4.

1.5.4 Equation of motion of a system

An important task for structural dynamics is to obtain the displacements that vary with time or other responses to prescribed loads. Approximate methods (such as the finite element method) considering a certain number of degrees of freedom can generally meet the accuracy requirements for most structures. Thus the problem boils down to solving the time history of these selected displacement variables. The mathematical expression of dynamic displacements is referred to as the equation of motion of a structural system. It is also known as the dynamic equilibrium equation once the inertial force is introduced. By solving the equation, the displacements and other responses can be obtained. The vibration characteristics of a multidegree-of-freedom system can be expressed by the following equation:

$$\boldsymbol{M}\ddot{\boldsymbol{q}} + \boldsymbol{C}\dot{\boldsymbol{q}} + \boldsymbol{K}\boldsymbol{q} = \boldsymbol{Q} \tag{1.1}$$

where $\boldsymbol{q}$ is the generalized displacement vector; $\dot{\boldsymbol{q}}$ is the generalized velocity vector; $\ddot{\boldsymbol{q}}$ is the generalized acceleration vector; $\boldsymbol{M}$ is the mass matrix; $\boldsymbol{C}$ is the damping matrix; $\boldsymbol{K}$ is the stiffness matrix; and $\boldsymbol{Q}$ is the generalized force vector.

1.5.5 Solution of equation of motion

The theory for the linear equation of motion of a system is comparatively mature. It can be categorized into the following two types:

1. Solution for linear equation of motion with constant coefficients: the main methods include numerical integration method (such as the Euler method or Runge–Kutta method), variational method, mode-superposition method, and weighted residual method.
2. Solution for linear equation of motion with variable coefficients: this is mainly tackled by the variational method, step-by-step integration method, and weighted residual method.

There is no general method available for solving a nonlinear equation of motion yet. The small parameter method, variational method, and weighted residual method are commonly applied to solve a nonlinear equation of motion. With the rapid development of computers, the step-by-step integration method has become the dominant algorithm.

1.5.6 Vibration tests

The main purpose of vibration tests is to validate the theoretical results, modify the theoretical model, and obtain the parameters required by the theoretical analysis. The natural frequencies, mode shapes, damping ratio, and seismic acceleration wave are among the test items. These parameters are the basis of the analysis of structural dynamics.

Problems

1.1 What are the main differences between the dynamic and static analysis of structures?

1.2 What are the main differences between prescribed and random dynamic loads? How should one express these two kinds of loads in mathematics?

1.3 What are the common problems related to engineering vibration analysis and what relationships do they have?

1.4 According to the characteristics of parametric vibration and self-excited vibration, which category does the motion of swing belong to?

References

[1] Zeng Q, Guo X. Theory of vibration analysis of train-bridge time-varying system and its application. Beijing: China Railway Press; 1999.

[2] Zeng Q, Xiang J, Zhou Z, Lou P. Theory of train derailment analysis and its application. Changsha: Central South University Press; 2006.

[3] Zhou Z, Wen Y, Zeng Q. Lectures on dynamics of structures. 2nd ed. Beijing: China Communications Press Co., Ltd; 2017.

[4] Clough RW, Penzien J. Dynamics of structures. 3rd ed. Berkeley, CA: Computers & Structures, Inc; 2003.

CHAPTER 2

Formulation of equations of motion of systems

The preliminary step for estimating structural response is to formulate the equation of motion of a structural system. This chapter focuses on the basic concepts of structural dynamics and several methods for formulating the equation of motion. These methods include (1) the direct equilibrium method, (2) the principle of virtual displacements, (3) Lagrange's equations, (4) Hamilton's principle, (5) the principle of total potential energy with a stationary value in elastic system dynamics, and (6) the "set-in-right-position" rule for assembling system matrices and the method of computer implementation in Matlab.

First, the concept of system constraint and the representation of the configuration of a system will be introduced in this chapter. Then, the principles and applications of the aforementioned six methods will be discussed.

2.1 System constraints

The earth is often selected as the reference frame in the vibration analysis of systems. The chosen Cartesian coordinate system is fixed on the Earth, as illustrated in Fig. 2.1. This kind of coordinate system is called a basic coordinate system. The notation O represents the origin of the coordinate

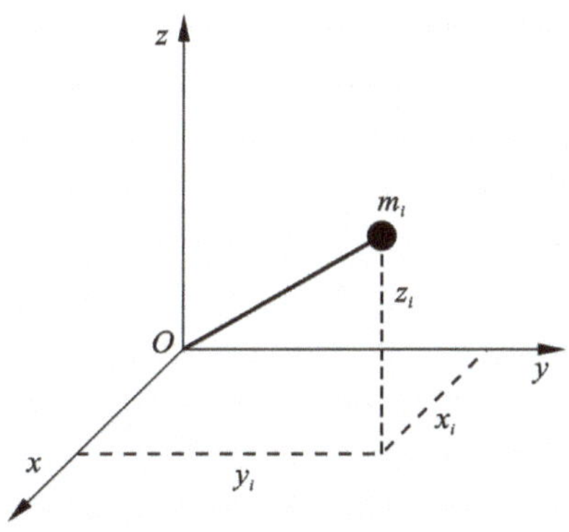

Figure 2.1 Position of a particle in the basic coordinate system.

Fundamentals of Structural Dynamics
DOI: https://doi.org/10.1016/B978-0-12-823704-5.00002-1

system. Bridges, buildings, and other infrastructures are considered to be fixed on the earth and are incapable of moving freely. The motions of these structures should satisfy external constraint conditions. Such kinds of systems are referred to as constrained systems of particles. In contrast, aircrafts, birds, etc., can move freely in all directions relative to the earth (i.e., the basic coordinate system). This kind of system is called a free system of particles. Each particle, which satisfies the requirements of internal constraints, can move freely in all directions relative to the basic coordinate system.

A constraint could be defined as a geometric or kinematic restriction imposed on the position and/or velocity of a particle. It is commonly expressed by a constraint equation. The boundary conditions of a structure are typical examples of constraint equations. The following is a brief introduction of constraint classifications.

1. Constraints can be categorized as either geometric or kinematic constraints according to the characteristics of state variables in constraint equations.

Geometric constraint: Only the positions of the particles of a system are restricted. For example, the coordinates of the particle m, x, y, z, as shown in Fig. 2.2, should satisfy the following equation

$$x^2 + y^2 + z^2 = l^2 \tag{2.1}$$

where l represents the length of the rigid rod. Eq. (2.1) is known as the geometric constraint equation. Therefore the position coordinates of the particle m at any instant of time t, $x(t)$, $y(t)$, $z(t)$, are not independent. Only two of them are independent.

Kinematic constraint: Both the position and velocity of the particles of a system are restricted. A cylinder moves along the positive direction of the x axis, as shown in Fig. 2.3. It should be noted that the position of the center of the cylinder C must satisfy the following relationship

$$z_C = R \tag{2.2}$$

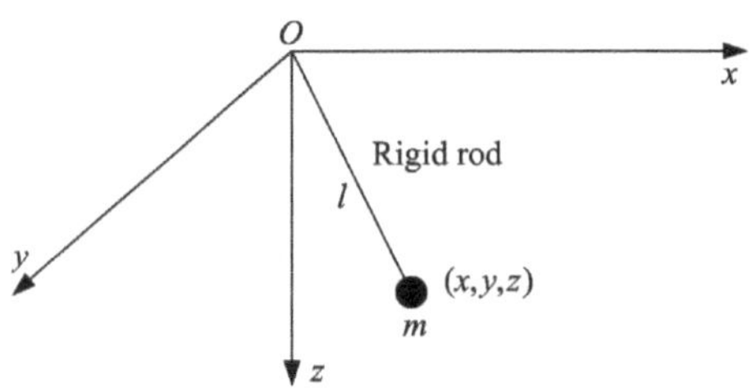

Figure 2.2 Particle constrained by a rigid rod.

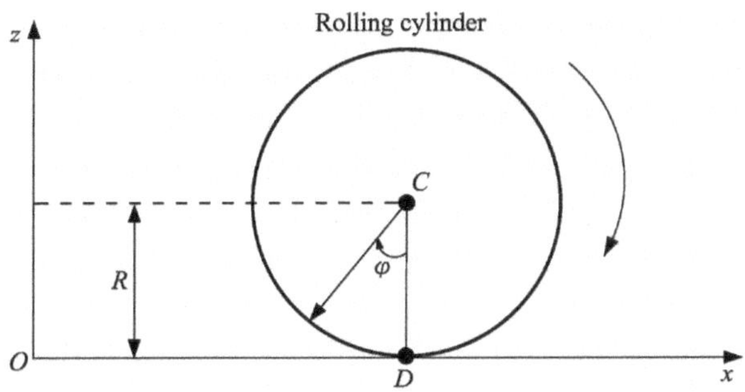

Figure 2.3 Cylinder rolling horizontally.

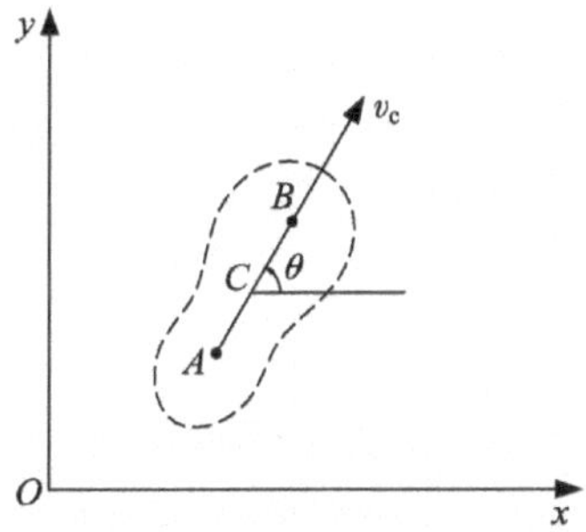

Figure 2.4 Motion of an ice skate in a plane.

where z_C is the position of the center of cylinder along the z axis, and R is the radius of the cylinder.

Eq. (2.2) is a geometric constraint equation. Once the cylinder can only roll without sliding, the velocity of the contact point D on the ground shall equal zero, which could be expressed as

$$\dot{x}_C - R\dot{\varphi} = 0 \tag{2.3}$$

where $\dot{x}_C$ is the velocity of the center of cylinder along the x axis, and $\dot{\varphi}$ is the angular velocity of the cylinder.

Eq. (2.3) is a kinematic constraint equation. Eq. (2.3) could be transformed into $x_C = R\varphi + c$ (c is an integral constant; φ is the rotation of the cylinder) by integration, which is a geometric constraint equation. The motion of an ice skate on the ground can be simplified to the motion of the rod AB in a plane, as shown in Fig. 2.4. The velocity v_C of the center of mass C is always along the direction of rod AB. Therefore the velocity components $\dot{x}_C$ and $\dot{y}_C$ along the direction of the x and y axes should satisfy the following relationship

$$\frac{\dot{y}_C}{\dot{x}_C} = \tan\theta \text{ or } \dot{x}_C \sin\theta - \dot{y}_C \cos\theta = 0$$

where θ is the rotation angle of rod AB measured from the x axis.

The above equation is a kinematic constraint equation. Due to the angle θ varying with the motion of the system, the above equation cannot be integrated to obtain a geometric constraint relation. More knowledge about transforming kinematic constraint equations into geometric constraint equations can be found in Ref. [1].

2. Constraints can be categorized as either steady or unsteady constraints according to whether the time variable is explicitly present in the constraint equation or not.

Steady constraint: Time variable t is not present in the constraint equation. Eqs. (2.1), (2.2), and (2.3) belong to the steady constraints.

Consider a system of l particles, the steady constraint equation could be expressed as follows:

$$\begin{aligned} &f_c(\boldsymbol{r}_1, \cdots, \boldsymbol{r}_l, \dot{\boldsymbol{r}}_1, \cdots, \dot{\boldsymbol{r}}_l) = 0 \\ &\text{or } f_c(x_1, y_1, z_1, \cdots, x_l, y_l, z_l, \dot{x}_1, \dot{y}_1, \dot{z}_1, \cdots, \dot{x}_l, \dot{y}_l, \dot{z}_l) = 0 \end{aligned} \tag{2.4}$$

where $\boldsymbol{r}_k$ is the position vector of the kth particle, $\dot{\boldsymbol{r}}_k$ is the velocity vector of the kth particle, x_k, y_k, z_k are the coordinate components of the kth particle in the basic coordinate system, and $\dot{x}_k, \dot{y}_k, \dot{z}_k$ are the velocity components of the kth particle in the basic coordinate system, where $k = 1, 2, \cdots, l$.

Unsteady constraint: Time t is an explicit variable in constraint equations. For example, Fig. 2.5 shows a planar pendulum dangled at point j. The point j moves in terms of sine function $y_0 = a \sin \omega t$ along

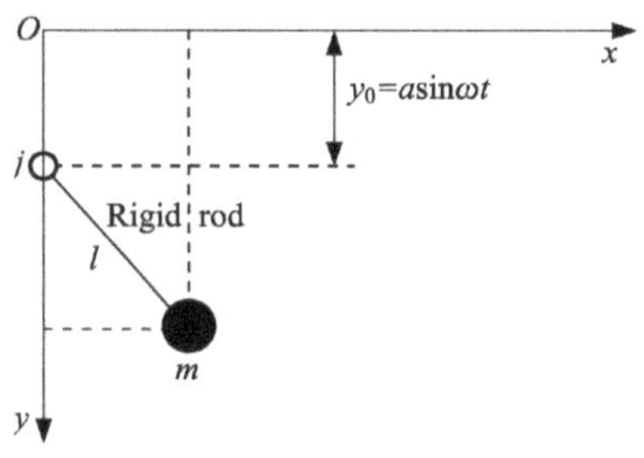

Figure 2.5 Motion of a planar pendulum.

the direction of the y axis. The constraint equation of particle m can be given as follows:

$$x^2 + (y - a \sin \omega t)^2 = l^2 \tag{2.5}$$

The general equation of an unsteady constraint can be expressed as:

$$\begin{aligned} &f_c(\boldsymbol{r}_1, \cdots, \boldsymbol{r}_l, \dot{\boldsymbol{r}}_1, \cdots, \dot{\boldsymbol{r}}_l, t) = 0 \\ &\text{or } f_c(x_1, y_1, z_1, \cdots, x_l, y_l, z_l, \dot{x}_1, \dot{y}_1, \dot{z}_1, \cdots, \dot{x}_l, \dot{y}_l, \dot{z}_l, t) = 0 \end{aligned} \tag{2.6}$$

3. Constraints can also be categorized as either holonomic or nonholonomic constraints, according to whether the terms of velocity are present in constraint equations or not.

Holonomic constraints: Geometric constraints and integrable kinematic constraints are called holonomic constraints. Holonomic constraints only depend on the coordinates and time t, and holonomic constraint equations exclude the terms of velocity. The general expression could be given as follows:

$$f_c(\boldsymbol{r}_1, \cdots, \boldsymbol{r}_l, t) = 0 \text{ or } f_c(x_1, y_1, z_1, \cdots, x_l, y_l, z_l, t) = 0 \tag{2.7}$$

Nonholonomic constraints: Kinematic constraints which cannot be integrated to get geometric constraints are called nonholonomic constraints. Nonholonomic constraint equations contain derivatives of coordinates with respect to time t. The general expression could be given as follows:

$$\begin{aligned} &f_c(\boldsymbol{r}_1, \cdots, \boldsymbol{r}_l, \dot{\boldsymbol{r}}_1, \cdots, \dot{\boldsymbol{r}}_l, t) = 0 \\ &\text{or } f_c(x_1, y_1, z_1, \cdots, x_l, y_l, z_l, \dot{x}_1, \dot{y}_1, \dot{z}_1, \cdots, \dot{x}_l, \dot{y}_l, \dot{z}_l, t) = 0 \end{aligned} \tag{2.8}$$

As discussed above, the constraints of the rolling cylinder, as shown in Fig. 2.3, can be considered to be holonomic. The constraint of the ice skate, as shown in Fig. 2.4, is nonholonomic due to its unintegrable kinematic constraint equation. For given constraint equations which contain the terms of velocity, integration transformations should be used to obtain constraint equations in the form of Eq. (2.7). Once these transformations are available, the corresponding constrains are holonomic. Otherwise, the constrains are nonholonomic.

Once all the constrains of a system are holonomic, the system can be defined as a holonomic system. Otherwise, the system is a nonholonomic system. The subsequent chapters of the book focus on holonomic systems. Detailed information about nonholonomic systems could be found in Ref. [1].

2.2 Representation of system configuration

The independent variables that can completely specify the configuration of a system are defined as the generalized coordinates. For the case of a holonomic system, the number of degrees of freedom (DOFs) of the system equals that of generalized coordinates, and n is used to represent the number of DOFs. However, the number of DOFs does not necessarily equal that of generalized coordinates, which will occur in the case of a nonholonomic system. More information can be found in Ref. [1]. Assuming a free system consisting of l particles (the system is assumed to contain l particles in this book, except that specific notes are addressed), the number of independent coordinates to determine the system configuration is therefore required to be $3l$. Due to some constraints in the constrained system of particles, the coordinates of particles in such a system are not independent and should satisfy some constraint conditions.

A constrained system of particles is set to be a holonomic system with s holonomic constraints. Then, $3l$ coordinates of the system should satisfy s constraint equations. This means that only $(3l - s)$ coordinates are independent. The remaining s coordinates are given as functions of these independent coordinates. Thus $(3l - s)$ independent coordinates are sufficient to determine the system configuration, that is, $n = 3l - s$.

For example, a free spatial particle with three DOFs is restricted to be in a plane, then the number of DOFs of the particle decreases from three to two. Once the particle is connected to a fixed point in the plane through a rigid rod, the particle would only have one DOF. Another example is the oscillation of a double pendulum, as shown in Fig. 2.6.

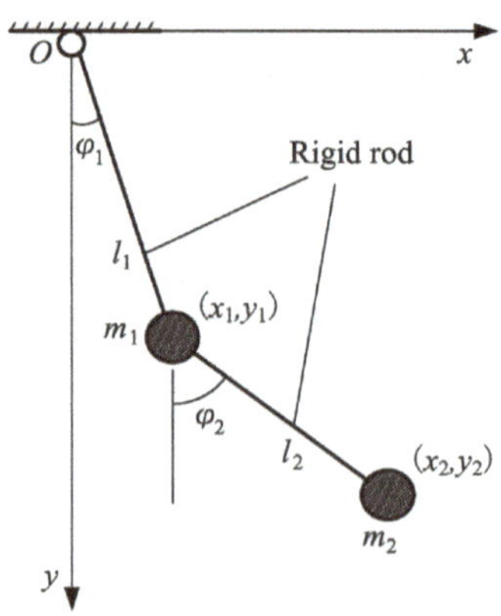

Figure 2.6 Motion of a double pendulum.

The coordinates x_1, y_1 of the mass m_1 and x_2, y_2 of the mass m_2 should satisfy the following constraint equations

$$x_1^2 + y_1^2 = l_1^2, (x_2 - x_1)^2 + (y_2 - y_1)^2 = l_2^2$$

In such a circumstance, only two independent coordinates are present in the system. This indicates that the system is a 2-DOF system.

Generally, it is not convenient to determine independent coordinates in the form of Cartesian coordinates. The uniqueness of independent coordinates may sometimes be damaged. For the example shown in Fig. 2.6, the independent coordinates x_1, x_2 (or y_1, y_2) correspond to the above or below positions (left or right positions). It is obvious that x_1 and x_2 (y_1 and y_2) are not appropriate for independent coordinates anymore. It is convenient to specify the system configuration completely by using the rotation angles φ_1 and φ_2 as the independent coordinates. The Cartesian coordinates of each mass can be expressed as continuous, single-valued functions of φ_1 and φ_2.

Actually, there are many options for the generalized coordinates for a given system. As shown in Fig. 2.7, the deflection of the simply supported beam could be expressed in the form of Fourier series by considering

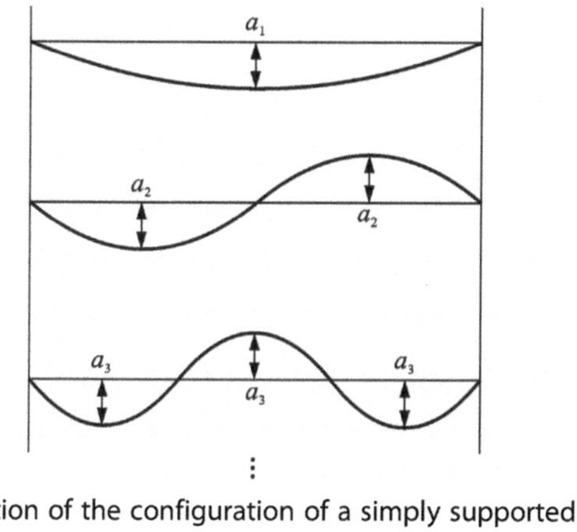

Figure 2.7 Description of the configuration of a simply supported beam.

the boundary constraints

$$v(x,t)=\sum_{i=1}^{\infty}a_i(t)\sin\frac{i\pi x}{L} \tag{2.9}$$

where $\sin(i\pi x/L)$ represents the ith shape function which is a prescribed function satisfying the boundary conditions, L represents the length of the beam, and $a_i(t)$ represents the ith generalized coordinate which is an unknown quantity. For dynamic problems, $a_i(t)$ is a function of time t. Therefore the deflection of the beam can be determined by using a set of generalized coordinates of $a_i(t)$, $i=1,2,\cdots,\infty$, and the number of DOFs of the system is infinite. Only the first few terms of the series are required to be retained in the actual analysis, which is similar to the truncation in a mathematical analysis. By considering the first n terms of the series, the deflection of the simply supported beam could be approximated as follows:

$$v(x,t)=\sum_{i=1}^{n}a_i(t)\sin\frac{i\pi x}{L} \tag{2.10}$$

Therefore a simply supported beam of infinite DOFs is simplified to a finite-DOF system. The generalized coordinates describe the amplitudes of shape functions. The generalized coordinates will have the dimension of the displacement if the shape functions are related to the displacement. However, the generalized coordinates are often not real physical quantities, and only the superposition of n terms of series represents the actual deflection. This kind of method, which is used to express system configuration, is called the generalized coordinate method.

In addition, the finite element method (FEM) may be considered to be an application of the generalized coordinate method, which has been widely used in structural analysis. The amplitudes of shape functions mentioned above are defined as the generalized coordinates, which are not physically meaningful. Meanwhile, the shape functions are defined throughout the entire structure. It is difficult to find a set of appropriate shape functions for complex structures. However, the variables adopted as generalized coordinates in the FEM have clear physical meanings. The shape functions in FEM can be expressed indirectly by means of the local functions throughout segments so that expressions are relatively simple. The simply supported beam as shown in Fig. 2.8A is used as an example to introduce the above method briefly.

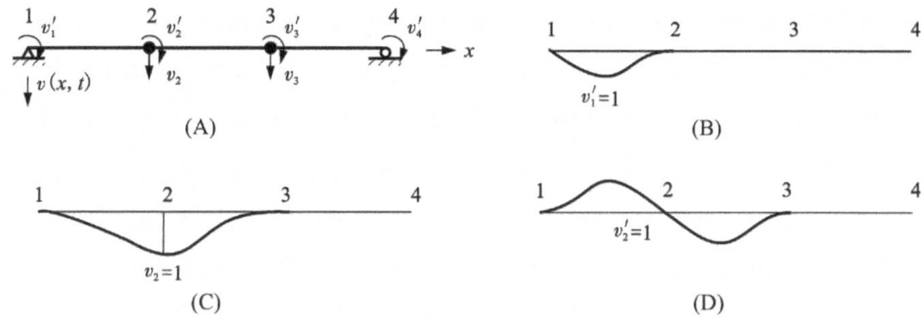

Figure 2.8 Discretization of a simply supported beam with FEM: (A) vertical translations and rotations of nodes; (B) shape function $\varphi_1(x)$; (C) shape function $\varphi_2(x)$; (D) shape function $\varphi_3(x)$.

The simply supported beam may be divided into three elements with four nodes. The vertical translation v and rotation $v'\left(v' = \partial v/\partial x\right)$ of all nodes, as shown in Fig. 2.8A, have been selected as the generalized coordinates. Taking account of the boundary conditions of nodes 1 and 4, the finite element model has six displacement coordinates, namely, v_1', v_2, v_2', v_3, v_3', and v_4'. The displacement coordinates of each node only affect the displacements of the adjacent elements. Fig. 2.8B, C, and D shows the shape functions $\varphi_1(x)$, $\varphi_2(x)$, and $\varphi_3(x)$ corresponding to node displacements v_1', v_2, and v_2', respectively, and other shape functions can be obtained similarly. Referring to Eq. (2.10), the configuration of the simply supported beam could be expressed in terms of six displacement coordinates and the corresponding shape functions as follows:

$$v(x, t) = v_1'\varphi_1(x) + v_2\varphi_2(x) + v_2'\varphi_3(x) + v_3\varphi_4(x) + v_3'\varphi_5(x) + v_4'\varphi_6(x)$$

Therefore a simply supported beam of infinite DOFs is simplified to a 6-DOF system by FEM.

The shape function $\varphi_i(x)$ in the present section is closely related to the element shape function N_i, which will be introduced in Section 2.11. However, there are some differences between them. Here, $\varphi_i(x)$ indicates a function of the entire region of the structure, and N_i only represents a function of the small region of an element. The $\varphi_i(x)$ can be determined by means of N_i. Therefore the shape functions of the structure can be expressed conveniently through this way. Generally, the coordinates used in the generalized coordinate method are the amplitudes of shape functions, which are not physically meaningful displacements. However, the displacement coordinates adopted by the FEM have physical meaning. These are the advantages of the FEM over the generalized coordinate method [2].

2.3 Real displacements, possible displacements, and virtual displacements

A constrained system with l particles starts to move under specified initial conditions. The position vectors of particles $\boldsymbol{r}_k$, $k=1,2,\cdots,l$, should satisfy initial conditions, dynamic differential equations, and all the constraint equations. This kind of motion is called real motion which occurs actually. The displacements of particles in the real motion are referred to as real displacements. The constraint equations of a holonomic system could be given as follows:

$$\begin{aligned}&f_c(\boldsymbol{r}_1,\cdots,\boldsymbol{r}_l,t)=0\\&\text{or}\ f_c(x_1,y_1,z_1,x_2,y_2,z_2,\cdots,x_l,y_l,z_l,t)=0,\ c=1,2,\cdots,s\end{aligned} \tag{2.11}$$

For simplicity, $x_1,y_1,z_1,x_2,y_2,z_2,\cdots,x_l,y_l,z_l$ are replaced by $x_1,x_2,x_3,x_4,x_5,x_6,\cdots,x_{3l-2},x_{3l-1},x_{3l}$, respectively, and the second expression of Eq. (2.11) is rewritten as

$$f_c(x_1,x_2,\cdots,x_{3l},t)=0,\ c=1,2,\cdots,s \tag{2.12}$$

The time t is assumed to vary from t to $t+dt$. The small displacements of particles could be expressed as $d\boldsymbol{r}_k$, $k=1,2,\cdots,l$, (dx_i, $i=1,2,\cdots,3l$, in the Cartesian coordinate system). When the displacements occur, the system should still satisfy Eq. (2.12), that is,

$$f_c(x_1+dx_1,x_2+dx_2,\cdots,x_{3l}+dx_{3l},t+dt)=0,\ c=1,2,\cdots,s$$

By expanding with Taylor series and ignoring the second and higher order terms, the above equations become

$$\begin{aligned}&f_c(x_1+dx_1,x_2+dx_2,\cdots,x_{3l}+dx_{3l},t+dt)\\&\quad=f_c(x_1,x_2,\cdots,x_{3l},t)+\frac{\partial f_c}{\partial x_1}dx_1+\frac{\partial f_c}{\partial x_2}dx_2+\cdots+\frac{\partial f_c}{\partial x_{3l}}dx_{3l}+\frac{\partial f_c}{\partial t}dt=0\\&\quad c=1,2,\cdots,s\end{aligned}$$

Considering Eq. (2.12), one obtains

$$\frac{\partial f_c}{\partial x_1}dx_1+\frac{\partial f_c}{\partial x_2}dx_2+\cdots+\frac{\partial f_c}{\partial x_{3l}}dx_{3l}+\frac{\partial f_c}{\partial t}dt=0$$

or after simplifying

$$\sum_{i=1}^{3l}\frac{\partial f_c}{\partial x_i}dx_i+\frac{\partial f_c}{\partial t}dt=0\ ,\ c=1,2,\cdots,s \tag{2.13}$$

For the case of steady constraints, f_c does not contain time t explicitly. Thus Eq. (2.13) becomes

$$\sum_{i=1}^{3l} \frac{\partial f_c}{\partial x_i} dx_i = 0, c = 1, 2, \cdots, s \tag{2.14}$$

Infinitesimal displacements that only satisfy Eq. (2.13) or Eq. (2.14) are called possible displacements. The possible displacements are not unique since they are only required to meet constraint equations rather than both initial conditions and equations of motion. It is obvious that the real displacements satisfy constraint equations. Therefore real displacements belong to one case of possible displacements. However, the real displacements also need to satisfy initial conditions and equations of motion. Thus there is only one solution for real displacements.

As shown in Fig. 2.9, the particle m is constrained to a spherical surface with constant radius R. The constraint equation could be given as: $x^2 + y^2 + z^2 = R^2$. At the instant of time $t + dt$, the particle m should satisfy $xdx + ydy + zdz = 0$, or $\boldsymbol{r} \cdot d\boldsymbol{r} = 0$.

There are infinite solutions for $d\boldsymbol{r}$ or dx, dy, and dz, which satisfy the above constraint equation. The solutions are arbitrary vectors $d\boldsymbol{r}$ which are located in the tangent plane at point M. Only five vectors in Fig. 2.9 are drawn for examples, and these vectors are possible displacements. The real displacement of particle m should satisfy initial conditions, equations of motion, and constraint equations simultaneously. Thus the real

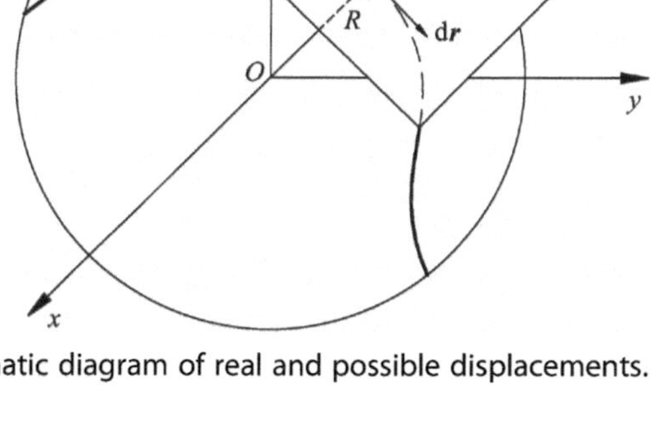

Figure 2.9 Schematic diagram of real and possible displacements.

displacement, which is unique, is located at the tangent plane at point M and along the actual trajectory. The solid line in Fig. 2.9 illustrates the real displacement. Obviously, it is only one case of possible displacements.

A virtual displacement is an arbitrary, infinitesimal, imaginary change of configuration, which is consistent with all displacement constraints on the system. The virtual displacements could be expressed in the form of $\delta \boldsymbol{r}_k$, $k = 1, 2, \cdots, l$, or δx_i, $i = 1, 2, \cdots, 3l$. According to the concepts of virtual displacement and DOF, the number of independent virtual displacements equals that of DOFs, as well as that of the independent equations of motion of the system. The system should also satisfy Eq. (2.12) at some time t with the virtual displacement δx_i, $i = 1, 2, \cdots, 3l$, that is,

$$f_c(x_1 + \delta x_1, x_2 + \delta x_2, \cdots, x_{3l} + \delta x_{3l}, t) = 0, c = 1, 2, \cdots, s$$

By expanding with Taylor series and ignoring the second and higher order terms, the above equation becomes

$$\begin{aligned} & f_c(x_1 + \delta x_1, x_2 + \delta x_2, \cdots, x_{3l} + \delta x_{3l}, t) \\ & \quad = f_c(x_1, x_2, \cdots, x_{3l}, t) + \frac{\partial f_c}{\partial x_1}\delta x_1 + \frac{\partial f_c}{\partial x_2}\delta x_2 + \cdots + \frac{\partial f_c}{\partial x_{3l}}\delta x_{3l} = 0 \end{aligned}$$

Considering Eq. (2.12), one obtains

$$\frac{\partial f_c}{\partial x_1}\delta x_1 + \frac{\partial f_c}{\partial x_2}\delta x_2 + \cdots + \frac{\partial f_c}{\partial x_{3l}}\delta x_{3l} = 0$$

or after simplifying

$$\sum_{i=1}^{3l} \frac{\partial f_c}{\partial x_i}\delta x_i = 0, c = 1, 2, \cdots, s \tag{2.15}$$

By comparing Eq. (2.15) with Eq. (2.13), the equations governing δx_i and dx_i are different. δx_i is time-independent, whereas dx_i depends on time. The constraints are time-varying in the circumstance of unsteady constraints. For this case, all the time-varying constraints can be "frozen" at some time, and the displacements compatible with the frozen constraints are the virtual displacements. Therefore the virtual displacements may not be possible displacements or real displacements.

As shown in Fig. 2.10A, the curvilinear motion of particle m in a plane is given. Its constraint would be steady if the plane is fixed. The real displacement of particle m is in the plane along the tangent line of point M, and its direction is determined, as illustrated in notation $d\boldsymbol{r}$ via the solid line.

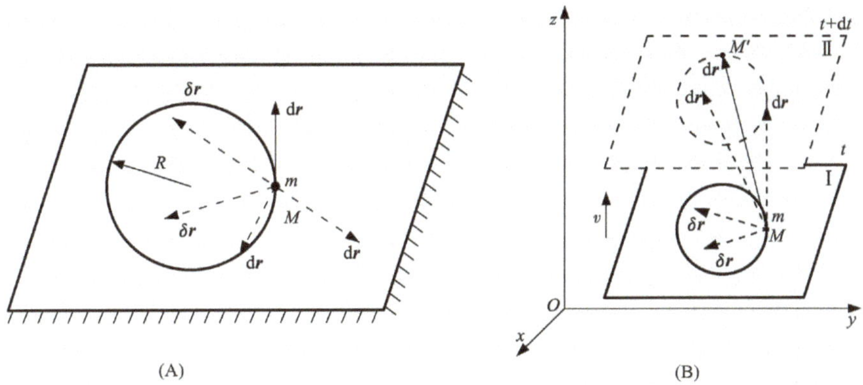

Figure 2.10 Schematic diagram of real, possible, and virtual displacements: (A) steady constraint; (B) unsteady constraint.

The possible displacements with arbitrary directions through point M are also located in the plane, as illustrated with notation $d\boldsymbol{r}$ via the dashed lines. Similarly, the virtual displacement $\delta\boldsymbol{r}$ with arbitrary directions through point M, is also located in the plane, as illustrated via the dashed lines. It should be noted that both the numbers of possible displacements and virtual displacements are infinite.

The constraint will become unsteady once the plane moves upward at a constant speed v, as illustrated in Fig. 2.10B. Then, the real displacement of the particle m is the vector represented by the solid line from the point M in the plane I at the time t to the point M' in the plane II at the time $t + dt$. The possible displacements of the particle m are the arbitrary vectors from point M in the plane I at the time t to any point in plane II at the time $t + dt$ (see dashed lines with notation $d\boldsymbol{r}$). However, the virtual displacements are arbitrary vectors staring from point M in plane I at the instant of time t (dashed lines with notation $\delta\boldsymbol{r}$).

2.4 Generalized force

Consider a system of particles having holonomic constraints. The numbers of particles and holonomic constraints are denoted as l and s, respectively. Therefore the number of DOFs of the system equals $n = 3l - s$. The position of the system could be determined from n generalized coordinates, denoted by $q_1, q_2, \cdots, q_n$. The spatial positions of the particle m_k can be expressed as the function of the generalized coordinates and the time t as follows:

$$\boldsymbol{r}_k = \boldsymbol{r}_k(q_1, q_2, \cdots, q_n, t) \tag{2.16}$$

Generalized coordinates that are adopted to describe the position of the system are independent. The variation of each generalized coordinate is identical to an independent virtual displacement of the system. Thus the virtual displacement of each particle can be described by the function of a set of independent virtual displacements, $\delta q_1, \delta q_2, \cdots, \delta q_n$. The time t corresponding to virtual displacements is stationary. Taking variation of Eq. (2.16) leads to

$$\delta \boldsymbol{r}_k = \sum_{i=1}^{n} \frac{\partial \boldsymbol{r}_k}{\partial q_i} \delta q_i \tag{2.17}$$

Suppose that a force $\boldsymbol{F}_k$ acts on particle m_k. The virtual work done by $\boldsymbol{F}_k$ under $\delta \boldsymbol{r}_k$ can be given by

$$\delta W_k = \boldsymbol{F}_k \cdot \delta \boldsymbol{r}_k \tag{2.18}$$

Substituting Eq. (2.17) into Eq. (2.18) yields

$$\delta W_k = \boldsymbol{F}_k \cdot \sum_{i=1}^{n} \frac{\partial \boldsymbol{r}_k}{\partial q_i} \delta q_i = \sum_{i=1}^{n} \boldsymbol{F}_k \cdot \frac{\partial \boldsymbol{r}_k}{\partial q_i} \delta q_i \tag{2.19}$$

Therefore the virtual work of all particles could be given as follows:

$$\delta W = \sum_{k=1}^{l} \sum_{i=1}^{n} \boldsymbol{F}_k \cdot \frac{\partial \boldsymbol{r}_k}{\partial q_i} \delta q_i = \sum_{i=1}^{n} \sum_{k=1}^{l} \boldsymbol{F}_k \cdot \frac{\partial \boldsymbol{r}_k}{\partial q_i} \delta q_i \equiv \sum_{i=1}^{n} Q_i \delta q_i \tag{2.20}$$

where

$$Q_i = \sum_{k=1}^{l} \boldsymbol{F}_k \cdot \frac{\partial \boldsymbol{r}_k}{\partial q_i} \tag{2.21}$$

Q_i is the generalized force corresponding to the generalized coordinate q_i. $\boldsymbol{F}_k$ represents all the external and internal forces which act on the system. If the virtual work done by the internal forces equals zero (such as the case of ideal constraints), only the virtual work done by the external forces needs to be considered. The generalized forces corresponding to each generalized coordinate can be obtained from Eq. (2.20). In addition, the generalized forces can be calculated by the following approaches [3]:

1. Eq. (2.21) can be rewritten in the form of projection as follows:

$$Q_i = \sum_{k=1}^{l} \left(F_{kx} \frac{\partial x_k}{\partial q_i} + F_{ky} \frac{\partial y_k}{\partial q_i} + F_{kz} \frac{\partial z_k}{\partial q_i} \right) \tag{2.22}$$

where F_{kx}, F_{ky}, and F_{kz} are projections of $\boldsymbol{F}_k$ onto the x, y, and z axes, respectively, and x_k, y_k, and z_k are the position coordinates of particle m_k. When x_k, y_k, and z_k can be easily expressed as the functions of generalized coordinates, it is convenient to obtain Q_i in accordance with Eq. (2.22).

2. All the generalized virtual displacements, except δq_i, can be set to be zero due to the independence of the generalized coordinates. Then, the virtual work of the system to δq_i could be given as δW_i. The generalized force corresponding to q_i could be obtained from the following equation

$$Q_i = \frac{\delta W_i}{\delta q_i} \tag{2.23}$$

When $\boldsymbol{F}_k$, $k = 1, 2, \cdots, l$, includes all the forces (both the external and internal forces) acting on the system, and Q_i, $i = 1, 2, \cdots, n$, are the generalized forces associated with all forces, then the equilibrium equations in the form of generalized forces can be expressed as follows:

$$Q_i = 0, i = 1, 2, \cdots, n \tag{2.24}$$

When $\boldsymbol{F}_k$, $k = 1, 2, \cdots, l$, only includes part of forces acting on the system, Q_i, $i = 1, 2, \cdots, n$, are the generalized forces associated with such part of forces. For example, the generalized force Q_i in the Lagrange's equation, as shown in Eq. (2.46) in Section 2.8 of this book, is the one associated with all forces except the inertial forces.

Example 2.1: Fig. 2.11 shows a double pendulum. P_1 and P_2 are the external forces acting on particles m_1 and m_2, respectively. Here, φ_1 and φ_2 are selected as the generalized coordinates. Determine the generalized forces associated with P_1 and P_2, respectively.

Solution (1):

$$F_{1x} = F_{2x} = F_{1z} = F_{2z} = 0, F_{1y} = P_1, F_{2y} = P_2$$

It can be observed that F_{1x}, F_{2x}, F_{1z}, and F_{2z} equal zero. Then, only y_1 and y_2 are required to be expressed as the functions of φ_1 and φ_2, given by

$$y_1 = l_1 \cos \varphi_1$$
$$y_2 = l_1 \cos \varphi_1 + l_2 \cos \varphi_2$$

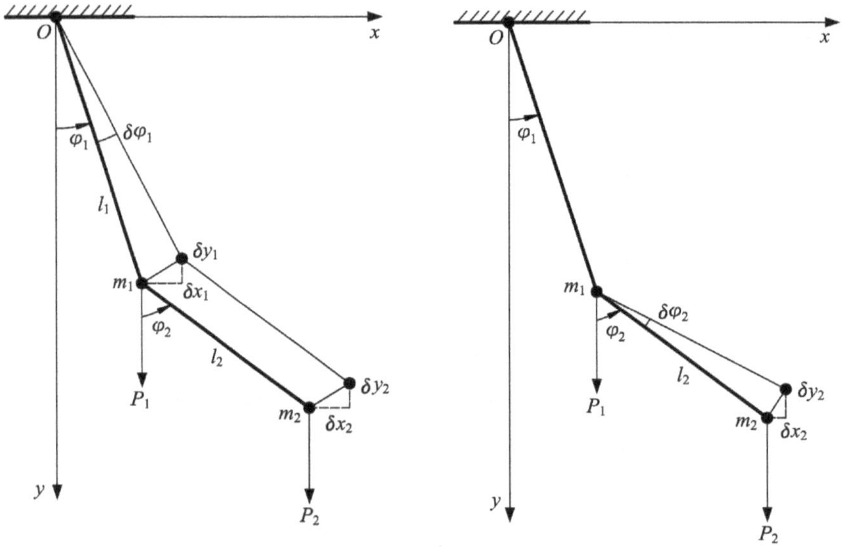

Figure 2.11 Analytical model of the generalized forces for a double pendulum.

Then,

$$\frac{\partial y_1}{\partial \varphi_1} = -\, l_1 \sin \varphi_1, \frac{\partial y_1}{\partial \varphi_2} = 0$$

$$\frac{\partial y_2}{\partial \varphi_1} = -\, l_1 \sin \varphi_1, \frac{\partial y_2}{\partial \varphi_2} = -\, l_2 \sin \varphi_2$$

Thus one obtains

$$Q_1 = F_{1y}\frac{\partial y_1}{\partial \varphi_1} + F_{2y}\frac{\partial y_2}{\partial \varphi_1} = -\,(P_1 + P_2) l_1 \sin \varphi_1$$

$$Q_2 = F_{1y}\frac{\partial y_1}{\partial \varphi_2} + F_{2y}\frac{\partial y_2}{\partial \varphi_2} = -\, P_2 l_2 \sin \varphi_2$$

Solution (2):

First, $\delta\varphi_1$ and $\delta\varphi_2$ are set to be nonzero and zero, respectively. Then, the virtual displacements in the Cartesian coordinate system are given as follows:

$$\delta x_1 = l_1 \delta\varphi_1 \cos \varphi_1, \delta y_1 = -\, l_1 \delta\varphi_1 \sin \varphi_1$$

$$\delta x_2 = l_1 \delta\varphi_1 \cos \varphi_1, \delta y_2 = -\, l_1 \delta\varphi_1 \sin \varphi_1$$

The virtual work by forces P_1 and P_2 to $\delta\varphi_1$ is given as follows:

$$\delta W_1 = P_1\delta y_1 + P_2\delta y_2 = -P_1 l_1 \delta\varphi_1 \sin\varphi_1 - P_2 l_1 \delta\varphi_1 \sin\varphi_1$$

Substituting the above equation into Eq. (2.23) yields

$$Q_1 = -(P_1 + P_2) l_1 \sin\varphi_1$$

Second, $\delta\varphi_1$ and $\delta\varphi_2$ are set to be zero and nonzero, respectively. The virtual displacements in the Cartesian coordinate system are given as follows:

$$\delta x_1 = 0, \delta y_1 = 0$$

$$\delta x_2 = l_2\delta\varphi_2 \cos\varphi_2, \delta y_2 = -l_2\delta\varphi_2 \sin\varphi_2$$

The virtual work by forces P_1 and P_2 to $\delta\varphi_2$ is given as follows:

$$\delta W_2 = P_1\delta y_1 + P_2\delta y_2 = -P_2 l_2 \delta\varphi_2 \sin\varphi_2$$

Finally, one obtains

$$Q_2 = -P_2 l_2 \sin\varphi_2$$

Example 2.2: Fig. 2.12 shows a mass–spring system. P_1 and P_2 are the external forces acting on masses m_1 and m_2, respectively. v_1 and v_2 are selected as the generalized coordinates. Determine the generalized forces associated with all the forces acting on the system.

Solution:

Suppose that the system is subjected to the virtual displacements δv_1 and δv_2. The virtual work done by the external and internal forces of the system can be given respectively as follows:

1. The virtual work done by external forces P_1 and P_2 is $P_1\delta v_1 + P_2\delta v_2$.
2. The forces acting on m_1 and m_2, which are induced by the spring k_1, are a pair of internal forces, which can be expressed as $-k_1(v_1 - v_2)$

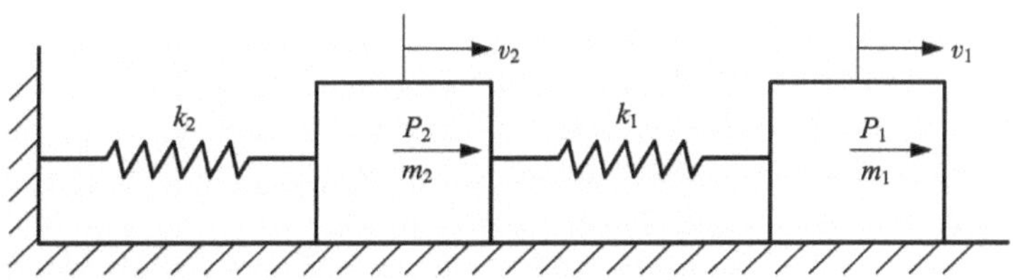

Figure 2.12 Schematic diagram of a mass–spring system.

and $k_1(v_1 - v_2)$, respectively. The virtual work by this pair of forces could be expressed as $-k_1(v_1 - v_2)\delta v_1 + k_1(v_1 - v_2)\delta v_2$.

3. Assume that the spring k_2 is removed and the elastic force of the spring k_2 acting on m_2 can be regarded as an external force expressed as $-k_2 v_2$. The virtual work done by this force could be given as $-k_2 v_2 \delta v_2$.

Finally, the total virtual work done by all the forces could be obtained as:

$$\begin{aligned} \delta W &= P_1\delta v_1 + P_2\delta v_2 - k_1(v_1 - v_2)\delta v_1 + k_1(v_1 - v_2)\delta v_2 - k_2 v_2 \delta v_2 \\ &= (P_1 - k_1 v_1 + k_1 v_2)\delta v_1 + (P_2 + k_1 v_1 - k_1 v_2 - k_2 v_2)\delta v_2 \end{aligned}$$

In accordance with Eq. (2.20), one obtains

$$Q_1 = P_1 - k_1 v_1 + k_1 v_2, \, Q_2 = P_2 + k_1 v_1 - k_1 v_2 - k_2 v_2$$

where Q_1 and Q_2 are generalized forces of the system associated with all the forces. The equilibrium equations in the form of generalized forces could be obtained as $Q_1 = 0$ and $Q_2 = 0$.

When the external forces P_1 and P_2 are time-varying, this means that the system is a dynamic system, and the generalized coordinates v_1 and v_2 vary with time. On the basis of the above deduction, the virtual work by the inertial forces should be added, and can be expressed as $-m_1\ddot{v}_1\delta v_1 - m_2\ddot{v}_2\delta v_2$. Then, the total virtual work by all the forces could be written as

$$\begin{aligned} \delta W &= P_1\delta v_1 + P_2\delta v_2 - k_1(v_1 - v_2)\delta v_1 + k_1(v_1 - v_2)\delta v_2 - k_2 v_2\delta v_2 - m_1\ddot{v}_1\delta v_1 - m_2\ddot{v}_2\delta v_2 \\ &= (P_1 - k_1 v_1 + k_1 v_2 - m_1\ddot{v}_1)\delta v_1 + (P_2 + k_1 v_1 - k_1 v_2 - k_2 v_2 - m_2\ddot{v}_2)\delta v_2 \end{aligned}$$

Similarly, one could also obtain the following

$$Q_1 = P_1 - k_1 v_1 + k_1 v_2 - m_1\ddot{v}_1, \, Q_2 = P_2 + k_1 v_1 - k_1 v_2 - k_2 v_2 - m_2\ddot{v}_2$$

where Q_1 and Q_2 are the generalized forces of the system associated with all the forces. It should be noted that the forces in this case include the inertial forces. The dynamic equilibrium equations in the form of generalized forces could be obtained as $Q_1 = 0$ and $Q_2 = 0$.

2.5 Conservative force and potential energy

According to the principle of the conservation of mechanical energy, the work done by the gravity when an object falls freely from a certain height is transformed into the kinetic energy of the object. This indicates that an object has certain energy at the initial height. This type of energy is known as gravitational potential energy. By considering the object falling from the height z to the reference plane, the work done by the gravity

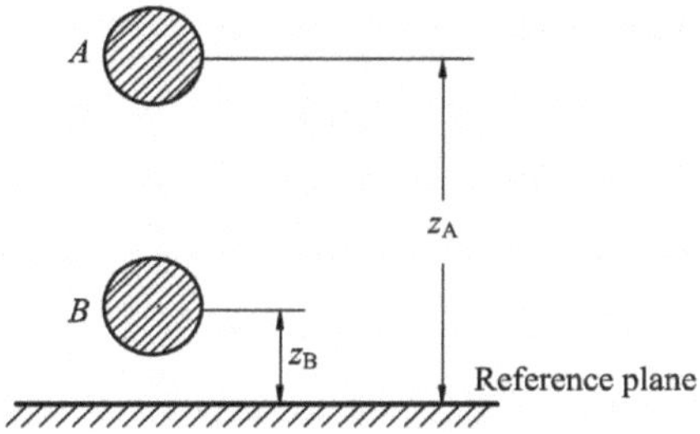

Figure 2.13 Gravitational potential energy of an object.

indicates the change of the object's potential energy. For instance, Fig. 2.13 shows the movement of an object from position B to A. Then, the work done by the gravity can be given as

$$W = -mg(z_A - z_B) = -(V_A - V_B) \tag{2.25}$$

where m is the mass of the object, g represents the acceleration of gravity, z_A and z_B are the heights at positions A and B, respectively, and V_A and V_B represent the potential energy of positions A and B, respectively.

It is shown from Eq.(2.25) that the change of the potential energy of the object equals the negative value of the work done by the gravity. Here, $z_B = 0$ and $V_B = 0$ could be obtained when the horizontal plane through position B is chosen to be the reference plane, thus

$$V_A = -(-mgz_A) \tag{2.26}$$

This indicates that the potential energy of an object at the position A equals the negative value of the work done by gravity when the object moves from the reference plane to position A. This is the criterion for evaluating the gravitational potential energy.

The aforementioned criterion for evaluating the gravitational potential energy is also applicable to the potential energy of an elastic system. As shown in Fig. 2.14, the stiffness of the spring is k. The potential energy of the spring at the positions x_2 and x_1 equals the negative value of the work done by the elastic internal force from zero (unstretched position) to x_2 and x_1, respectively,

$$V_2 = -W_2 = -\left(-\frac{1}{2}kx_2^2\right), V_1 = -W_1 = -\left(-\frac{1}{2}kx_1^2\right) \tag{2.27}$$

The direction of the elastic internal force is opposite to that of the spring's displacement, which leads to the negative sign in the bracket of

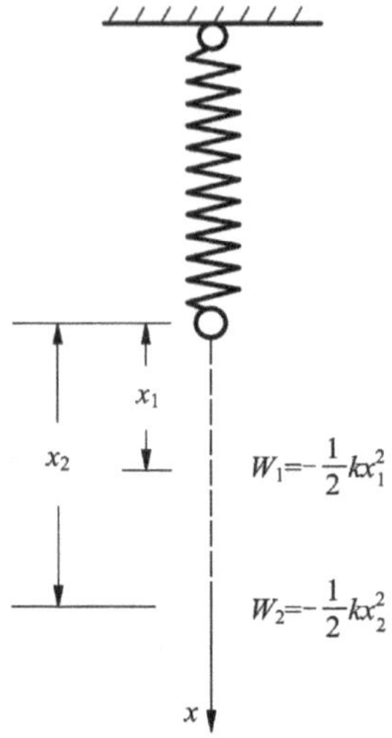

Figure 2.14 Work by the spring's elastic force.

Eq. (2.27). The change of the spring's potential energy equals the negative value of work done by the internal force of the spring when moving from x_2 to x_1, that is,

$$V_2 - V_1 = -\left[-\frac{k}{2}\left(x_2^2 - x_1^2\right)\right] \tag{2.28}$$

It should be noted that the elastic force is assumed to be a linear function of displacement. Therefore a coefficient 1/2 is present in Eq. (2.28). Since the displacement of an object to the gravity is negligible in comparison with the distance between the object and the earth's center, the gravity can be regarded as a constant. Thus the coefficient 1/2 is not present in Eq. (2.26).

The common characteristics of the gravitational and elastic forces can be concluded as follows:

1. The magnitude and direction of forces are entirely determined from the position of the object.
2. As shown in Fig. 2.15, the object moves from position B to A. The work done by the force only depends on the initial and final positions. It is independent of the movement path of the object.

The force with the above characteristics is defined as the conservative force. Choosing the position B as the zero position of potential energy, the potential energy at an arbitrary position A is defined as the sum of

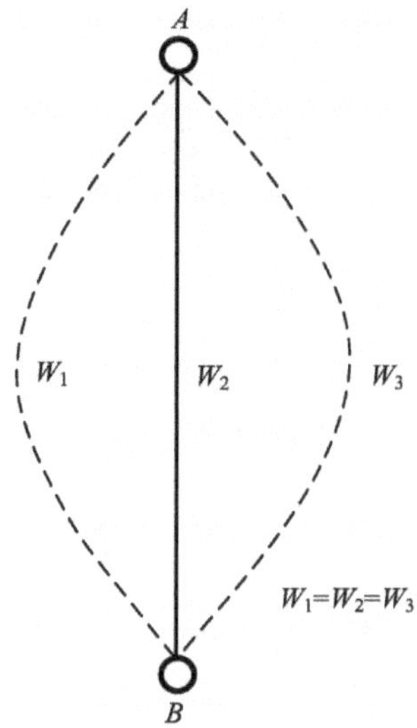

Figure 2.15 Effects of different paths on the work.

negative work by all the conservative forces from B to A. The work by the conservative forces only depends on the initial and final positions. Different paths from B to A have no effect on the work by conservative forces. Therefore the potential energy of an object is the function of its position once the zero position has been determined, which could be expressed as

$$V = V(x, y, z) \tag{2.29}$$

which is also called the potential energy function. The potential energy function at zero position is zero. By considering a small change of the position of an object, the change of the potential energy could be given as

$$dV = -dW = -(f_x dx + f_y dy + f_z dz) \tag{2.30}$$

where f_x, f_y, and f_z are the three components of the conservative force $\boldsymbol{f}$.

From Eq. (2.30), one obtains

$$f_x = -\frac{\partial V}{\partial x}, f_y = -\frac{\partial V}{\partial y}, f_z = -\frac{\partial V}{\partial z} \tag{2.31}$$

Then, the conservative force could be written as

$$\boldsymbol{f} = -\nabla V \tag{2.32}$$

where ∇ represents the gradient function and can be expressed as

$$\nabla = \left\{ \frac{\partial}{\partial x} \quad \frac{\partial}{\partial y} \quad \frac{\partial}{\partial z} \right\}$$

2.6 Direct equilibrium method

Inertia is the ability of an object to maintain its original state of motion. The characteristics of inertia could be described as the resistance of an object to changes in motion. Inertia provides a force against the changes of an object's motion state. This type of force is known as inertial force and denoted as f_I. The inertial force equals the product of the object's mass and its acceleration, and the direction of the inertial force is opposite to that of acceleration.

In general, the forces acting on a system of particles could be classified into active forces, reactive forces of constraint, and inertial forces. However, there is no strict distinction between active forces and reactive forces of constraint, and the corresponding classification depends on specific problems. For instance, the force of a support can be considered as an active force or reactive force in different analysis. The d'Alembert's principle could be expressed as: when a system of particles is subjected to the actual active forces, reactive forces as well as the imaginary inertial forces at an instant of time, the system is said to be in dynamic equilibrium.

For a system consisting of l particles, $\boldsymbol{F}_k$, $\boldsymbol{R}_k$, and $\boldsymbol{f}_{Ik}$ represent the active force, the reactive force, and the inertial force acting on the particle m_k, respectively, and the d'Alembert's principle can be expressed as follows:

$$\boldsymbol{F}_k + \boldsymbol{R}_k + \boldsymbol{f}_{Ik} = \boldsymbol{0}, k = 1, 2, \cdots, l \tag{2.33}$$

Generally, the active force $\boldsymbol{F}_k$ includes the external load $\boldsymbol{P}(t)$, the damping force $\boldsymbol{f}_D$, and the elastic force $\boldsymbol{f}_S$. Eq. (2.33) is also known as the dynamic equilibrium equation.

Example 2.3: A dynamic system consisting of two particles is shown in Fig. 2.16. Formulate its equations of motion.

Solution:

The active force acting on the mass m_1 is $F_1 = P_1 + k_2(x_2 - x_1) + c_2(\dot{x}_2 - \dot{x}_1) - k_1 x_1 - c_1 \dot{x}_1$. The inertial force of the mass m_1 is $f_{I1} = - m_1 \ddot{x}_1$.

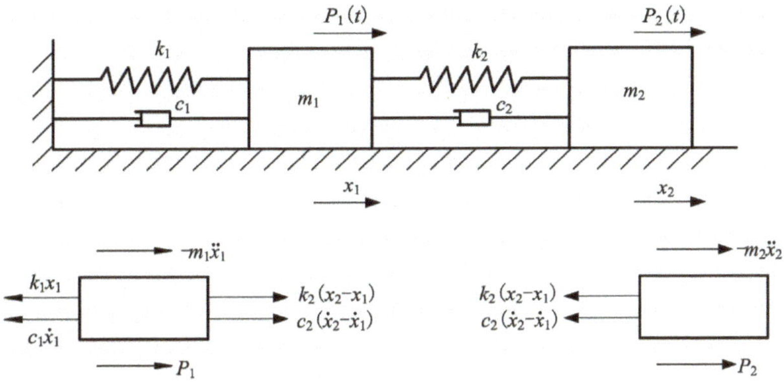

Figure 2.16 Schematic diagram of a 2-DOF mass–spring–damper system. *DOF*, Degree of freedom.

The active force acting on the mass m_2 is $F_2 = P_2 - k_2(x_2 - x_1) - c_2(\dot{x}_2 - \dot{x}_1)$. The inertial force of the mass m_2 is $f_{I2} = -m_2\ddot{x}_2$.

Substituting the above expressions of forces into the dynamic equilibrium equation yields

$$P_1 + k_2(x_2 - x_1) + c_2(\dot{x}_2 - \dot{x}_1) - k_1x_1 - c_1\dot{x}_1 - m_1\ddot{x}_1 = 0$$
$$P_2 - k_2(x_2 - x_1) - c_2(\dot{x}_2 - \dot{x}_1) - m_2\ddot{x}_2 = 0$$

After some rearrangement, one obtains the equations of motion in the form of matrix as follows:

$$\begin{bmatrix} m_1 & 0 \\ 0 & m_2 \end{bmatrix}\begin{Bmatrix} \ddot{x}_1 \\ \ddot{x}_2 \end{Bmatrix} + \begin{bmatrix} c_1 + c_2 & -c_2 \\ -c_2 & c_2 \end{bmatrix}\begin{Bmatrix} \dot{x}_1 \\ \dot{x}_2 \end{Bmatrix} + \begin{bmatrix} k_1 + k_2 & -k_2 \\ -k_2 & k_2 \end{bmatrix}\begin{Bmatrix} x_1 \\ x_2 \end{Bmatrix} = \begin{Bmatrix} P_1 \\ P_2 \end{Bmatrix}$$

2.7 Principle of virtual displacements

The principle of virtual displacements may be expressed as follows: if a system of particles which is in dynamic equilibrium under the action of a set of forces (including the active forces $\boldsymbol{F}_k$, reactive forces of constraint $\boldsymbol{R}_k$, and inertial forces $\boldsymbol{f}_{Ik} = -m_k\ddot{\boldsymbol{r}}_k$), is subjected to virtual displacements compatible with the system's constraints, the total virtual work done by this set of forces will be zero, that is,

$$\sum_{k=1}^{l}(\boldsymbol{F}_k \cdot \delta\boldsymbol{r}_k + \boldsymbol{R}_k \cdot \delta\boldsymbol{r}_k - m_k\ddot{\boldsymbol{r}}_k \cdot \delta\boldsymbol{r}_k) = 0 \tag{2.34}$$

For the system of particles with ideal constraints, the total work done by the reactive forces vanishes. This indicates that $\sum_{k=1}^{l} \boldsymbol{R}_k \cdot \delta\boldsymbol{r}_k = 0$.

Then, the total virtual work done by both the active forces and inertial forces equals zero at any instant of time, that is,

$$\sum_{k=1}^{l}(\boldsymbol{F}_k\cdot\delta\boldsymbol{r}_k - m_k\ddot{\boldsymbol{r}}_k\cdot\delta\boldsymbol{r}_k) = 0 \tag{2.35}$$

Once the constraints of the system are not ideal, the reactive forces could be categorized as either reactive forces of ideal constraints or reactive forces of nonideal constraints. The total virtual work done by the reactive forces of ideal constraints still equals zero, and the reactive forces of nonideal constraints could be regarded as active forces and combined into the active forces $\boldsymbol{F}_k$. Therefore the principle of virtual displacements can still be expressed as Eq. (2.35).

When this principle is applied, the first step is to identify all the forces acting on the system, including the inertial forces defined in accordance with d'Alembert's principle. Then, the equations of motion are obtained by separately introducing a virtual displacement corresponding to each DOF and equating the total virtual work to zero. A major advantage of this approach is that the virtual work contributions are scalar quantities and can be added algebraically, whereas the forces acting on the structure are vectorial, which can only be superposed vectorially. Therefore the virtual displacement method is more convenient than the direct equilibrium method, especially for complex systems.

Example 2.4: Fig. 2.16 shows a dynamic system. Formulate its equations of motion using the principle of virtual displacements.

Solution: The active force and inertial force acting on the mass m_1 can be given as: $F_1 = P_1 + k_2(x_2 - x_1) + c_2(\dot{x}_2 - \dot{x}_1) - k_1x_1 - c_1\dot{x}_1$, and $f_{I1} = -m_1\ddot{x}_1$, respectively.

The active force and inertial force acting on the mass m_2 can be given as: $F_2 = P_2 - k_2(x_2 - x_1) - c_2(\dot{x}_2 - \dot{x}_1)$, and $f_{I2} = -m_2\ddot{x}_2$, respectively.

Once the system is subjected to the virtual displacements δx_1 and δx_2, the equation of virtual work can be expressed as follows:

$$[P_1 + k_2(x_2 - x_1) - k_1x_1 - m_1\ddot{x}_1]\delta x_1 + [P_2 - k_2(x_2 - x_1) - m_2\ddot{x}_2]\delta x_2 = 0$$

For δx_1 and δx_2 are arbitrary, one obtains

$$\begin{cases} P_1 + k_2(x_2 - x_1) - k_1x_1 - m_1\ddot{x}_1 = 0 \\ P_2 - k_2(x_2 - x_1) - m_2\ddot{x}_2 = 0 \end{cases}$$

The equations of motion in matrix form can be obtained by rearrangement, which are the same as the equations of motion in Example 2.3.

Example 2.5: Fig. 2.17A shows a system consisting of two rigid rods denoted by *AB* and *BC*, respectively. Points *A* and *C* are in rigid constraints. The vertical spring with stiffness coefficient *k* is located at point *B*. There is a vertical damper with damping coefficient *c* at the point *E*. The rigid rod *AB* is a uniform rod of mass $\overline{m}$ per unit length. The rod *BC* is regarded as a weightless rod, and its mass *m* is concentrated at its center point *E*. A concentrated dynamic load *P*(*t*) is applied at point *E*. A uniformly distributed load *p*(*t*) is applied at the rigid rod *AB*. A constant axial force *N* acts on the point *C*. Formulate the equation of motion of the system using the principle of virtual displacements.

Solution: Although the system shown in Fig. 2.17 includes both a concentrated-mass rod and a uniformly-distributed-mass rod, the displacement of the system can be determined through only one displacement variable due to two rods being rigid. Therefore it is a single-DOF (SDOF) system.

The vertical displacement *y*(*t*) of the point *B* is selected as the primary displacement variable. The configuration of the system could be expressed in terms of *y*(*t*), as illustrated in Fig. 2.17B.

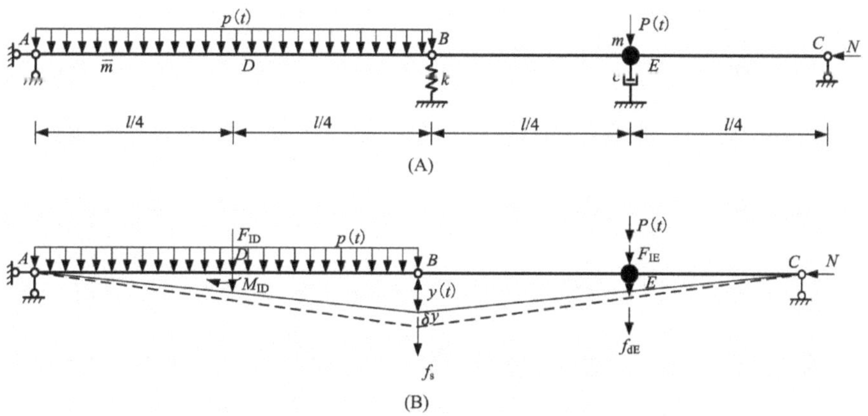

Figure 2.17 Single-DOF system consisting of two rigid rods: (A) schematic diagram of the system consisting of two rigid rods; (B) deflection and loads of the system. *DOF*, Degree of freedom.

The equation of motion of the system could be formulated by the direct equilibrium method. However, the virtual displacement method is more convenient here. The procedures of the virtual displacement method are as follows:

The active forces and inertial forces of the system could be given as follows:

1. The elastic force is $f_s = -ky$;
2. The damping force is $f_{dE} = -c(\dot{y}/2)$;
3. The inertial force of the concentrated mass m is $F_{IE} = -m\ddot{y}/2$;
4. The translational inertial force of the rod AB with uniformly distributed mass is $F_{ID} = -\overline{m}(l/2) \times (\ddot{y}/2)$;
5. The moment of inertia of the rod AB about the center of mass D is

$$I_D = \overline{m}(l/2)[(l/2)^2/12] = \overline{m}l^3/96;$$

and the corresponding inertial moment is

$$M_{ID} = -I_D \times [\ddot{y}/(l/2)] = -(\overline{m}l^2/48)\ddot{y};$$

6. The applied loads are $p(t)$ and $P(t)$.

The system is subjected to the virtual displacement δy, as illustrated in Fig. 2.17B. The virtual work by the reactive forces of constraints equals zero. The total virtual work by the active forces and inertial forces could be given as follows:

$$\delta W = -ky\delta y - c\frac{\dot{y}}{2}\frac{\delta y}{2} - m\frac{\ddot{y}}{2}\frac{\delta y}{2} - \frac{\overline{m}l}{4}\ddot{y}\frac{\delta y}{2} - \frac{\overline{m}l^2}{48}\ddot{y}\frac{\delta y}{(l/2)} + P\frac{\delta y}{2} + p\frac{l}{4}\delta y$$

According to $\delta W = 0$, the equation of motion of the system can be obtained as

$$\left(\frac{m}{4} + \frac{\overline{m}l}{6}\right)\ddot{y} + \frac{c}{4}\dot{y} + ky = \frac{P}{2} + \frac{pl}{4}$$

Once an axial force N is applied at the point C, the virtual work done by the force N should be added to the equation of virtual work. Both the rods AB and BC will have a horizontal virtual displacement $\left[y(t)/(l/2)\right]\delta y$ due to the rotation of the rods AB and BC, when a vertical virtual displacement δy occurs at point B. Therefore the virtual work δW_N due to the force N could be written as

$$\delta W_N = N\left(\frac{y}{l/2} + \frac{y}{l/2}\right)\delta y = \frac{4N}{l}y\delta y$$

After considering the effect of the axial force N, the total virtual work of the system becomes

$$\delta W = -k y \delta y - c\frac{\dot{y}}{2}\frac{\delta y}{2} - m\frac{\ddot{y}}{2}\frac{\delta y}{2} - \frac{\overline{m}l}{4}\ddot{y}\frac{\delta y}{2} - \frac{\overline{m}l^2}{48}\ddot{y}\frac{\delta y}{(l/2)} + P\frac{\delta y}{2} + p\frac{l}{4}\delta y + \frac{4N}{l}y\delta y$$

According to $\delta W = 0$, the equation of motion of the system with the axial force N can be obtained as

$$\left(\frac{m}{4} + \frac{\overline{m}l}{6}\right)\ddot{y} + \frac{c}{4}\dot{y} + \left(k - \frac{4N}{l}\right)y = \frac{P}{2} + \frac{pl}{4}$$

It should be noted that the added axial force only has an effect on the generalized stiffness term. The axial pressure reduces the stiffness of the system, whereas the axial tension increases the stiffness of the system.

2.8 Lagrange's equation

It is assumed that there are s constraint equations for a system of particles (l particles), given by

$$f_c(x_1, y_1, z_1, \cdots, x_l, y_l, z_l, t) = 0, c = 1, 2, \cdots, s \tag{2.36}$$

where $x_k = x_k(q_1, q_2, \cdots, q_n, t)$, $y_k = y_k(q_1, q_2, \cdots, q_n, t)$, and $z_k = z_k(q_1, q_2, \cdots, q_n, t)$, $k = 1, 2, \cdots, l$.

The system configuration can be described by n generalized coordinates, that is, $q_1, q_2, \cdots, q_n$.The position vector $\boldsymbol{r}_k$ of all particles could be expressed as the function of generalized coordinates, that is,

$$\boldsymbol{r}_k = \boldsymbol{r}_k(q_1, q_2, \cdots, q_n, t), k = 1, 2, \cdots, l$$

Thus one obtains

$$\dot{\boldsymbol{r}}_k = \frac{d\boldsymbol{r}_k}{dt} = \sum_{m=1}^{n}\frac{\partial \boldsymbol{r}_k}{\partial q_m}\dot{q}_m + \frac{\partial \boldsymbol{r}_k}{\partial t} \tag{2.37}$$

$$\frac{\partial \dot{\boldsymbol{r}}_k}{\partial \dot{q}_i} = \frac{\partial}{\partial \dot{q}_i}\left[\sum_{m=1}^{n}\frac{\partial \boldsymbol{r}_k}{\partial q_m}\dot{q}_m + \frac{\partial \boldsymbol{r}_k}{\partial t}\right] = \frac{\partial}{\partial \dot{q}_i}\left[\sum_{m=1}^{n}\frac{\partial \boldsymbol{r}_k}{\partial q_m}\dot{q}_m\right] = \frac{\partial \boldsymbol{r}_k}{\partial q_i} \tag{2.38}$$

The forces acting on the kth particle are illustrated in Fig. 2.18. According to Newton's second law of motion, the equation of motion of the particle can be expressed as

$$\boldsymbol{F}_k + \boldsymbol{R}_k = m_k\ddot{\boldsymbol{r}}_k, k = 1, 2, \cdots, l \tag{2.39}$$

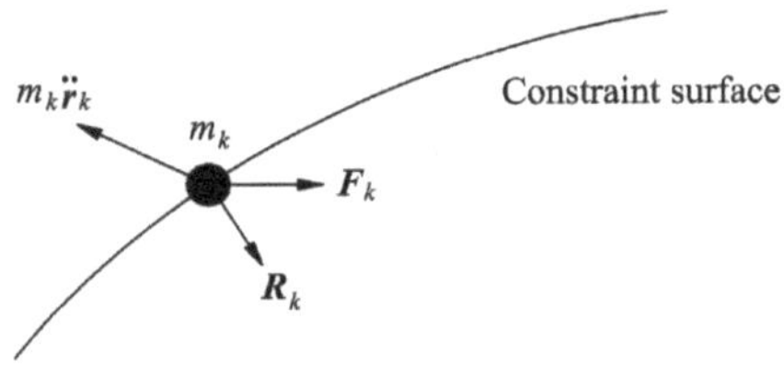

Figure 2.18 Dynamic equilibrium of the *k*th particle.

where m_k is the mass of the *k*th particle, $\ddot{\boldsymbol{r}}_k$ is the acceleration of the *k*th particle, $\boldsymbol{F}_k$ is the active force acting on the *k*th particle, and $\boldsymbol{R}_k$ is the reactive force of constrains acting on the *k*th particle.

The virtual work by the reactive forces equals zero when the constraints of the system are ideal (i.e., $\sum_{k=1}^{l}(\boldsymbol{R}_k \cdot \delta\boldsymbol{r}_k) = 0$). If the constraints of the system are nonideal, the equation of motion of the particle can still be expressed as Eq. (2.39) according to the method introduced in Section 2.7.

The equation of the virtual work of the system could be given as

$$\sum_{k=1}^{l}(\boldsymbol{F}_k \cdot \delta\boldsymbol{r}_k - m_k\ddot{\boldsymbol{r}}_k \cdot \delta\boldsymbol{r}_k) = -\sum_{k=1}^{l}(\boldsymbol{R}_k \cdot \delta\boldsymbol{r}_k) = 0 \tag{2.40}$$

According to the concept of virtual displacements, as mentioned above, $\delta\boldsymbol{r}_k$ is the variation of position of the *k*th particle at the instant of time t, that is,

$$\delta\boldsymbol{r}_k = \sum_{i=1}^{n}\frac{\partial\boldsymbol{r}_k}{\partial q_i}\delta q_i \tag{2.41}$$

Then, one can obtain

$$\sum_{k=1}^{l}\left[(\boldsymbol{F}_k - m_k\ddot{\boldsymbol{r}}_k)\cdot\left(\sum_{i=1}^{n}\frac{\partial\boldsymbol{r}_k}{\partial q_i}\delta q_i\right)\right] = 0$$

Exchanging the order of summation, the above equation can be rewritten as

$$\sum_{i=1}^{n}\left[\delta q_i\sum_{k=1}^{l}(\boldsymbol{F}_k - m_k\ddot{\boldsymbol{r}}_k)\cdot\frac{\partial\boldsymbol{r}_k}{\partial q_i}\right] = 0$$

Since δq_i is independent and arbitrary, the coefficients of each δq_i should equal zero, giving

$$\sum_{k=1}^{l}(\boldsymbol{F}_k - m_k\ddot{\boldsymbol{r}}_k)\cdot\frac{\partial\boldsymbol{r}_k}{\partial q_i} = 0 \tag{2.42}$$

or

$$\sum_{k=1}^{l}\left(m_k\ddot{\boldsymbol{r}}_k\cdot\frac{\partial \boldsymbol{r}_k}{\partial q_i}\right)=\sum_{k=1}^{l}\left(\boldsymbol{F}_k\cdot\frac{\partial \boldsymbol{r}_k}{\partial q_i}\right) \tag{2.43}$$

The kinetic energy of the system can be expressed as $T=(1/2)\sum_{k=1}^{l}(m_k\dot{\boldsymbol{r}}_k\cdot\dot{\boldsymbol{r}}_k)$. Combining Eqs. (2.37) and (2.38) yields

$$\frac{\partial T}{\partial q_i}=\sum_{k=1}^{l}\left(m_k\dot{\boldsymbol{r}}_k\cdot\frac{\partial \dot{\boldsymbol{r}}_k}{\partial q_i}\right)=\sum_{k=1}^{l}\left[m_k\dot{\boldsymbol{r}}_k\cdot\left(\sum_{m=1}^{n}\frac{\partial^2 \boldsymbol{r}_k}{\partial q_i\partial q_m}\dot{q}_m+\frac{\partial^2 \boldsymbol{r}_k}{\partial q_i\partial t}\right)\right]$$

$$\frac{\partial T}{\partial \dot{q}_i}=\sum_{k=1}^{l}\left(m_k\dot{\boldsymbol{r}}_k\cdot\frac{\partial \dot{\boldsymbol{r}}_k}{\partial \dot{q}_i}\right)=\sum_{k=1}^{l}\left(m_k\dot{\boldsymbol{r}}_k\cdot\frac{\partial \boldsymbol{r}_k}{\partial q_i}\right)$$

$$\begin{aligned}\frac{d}{dt}\left(\frac{\partial T}{\partial \dot{q}_i}\right)&=\frac{d}{dt}\left[\sum_{k=1}^{l}\left(m_k\dot{\boldsymbol{r}}_k\cdot\frac{\partial \boldsymbol{r}_k}{\partial q_i}\right)\right]\\&=\sum_{k=1}^{l}\left(m_k\ddot{\boldsymbol{r}}_k\cdot\frac{\partial \boldsymbol{r}_k}{\partial q_i}\right)+\sum_{k=1}^{l}\left[m_k\dot{\boldsymbol{r}}_k\cdot\frac{d}{dt}\left(\frac{\partial \boldsymbol{r}_k}{\partial q_i}\right)\right]\\&=\sum_{k=1}^{l}\left(m_k\ddot{\boldsymbol{r}}_k\cdot\frac{\partial \boldsymbol{r}_k}{\partial q_i}\right)+\sum_{k=1}^{l}\left[m_k\dot{\boldsymbol{r}}_k\cdot\left(\sum_{m=1}^{l}\frac{\partial^2 \boldsymbol{r}_k}{\partial q_i\partial q_m}\dot{q}_m+\frac{\partial^2 \boldsymbol{r}_k}{\partial q_i\partial t}\right)\right]\end{aligned}$$

Thus

$$\sum_{k=1}^{l}\left(m_k\ddot{\boldsymbol{r}}_k\cdot\frac{\partial \boldsymbol{r}_k}{\partial q_i}\right)=\frac{d}{dt}\left(\frac{\partial T}{\partial \dot{q}_i}\right)-\frac{\partial T}{\partial q_i} \tag{2.44}$$

Letting $\sum_{k=1}^{l}(\boldsymbol{F}_k\cdot\delta\boldsymbol{r}_k)=\sum_{i=1}^{n}Q_i\delta q_i$, setting the virtual displacement δq_i to be nonzero and the other virtual displacements to be zero, and considering Eq. (2.41) gives

$$Q_i=\sum_{k=1}^{l}\left(\boldsymbol{F}_k\cdot\frac{\partial \boldsymbol{r}_k}{\partial q_i}\right) \tag{2.45}$$

Substituting Eqs. (2.44) and (2.45) into Eq. (2.43), one gets the Lagrange's equation as follows:

$$\frac{d}{dt}\left(\frac{\partial T}{\partial \dot{q}_i}\right)-\frac{\partial T}{\partial q_i}=Q_i, i=1,2,\cdots,n \tag{2.46}$$

where Q_i is the generalized force associated with the active forces (including the reactive forces of nonideal constraints). The virtual work done by the reactive forces (including the reactive forces of ideal constraints only) is zero, and the corresponding generalized forces are also equal to zero. Q_i can be considered as the generalized force associated with the active and reactive forces, rather than inertial forces.

It is also shown from Eq. (2.44) that $\frac{d}{dt}\left(\frac{\partial T}{\partial \dot{q}_i}\right) - \frac{\partial T}{\partial q_i}$ is the negative value of the generalized force associated with the inertial forces. $-\left[\frac{d}{dt}\left(\frac{\partial T}{\partial \dot{q}_i}\right) - \frac{\partial T}{\partial q_i}\right]$ is called generalized inertial force. The Lagrange's equation, Eq. (2.46), indicates that the generalized forces associated with all forces should be equal to zero. Lagrange's equation is essentially consistent with Eq. (2.24).

Generally, the active forces $\boldsymbol{F}_k$ acting on the kth particle can be written as the sum of the conservative forces and nonconservative forces, that is,

$$\boldsymbol{F}_k = \boldsymbol{f}_k + \boldsymbol{\varphi}_k = -\nabla_k V + \boldsymbol{\varphi}_k \tag{2.47}$$

where V is the total potential energy of the system including the external potential energy of conservative forces and internal potential energy. The external potential energy may contain the time t explicitly. $\nabla_k = \{\partial/\partial x_k \quad \partial/\partial y_k \quad \partial/\partial z_k\}$. Here, $\boldsymbol{\varphi}_k$ is the nonconservative forces acting on the kth particle, for instance, the resistance of the medium, etc.

Substituting Eq. (2.47) into Eq. (2.45) yields

$$Q_i = \sum_{k=1}^{l}\left(\boldsymbol{F}_k \cdot \frac{\partial \boldsymbol{r}_k}{\partial q_i}\right) = -\sum_{k=1}^{l}\left(\nabla_k V \cdot \frac{\partial \boldsymbol{r}_k}{\partial q_i}\right) + \sum_{k=1}^{l}\left(\boldsymbol{\varphi}_k \cdot \frac{\partial \boldsymbol{r}_k}{\partial q_i}\right)$$

Considering

$$\sum_{k=1}^{l}\left(\nabla_k V \cdot \frac{\partial \boldsymbol{r}_k}{\partial q_i}\right) = \sum_{k=1}^{l}\left(\frac{\partial V}{\partial x_k}\frac{\partial x_k}{\partial q_i} + \frac{\partial V}{\partial y_k}\frac{\partial y_k}{\partial q_i} + \frac{\partial V}{\partial z_k}\frac{\partial z_k}{\partial q_i}\right) = \frac{\partial V}{\partial q_i}$$

and letting

$$Q_i' = \sum_{k=1}^{l}\left(\boldsymbol{\varphi}_k \cdot \frac{\partial \boldsymbol{r}_k}{\partial q_i}\right) \tag{2.48}$$

gives

$$Q_i = -\frac{\partial V}{\partial q_i} + Q_i' \tag{2.49}$$

It is shown that the generalized force Q_i can be expressed as the sum of the generalized conservative force and nonconservative force, thus

$$\frac{d}{dt}\left(\frac{\partial T}{\partial \dot{q}_i}\right) - \frac{\partial (T - V)}{\partial q_i} = Q_i' \tag{2.50}$$

The difference between a system's kinetic energy and its potential energy is called the Lagrange function, which is denoted by L, that is, $L = T - V$. The potential energy only depends on the generalized coordinates q_i and t, rather than $\dot{q}_i$. Therefore one obtains $\partial L/\partial \dot{q}_i = \partial T/\partial \dot{q}_i$, and thus

$$\frac{d}{dt}\left(\frac{\partial L}{\partial \dot{q}_i}\right) - \frac{\partial L}{\partial q_i} = Q_i' \tag{2.51}$$

Eq. (2.51) is the final expression of Lagrange's equation. If the constraints are unsteady, the Lagrange function may contain the time t explicitly, that is, $L = L(q_1, q_2, \cdots, q_n, \dot{q}_1, \dot{q}_2, \cdots, \dot{q}_n, t)$. The variables q_i and $\dot{q}_i$ are called Lagrange variables.

Example 2.6: Formulate the equation of motion of the pendulum shown in Fig. 2.19 using Lagrange's equation.

Solution: Here, point O is the origin of the coordinate system, and the position of zero potential energy. The generalized coordinate of the system is

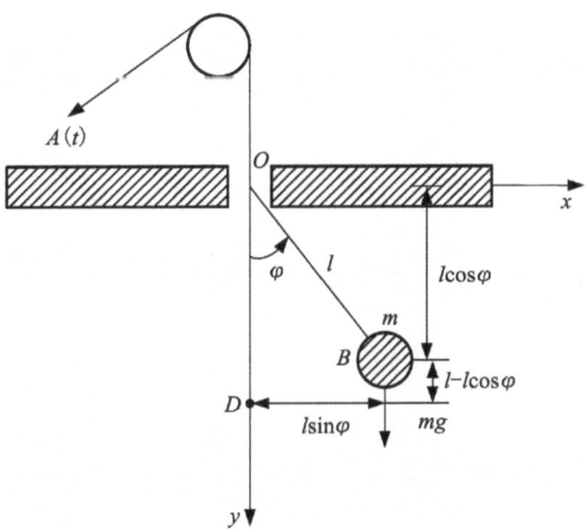

Figure 2.19 Pendulum with time-varying length.

selected as φ. The conservative forces acting on the system is the gravity mg. No nonconservative forces act on this system, therefore the generalized force Q' is equal to zero. Cartesian coordinates of the particle m can be expressed as the functions of φ, that is, $x = l \sin \varphi$, and $y = l \cos \varphi$. Thus

$$\dot{x} = \frac{d}{dt}(l \sin \varphi) = \dot{l} \sin \varphi + l \cos \varphi\dot{\varphi}$$

$$\dot{y} = \frac{d}{dt}(l \cos \varphi) = \dot{l} \cos \varphi - l \sin \varphi\dot{\varphi}$$

The kinetic energy could be given as follows:

$$\begin{aligned} T &= \frac{1}{2}m(\dot{x}^2 + \dot{y}^2) \\ &= \frac{1}{2}m\left[(\dot{l} \sin \varphi + l \cos \varphi\dot{\varphi})^2 + (\dot{l} \cos \varphi - l \sin \varphi\dot{\varphi})^2\right] \\ &= \frac{1}{2}m(\dot{l}^2 + l^2\dot{\varphi}^2) \end{aligned}$$

The negative value of the work done by the gravity during the movement of the particle from point O to B is its potential energy V, that is,

$$V = -mgl \cos \varphi$$

Considering $L = T - V$ and introducing it into Lagrange's equation leads to

$$\frac{d}{dt}(ml^2\dot{\varphi}) + mgl \sin \varphi = 0$$

Then, the nonlinear equation of motion can be obtained as

$$l^2\ddot{\varphi} + 2l\dot{l}\dot{\varphi} + gl \sin \varphi = 0$$

For the small-amplitude oscillation, φ is small enough, $\sin \varphi \approx \varphi$, and $\ddot{\varphi} + 2(\dot{l}/l)\dot{\varphi} + (g/l)\varphi = 0$. Thus a linear equation of motion with variable coefficients is obtained. The second term on the left-hand side of the linear equation is equivalent to the damping term. When $\dot{l}$ is positive, positive damping is present in the system and the amplitude will decay with time. In the contrast, the negative value of $\dot{l}$ would cause negative damping and the amplitude will increase with time continuously (the detailed explanation can be found in Section 3.1.3).

2.9 Hamilton's principle

Lagrange's equations are applicable to the formulation of the equations of motion of discrete systems (finite-DOF systems). For continuous systems (infinite-DOF systems), Hamilton's principle is more appropriate. These two methods are based on the principle of virtual displacements. Therefore they could be deduced from each other. Hamilton's principle belongs to the variation principles in dynamics and can also be derived by means of the variational method. However, it is more convenient to derive Hamilton's principle from Lagrange's equation, and the detailed process is introduced below.

Multiplying Eq. (2.51) by the virtual displacement δq_i of each generalized coordinate, summating them, and integrating the summation expression from time t_1 to t_2 yields

$$\int_{t_1}^{t_2}\sum_{i=1}^{n}\frac{d}{dt}\left(\frac{\partial L}{\partial \dot{q}_i}\right)\delta q_i dt-\int_{t_1}^{t_2}\sum_{i=1}^{n}\frac{\partial L}{\partial q_i}\delta q_i dt=\int_{t_1}^{t_2}\sum_{i=1}^{n}Q_i'\delta q_i dt \tag{2.52}$$

Considering

$$\int_{t_1}^{t_2}\sum_{i=1}^{n}\frac{d}{dt}\left(\frac{\partial L}{\partial \dot{q}_i}\right)\delta q_i dt=\int_{t_1}^{t_2}\sum_{i=1}^{n}\delta q_i d\left(\frac{\partial L}{\partial \dot{q}_i}\right)=\sum_{i=1}^{n}\left[\frac{\partial L}{\partial \dot{q}_i}\delta q_i\right]_{t_1}^{t_2}-\int_{t_1}^{t_2}\sum_{i=1}^{n}\left(\frac{\partial L}{\partial \dot{q}_i}\delta \dot{q}_i\right)dt$$

and substituting the above equation into Eq. (2.52) gives

$$\sum_{i=1}^{n}\left[\frac{\partial L}{\partial \dot{q}_i}\delta q_i\right]_{t_1}^{t_2}-\int_{t_1}^{t_2}\sum_{i=1}^{n}\left(\frac{\partial L}{\partial \dot{q}_i}\delta \dot{q}_i+\frac{\partial L}{\partial q_i}\delta q_i\right)dt=\int_{t_1}^{t_2}\sum_{i=1}^{n}Q_i'\delta q_i dt \tag{2.53}$$

$\delta q_i=0$ holds at both instants of time t_1 and t_2 since the initial and final positions of the system have been given. Thus

$$\sum_{i=1}^{n}\left[\frac{\partial L}{\partial \dot{q}_i}\delta q_i\right]_{t_1}^{t_2}=0 \tag{2.54}$$

Considering that the variable t remains unchanged (the essence of the principle of virtual displacements), from $L=L(q_1,q_2,\cdots,q_n,\dot{q}_1,\dot{q}_2,\cdots,\dot{q}_n,t)$, one obtains

$$\delta L=\sum_{i=1}^{n}\frac{\partial L}{\partial \dot{q}_i}\delta \dot{q}_i+\frac{\partial L}{\partial q_i}\delta q_i \tag{2.55}$$

Letting $\delta W_{nc} = \sum_{i=1}^{n} Q_i'\delta q_i$ and substituting Eqs. (2.54) and (2.55) into Eq. (2.53), Hamilton's principle can be obtained as follows:

$$\int_{t_1}^{t_2} \delta L dt + \int_{t_1}^{t_2} \delta W_{nc} dt = 0 \text{ or } \int_{t_1}^{t_2} \delta(T - V) dt + \int_{t_1}^{t_2} \delta W_{nc} dt = 0 \qquad (2.56)$$

where W_{nc} is the work by all the nonconservative forces acting on a system, and δ represents the variation during the time interval from t_1 to t_2.

Once nonconservative forces are not present, $\delta W_{nc} = 0$, one obtains

$$\int_{t_1}^{t_2} \delta L dt = 0 \text{ or } \int_{t_1}^{t_2} \delta(T - V) dt = 0 \qquad (2.57)$$

Eq. (2.57) is the expression of Hamilton's principle, which is applicable to any linear or nonlinear system. Inertial forces and elastic forces are not explicitly adopted in Hamilton's principle. Variations of kinetic energy and potential energy are used instead. For these two terms regarding energy, only scalar quantities are required to be dealt with. Although the virtual work is scalar in the principle of virtual displacements, the forces and displacements involved in virtual work are still vectors. These are the major advantages of Hamilton's principle.

Example 2.7: Use Hamilton's principle to formulate the equation of motion of the particle m shown in Fig. 2.20.

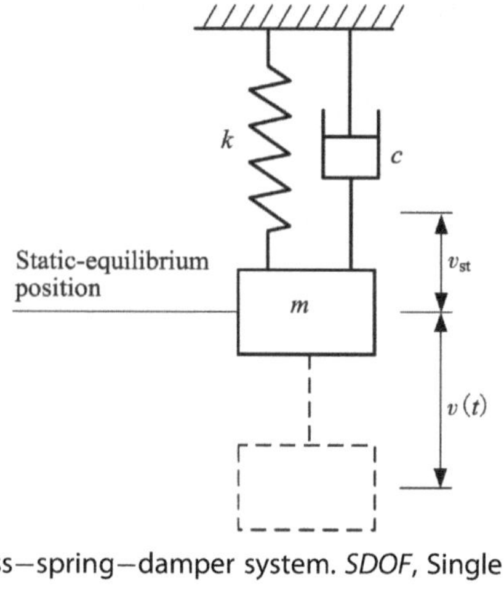

Figure 2.20 SDOF mass–spring–damper system. *SDOF*, Single degree of freedom.

Solution: A SDOF system is shown in Fig. 2.20. The static displacement is v_{st}, and its dynamic displacement $v(t)$ is measured from the static equilibrium position. The gravity mg can be equilibrated by elastic force kv_{st}.

Kinetic energy of the particle m is $T = 1/2m\dot{v}^2$.

Potential energy of system is $V = 1/2kv^2$.

The variation of the work by nonconservative forces equals the work by nonconservative forces during virtual displacement. Thus

$$\delta W_{nc} = P\delta v - c\dot{v}\delta v$$

Substituting the above equation into Eq. (2.56) gives

$$\int_{t_1}^{t_2} (m\dot{v}\delta\dot{v} - kv\delta v + P\delta v - c\dot{v}\delta v)dt = 0$$

Then,

$$\begin{aligned}\int_{t_1}^{t_2} m\dot{v}\delta\dot{v}dt &= \int_{t_1}^{t_2} m\dot{v}\frac{d(\delta v)}{dt}dt = [m\dot{v}\delta v]_{t_1}^{t_2} - \int_{t_1}^{t_2} m\ddot{v}\,\delta v dt \\ &= m\dot{v}\delta v|_{t=t_2} - m\dot{v}\delta v|_{t=t_1} - \int_{t_1}^{t_2} m\ddot{v}\,\delta v dt \\ &= -\int_{t_1}^{t_2} m\ddot{v}\,\delta v dt\end{aligned}$$

Thus

$$\int_{t_1}^{t_2} (-m\ddot{v} - c\dot{v} - kv + P)\delta v dt = 0$$

Since δv is arbitrary, one obtains the equation of motion of the system as

$$m\ddot{v} + c\dot{v} + kv = P$$

Example 2.8: A nonuniform straight beam is shown in Fig. 2.21. The length of the beam is L, and the neutral axis of the beam is the ox axis. The origin of coordinate system is located at the left end of the beam. The mass per unit length of the beam is $m(x)$. The transverse load per unit length

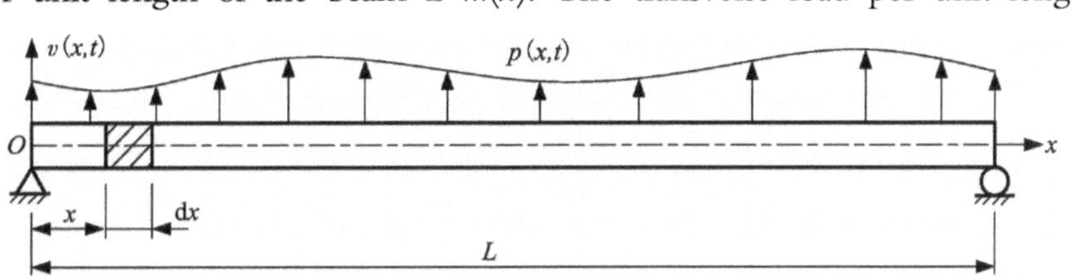

Figure 2.21 Schematic diagram of the transverse vibration of a simply supported beam.

acting on the beam is $p(x,t)$. The transverse displacement on the neutral axis of the beam is expressed as $v(x,t)$. The flexural stiffness of the beam is $EI(x)$. The static equilibrium position of the beam under its self-weight is determined as its initial position. Therefore the influence of the self-weight can be ignored in the analysis of the structural dynamic response. Hamilton's principle is adopted here to formulate the equation of motion of the nonuniform beam.

Solution:

The kinetic energy of the beam is given as follows:

$$T = \frac{1}{2}\int_0^L m(x)\left(\frac{\partial v}{\partial t}\right)^2 dx \tag{2.58}$$

The potential energy of the beam is given as

$$V = \frac{1}{2}\int_0^L EI(x)\left(\frac{\partial^2 v}{\partial x^2}\right)^2 dx \tag{2.59}$$

The work W_{nc} by external loads $p(x,t)$ (nonconservative force) can be written as

$$W_{nc} = \int_0^L p(x,t)v(x,t)dx \tag{2.60}$$

According to Hamilton's principle, one has

$$\int_{t_1}^{t_2} \delta(T - V)dt + \int_{t_1}^{t_2} \delta W_{nc}dt = 0 \tag{2.61}$$

Taking the variation of T, V, and W_{nc}, respectively, and integrating by parts yields

$$\begin{aligned}
\int_{t_1}^{t_2} \delta T dt &= \int_{t_1}^{t_2} \frac{1}{2}\int_0^L m(x)\delta\left(\frac{\partial v}{\partial t}\right)^2 dxdt \\
&= \int_{t_1}^{t_2}\int_0^L m(x)\dot{v}\delta\dot{v}dxdt \\
&= \int_0^L m(x)\dot{v}[\delta v]_{t_1}^{t_2} dx - \int_{t_1}^{t_2}\int_0^L m(x)\,\ddot{v}\,\delta v dxdt \\
&= -\int_{t_1}^{t_2}\int_0^L m(x)\,\ddot{v}\delta v dxdt \\
&= -\int_{t_1}^{t_2}\int_0^L m(x)\frac{\partial^2 v}{\partial t^2}\delta v dxdt
\end{aligned} \tag{2.62}$$

$$\begin{aligned}\int_{t_1}^{t_2} \delta V dt &= \int_{t_1}^{t_2} \frac{1}{2}\int_0^L EI(x)\delta\left(\frac{\partial^2 v}{\partial x^2}\right)^2 dxdt = \int_{t_1}^{t_2}\int_0^L EI(x)v''\delta v'' dxdt \\ &= \int_{t_1}^{t_2}\left\{EI(x)v''[\delta v']_0^L - \int_0^L \frac{\partial}{\partial x}[EI(x)v'']\delta v' dx\right\}dt \\ &= \int_{t_1}^{t_2}\left\{EI(x)v''[\delta v']_0^L - \frac{\partial}{\partial x}[EI(x)v''][\delta v]_0^L + \int_0^L \frac{\partial^2}{\partial x^2}[EI(x)v'']\delta v dx\right\}dt \\ &= \int_{t_1}^{t_2}\int_0^L \frac{\partial^2}{\partial x^2}[EI(x)v'']\delta v dxdt = \int_{t_1}^{t_2}\int_0^L \frac{\partial^2}{\partial x^2}\left[EI(x)\frac{\partial^2 v}{\partial x^2}\right]\delta v dxdt\end{aligned} \tag{2.63}$$

$$\int_{t_1}^{t_2} \delta W_{nc} dt = \int_{t_1}^{t_2}\int_0^L p(x,t)\delta v dxdt \tag{2.64}$$

Then Eq. (2.61) can be written as

$$\begin{aligned}&-\int_{t_1}^{t_2}\int_0^L m(x)\frac{\partial^2 v}{\partial t^2}\delta v dxdt - \int_{t_1}^{t_2}\int_0^L \frac{\partial^2}{\partial x^2}\left[EI(x)\frac{\partial^2 v}{\partial x^2}\right]\delta v dxdt \\ &+ \int_{t_1}^{t_2}\int_0^L p(x,t)\delta v dxdt = 0\end{aligned} \tag{2.65}$$

Rearranging Eq. (2.65) gives

$$\int_{t_1}^{t_2}\int_0^L \left\{-m(x)\frac{\partial^2 v}{\partial t^2} - \frac{\partial^2}{\partial x^2}\left[EI(x)\frac{\partial^2 v}{\partial x^2}\right] + p(x,t)\right\}\delta v dxdt = 0 \tag{2.66}$$

Since δv is arbitrary, one obtains

$$-m(x)\frac{\partial^2 v}{\partial t^2} - \frac{\partial^2}{\partial x^2}\left[EI(x)\frac{\partial^2 v}{\partial x^2}\right] + p(x,t) = 0 \tag{2.67}$$

Then, the differential equation of the transverse vibration of the simply supported nonuniform beam can be obtained as

$$\frac{\partial^2}{\partial x^2}\left[EI(x)\frac{\partial^2 v}{\partial x^2}\right] + m(x)\frac{\partial^2 v}{\partial t^2} = p(x,t) \tag{2.68}$$

2.10 Principle of total potential energy with a stationary value in elastic system dynamics

Prof. Qingyuan Zeng transformed the dynamic problem into a dynamic equilibrium problem in accordance with d'Alembert's principle. By analogy with the principle of total potential energy with a stationary value in statics, he proposed the principle of total potential energy with a stationary value in elastic system dynamics [4]. In order to clarify this principle, the principle of total potential energy with a stationary value in statics is introduced first, the derivation of this principle is then presented, and finally three examples are used to illustrate its application.

2.10.1 Principle of virtual work and principle of total potential energy with a stationary value in statics

The principle of virtual work could be expressed as: the total virtual work done by forces in equilibrium during the system's virtual displacements is equal to zero. The virtual displacements are arbitrary and infinitesimal displacements compatible with geometric constraint conditions. The virtual displacements are independent of the actual forces acting on the system. In Ref. [5], three equations of elastic mechanics are multiplied by the virtual displacements, and the equation of virtual work could be obtained through mathematical derivation as follows:

$$\int_{s} \boldsymbol{\varphi} \cdot \delta \boldsymbol{u} ds + \int_{v} \boldsymbol{X} \cdot \delta \boldsymbol{u} dv = \int_{v} \boldsymbol{\sigma} \cdot \delta \boldsymbol{\varepsilon} dv \tag{2.69}$$

where $\boldsymbol{\varphi}$ is the surface force, $\boldsymbol{X}$ is the body force, $\delta \boldsymbol{u}$ is the virtual displacement, $\boldsymbol{\sigma}$ is the stress, and $\delta \boldsymbol{\varepsilon}$ is the virtual strain. The left-hand side of Eq. (2.69) can be denoted by δW, which is the sum of the virtual work by the surface and body forces. The right-hand side of the equation is the virtual strain energy of the system, and could be denoted by δU_i. Therefore Eq. (2.69) can be rewritten as

$$\delta W = \delta U_i \tag{2.70}$$

The relation of $\delta W = -\delta U_e$ holds since the negative value of the work W by a force is equal to its potential energy U_e. Therefore a more concise expression for the principle of virtual work can be achieved by substituting $\delta W = -\delta U_e$ into Eq. (2.70) as follows:

$$\delta_{\varepsilon} U = \delta_{\varepsilon}(U_i + U_e) = 0 \tag{2.71}$$

Eq. (2.71) is known as the principle of total potential energy with a stationary value in statics. $U = U_i + U_e$ is the total potential energy of the system, and U_e is the potential energy of external forces. In Ref. [6], Eq. (2.71) is regarded as the mathematical condition for $U = U_i + U_e$ with a stationary value, and "$U = U_i + U_e$ = constant" is called the principle of potential energy with a stationary value of a system. It is pointed out in Ref. [5] that "The subscript ε added to the variational sign δ indicates that variation is taken with respect to elastic strains and displacements only. In the calculation of the potential energy of external forces, U_e, all displacements are variable and all forces are definite (i.e., remain unchanged)." The aim of this paragraph is to emphasize the nature of the principle of virtual work of Eq. (2.71) in the variation of the potential energy of a system. The external forces and stresses should not change, while the variation of the potential energy is taken with respect to displacements and strains. The sign δ in Eq. (2.71) maintains the effects of the variation taken with respect to displacements and strains in Eq. (2.69). Although Eq. (2.69), the expression of the principle of virtual work, is expressed as Eq. (2.71) corresponding to the principle of total potential energy with a stationary value, the variational sign δ should keep its nature to represent virtual displacements and virtual strains, and it cannot be regarded as the variation of total potential energy in mathematics.

For example, the potential energy of the external load P at the static equilibrium position B is assumed to be V_B, as shown in Fig. 2.22A. The virtual displacement $\delta\Delta$ occurs at the static equilibrium position, and the virtual work done by the external load P is $\delta W = P\delta\Delta$. According to the definition of the potential energy, $\delta V_B = -\delta W = -P\delta\Delta$ could be obtained. In accordance with the physical concept of the principle of virtual work, the magnitude and direction of the forces acting on the structure remain unchanged during the virtual displacements. Therefore one gets $\delta V_B = \delta(-P\Delta)$ and $V_B = -P\Delta + V_0$ (where V_0 represents the potential energy of the beam in initial state, and it is set to be zero in

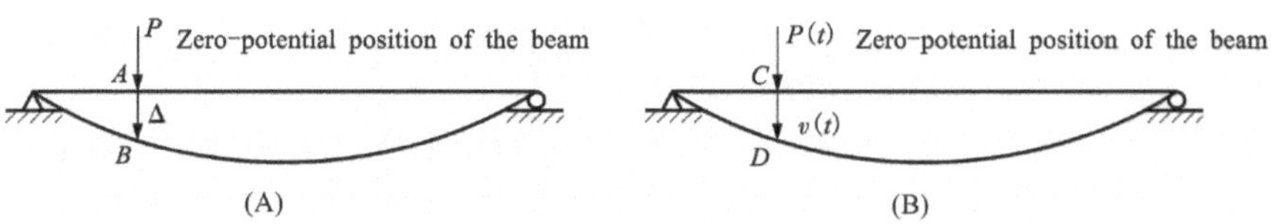

Figure 2.22 Potential energy of different external loads: (A) beam subjected to static forces; (B) beam subjected to dynamic loads.

this example). Therefore the potential energy of the external load P is $V_B = -P\Delta$ (Δ represents the displacement of the beam at the acting point in the direction of external load P).

For the vibration analysis of the planar beam, the direction and magnitude of the dynamic load $P(t)$ are unchangeable at the instant of time t. Such an idea could be applied to define potential energy of a dynamic load. As shown in Fig. 2.22B, the beam is subjected to a dynamic load $P(t)$ at point C, and the displacement from point C to D is $v(t)$. The potential energy of $P(t)$ at the instant of time t could be given as $V = -P(t)v(t)$.

According to the aforementioned discussions, conclusions could be drawn as follows. (1) The virtual displacement is an arbitrary, infinitesimal and imaginary displacement compatible with geometric constraint conditions for a system in equilibrium. It is independent of the external forces and stresses. It is not the actual displacement and will not change the force equilibrium. Therefore the external forces and the internal forces will not change during the virtual displacement. (2) The principle of total potential energy with a stationary value in statics could be obtained by putting the variational sign δ of the virtual displacements and virtual strains in Eq. (2.69) to the outside of the integral sign. The external forces and stresses of the system remain unchanged during the manipulation. Thus the first-order variation of the potential energy of the system U only involves the variation with respect to the displacements $\boldsymbol{u}$ and the strains ε, rather than the external forces and stresses, so that the recovering from Eq. (2.71) to Eq. (2.69) is feasible, and the requirement of the principle of virtual work is satisfied exactly.

2.10.2 Derivation of the principle of total potential energy with a stationary value in elastic system dynamics

According to the d'Alembert's principle, the dynamic problem of an elastic system can be transformed into a problem of dynamic equilibrium. The general form of dynamic equilibrium equations is given as follows:

$$\boldsymbol{f}_m + \boldsymbol{f}_c + \boldsymbol{f}_s + \boldsymbol{P}(t) = \boldsymbol{0} \tag{2.72}$$

where $\boldsymbol{f}_s$ is the elastic force, $\boldsymbol{f}_m$ is the inertial force, $\boldsymbol{f}_c$ is the damping force, and $\boldsymbol{P}(t)$ is the applied load (including the gravity).

Imagining any instant of time t to be fixed transiently, multiplying both sides of Eq. (2.72) by the virtual displacement $\delta\boldsymbol{u}$, and noting that

$-\delta \boldsymbol{u} \cdot \boldsymbol{f}_s$ equals the virtual strain energy of the system δU_i, one obtains the following equation

$$\delta U_i = \delta \boldsymbol{u} \cdot \boldsymbol{f}_m + \delta \boldsymbol{u} \cdot \boldsymbol{f}_c + \delta \boldsymbol{u} \cdot \boldsymbol{P}(t) \tag{2.73}$$

Eq. (2.73) is the general expression of the principle of virtual work in elastic system dynamics. Similar to the derivation of Eq. (2.71) from Eq. (2.69), Eq. (2.73) can be written in concise form as follows:

$$\delta_\varepsilon (U_i + V_m + V_c + V_P) = 0 \tag{2.74}$$

where U_i is the strain energy of the elastic system, $V_m = -\boldsymbol{u} \cdot \boldsymbol{f}_m$ is the negative value of work done by inertial forces, $V_c = -\boldsymbol{u} \cdot \boldsymbol{f}_c$ is the negative value of work done by damping forces, and $V_P = -\boldsymbol{u} \cdot \boldsymbol{P}(t)$ is the negative value of work done by applied loads.

When an equilibrium system is subjected to virtual displacements, all the forces acting on the system remain unchanged due to the time t being fixed transiently. The virtual work done by these forces is only related to the initial and final positions, rather than the path of the acting point of forces. Therefore these forces can be regarded as conservative forces in accordance with the definition of conservative forces. V_m, V_c, and V_P are called the potential energy of the inertial force, damping force, and applied load, respectively. Thus we call Π_d defined in Eq.(2.75) as the total potential energy of an elastic dynamic system, following the concept of the total potential energy in elastic static system

$$\Pi_d = U_i + U_m + V_c + V_P \tag{2.75}$$

By analogy with Eq. (2.71), Eq. (2.74) is named as the principle of the total potential energy with a stationary value in elastic system dynamics. The subscript ε of the variational sign δ highlights that in order to reserve the nature of the principle of virtual work in Eq. (2.74), the variation of Π_d can only be taken with respect to the displacements and the elastic strains, while the inertial force, damping force, applied load, etc., are all regarded as constants. In other words, all the forces involved are invariant in the calculation of V_m, V_c, and V_P, but all the displacements are regarded as variables. The physical meaning of Eq. (2.74) is as follows: when the d'Alembert's principle is applied and the time t is fixed transiently, the first-order variation of Π_d (the total potential energy in elastic dynamic system) must be zero, that is, Π_d possesses a stationary value. Generally, the stationary value is obtained in accordance with the functional variation principle. However, the stationary value of Π_d is obtained herein from the principle of virtual work.

Obviously, it is not reasonable to obtain the first-order variation of Π_d according to the variational method. Based on the physical concept of the principle of virtual work, only the variations of Π_d with respect to the displacements and strains are required.

The basic procedure of formulating the equations of motion of a system using the principle of total potential energy with a stationary value in elastic system dynamics could be summarized as follows: (1) the potential energy of each force acting on the system is calculated, as well as the elastic strain energy, and then the total potential energy of the system Π_d is obtained; and (2) the variation of Π_d with respect to the displacements and strains is taken to formulate the equations of motion of the system in accordance with $\delta_\varepsilon \Pi_d = 0$. The major advantage of this principle is that it is not required to distinguish between the conservative and nonconservative forces, as well as the steady and unsteady constraints. In order to illustrate the validity and simplicity of this principle, three examples are given below.

Example 2.9: Use the principle of total potential energy with a stationary value in elastic system dynamics to formulate the equation of motion of the particle m, as shown in Fig. 2.23.

Solution: The coordinate system used here is the same as that in Example 2.6. Point O is selected as the position of zero potential energy. The generalized coordinate of the particle is selected as φ, as illustrated in Fig. 2.23.

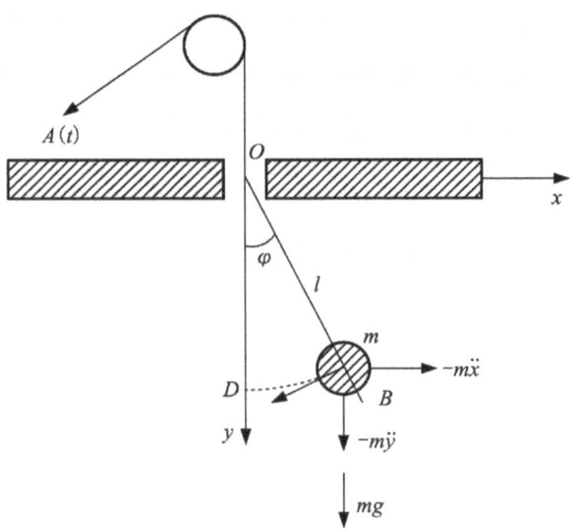

Figure 2.23 Dynamic equilibrium of a single pendulum with time-varying length.

From Fig. 2.23, one obtains the Cartesian coordinates of particle m as follows:

$$x = l \sin \varphi, y = l \cos \varphi$$

then

$$\dot{x} = \dot{l} \sin \varphi + l \cos \varphi \dot{\varphi},\ \ddot{x} = \frac{d\dot{x}}{dt} = \frac{d}{dt}(\dot{l} \sin \varphi + l \cos \varphi \dot{\varphi})$$

$$\dot{y} = \dot{l} \cos \varphi - l \sin \varphi \dot{\varphi}, \ddot{y} = \frac{d}{dt}(\dot{l} \cos \varphi - l \sin \varphi \dot{\varphi})$$

Therefore the total potential energy of the system could be obtained as

$$\begin{aligned} \Pi_d &= -mgl \cos \varphi - (-m\ddot{x}x) - (-m\ddot{y}y) \\ &= -mgl \cos \varphi + m\frac{d}{dt}(\dot{l} \sin \varphi + l \cos \varphi \dot{\varphi})l \sin \varphi \\ &\quad + m\frac{d}{dt}(\dot{l} \cos \varphi - l \sin \varphi \dot{\varphi})l \cos \varphi \end{aligned}$$

Then

$$\begin{aligned} \delta_\varepsilon \Pi_d &= mgl \sin \varphi \delta\varphi + m\frac{d}{dt}(\dot{l} \sin \varphi + l \cos \varphi \dot{\varphi})l \cos \varphi \delta\varphi \\ &\quad + m\frac{d}{dt}(\dot{l} \cos \varphi - l \sin \varphi \dot{\varphi})(-l \sin \varphi \delta\varphi) \\ &= \delta\varphi\left(ml^2\ddot{\varphi} + 2ml\dot{l}\dot{\varphi} + mgl \sin \varphi\right) = 0 \end{aligned}$$

Considering $\delta\varphi \neq 0$, one obtains

$$ml^2\ddot{\varphi} + 2ml\dot{l}\dot{\varphi} + mgl \sin \varphi = 0$$

which can be rewritten as

$$l^2\ddot{\varphi} + 2l\dot{l}\dot{\varphi} + gl \sin \varphi = 0$$

Example 2.10: Mass M is connected to a moveable point O by a spring of stiffness k. Both mass M and point O can only move along the x axis. The motion of point O is known as $x_0(t)$. In addition, a pendulum is suspended on mass M. The mass of the pendulum is m. The distance between the center of gravity C and the suspension point is l. The radius of gyration about the center of gravity is ρ. The system is shown in Fig. 2.24. Formulate the equations of motion of this system.

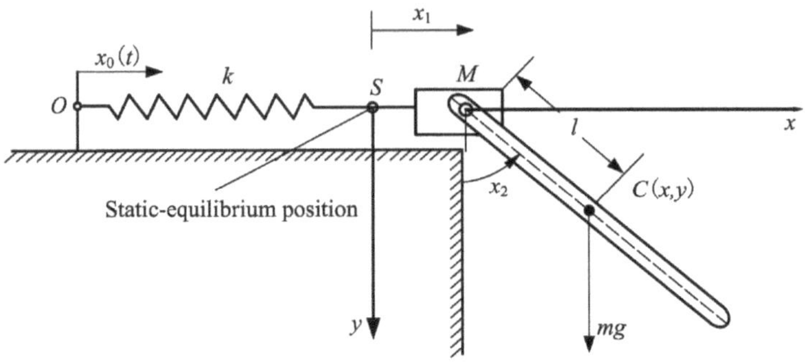

Figure 2.24 Motion of a multi-rigid-body system.

Solution: The configuration of the system is expressed by the generalized coordinates x_1 and x_2, as illustrated in Fig. 2.24. Point S is the unstretched position for the spring at the initial time. x_1 represents the horizontal displacement of mass M relative to point S, and x_2 represents the rotation of the pendulum measured counterclockwise from the vertical axis. The Cartesian coordinates of point C could be given as $x = l\sin x_2 + x_1$ and $y = l\cos x_2$. Therefore one obtains

$$\dot{x} = l\cos x_2\dot{x}_2 + \dot{x}_1,\ \ddot{x} = l\cos x_2\ddot{x}_2 - l\sin x_2\dot{x}_2^2 + \ddot{x}_1$$
$$\dot{y} = -l\sin x_2\dot{x}_2,\ \ddot{y} = -l\sin x_2\ddot{x}_2 - l\dot{x}_2^2\cos x_2$$

The potential energy of inertial forces could be given as

$$V_m = -(-m\ddot{x}\,x - m\ddot{y}\,y - M\ddot{x}_1x_1 - m\rho^2\ddot{x}_2x_2)$$

$$= m(l\cos x_2\ddot{x}_2 - l\sin x_2\dot{x}_2^2 + \ddot{x}_1)(l\sin x_2 + x_1)$$

$$+ m(-l\sin x_2\ddot{x}_2 - l\dot{x}_2^2\cos x_2)l\cos x_2 + M\ddot{x}_1x_1 + m\rho^2\ddot{x}_2x_2$$

The potential energy of the gravity (point S is chosen as the position of zero potential energy) is given as

$$V_P = -mgl\cos x_2$$

The strain energy of the spring is given as follows:

$$U_i = \frac{1}{2}k(x_1 - x_0)^2$$

Therefore the total potential energy of the system is given as follows:

$$\Pi_d = V_m + V_P + U_i$$

According to $\delta_\varepsilon \Pi_d = 0$, one obtains

$$m(l\cos x_2\ddot{x}_2 - l\sin x_2\dot{x}_2^2 + \ddot{x}_1)(l\cos x_2\delta x_2 + \delta x_1)$$

$$+ m(-l\sin x_2\ddot{x}_2 - l\dot{x}_2^2\cos x_2)(-l\sin x_2\delta x_2) + M\ddot{x}_1\delta x_1$$

$$+ m\rho^2\ddot{x}_2\delta x_2 + mgl\sin x_2\delta x_2 + k(x_1 - x_0)\delta x_1 = 0$$

Factoring out δx_1 and δx_2 leads to

$$\delta x_1\left[(m+M)\ddot{x}_1 + ml\cos x_2\ddot{x}_2 - ml\sin x_2\dot{x}_2^2 + k(x_1 - x_0)\right]$$

$$+ \delta x_2\left[m(l^2\cos^2 x_2\ddot{x}_2 - l^2\sin x_2\cos x_2\dot{x}_2^2 + \ddot{x}_1 l\cos x_2\right.$$

$$\left.+ l^2\sin^2 x_2\ddot{x}_2 + l^2\sin x_2\cos x_2\dot{x}_2^2) + m\rho^2\ddot{x}_2 + mgl\sin x_2\right] = 0$$

Considering $\delta x_1 \neq 0$ and $\delta x_2 \neq 0$, one obtains

$$\begin{cases} (m+M)\ddot{x}_1 + ml\,\ddot{x}_2\cos x_2 - ml\,\dot{x}_2^2\sin x_2 + k(x_1 - x_0) = 0 \\ ml\ddot{x}_1\cos x_2 + (ml^2 + m\rho^2)\ddot{x}_2 + mgl\sin x_2 = 0 \end{cases}$$

Once only small-amplitude oscillations are considered, both x_1 and x_2 are relatively small. Therefore $\cos x_2 \approx 1$ and $\sin x_2 \approx x_2$ could be obtained. By ignoring all the nonlinear terms, the above equations can be approximated as

$$\begin{cases} (m+M)\ddot{x}_1 + ml\ddot{x}_2 + k(x_1 - x_0) = 0 \\ ml\ddot{x}_1 + (ml^2 + m\rho^2)\ddot{x}_2 + mglx_2 = 0 \end{cases}$$

It is shown from Examples 2.6 and 2.10 that the equations of motion can usually be approximated as linear equations by only considering small-amplitude vibrations. The following chapters in this book mainly focus on the vibration analysis of linear systems. Only a little content related to nonlinear vibration will be involved in the step-by-step integration method.

Example 2.11: Use the principle of total potential energy with a stationary value in elastic system dynamics to derive the equation of motion of a simply supported nonuniform beam, as shown in Fig. 2.21. The detailed description of the system is given by Example 2.8.

Solution: The bending strain energy of the beam is given as

$$U_i = \frac{1}{2}\int_0^L EI(x)\left(\frac{\partial^2 v}{\partial x^2}\right)^2 dx$$

The potential energy of external forces (the work done by reactive forces of the supports at the ends of the beam equals zero, and only the applied load $p(x, t)$ is considered here) is expressed as

$$V_P = -\int_0^L p(x, t)v dx$$

The potential energy of inertial forces is given as

$$V_m = -\int_0^L -m(x)\frac{\partial^2 v}{\partial t^2} v dx = \int_0^L m(x)\frac{\partial^2 v}{\partial t^2} v dx$$

Taking the variation of U_i with respect to v and integrating the variation by parts leads to

$$\begin{aligned}
\delta U_i &= \int_0^L EI(x)\frac{\partial^2 v}{\partial x^2}\delta\left(\frac{\partial^2 v}{\partial x^2}\right) dx \\
&= EI(x)\frac{\partial^2 v}{\partial x^2}\delta\left(\frac{\partial v}{\partial x}\right)\Bigg|_0^L - \int_0^L \frac{\partial}{\partial x}\left[EI(x)\frac{\partial^2 v}{\partial x^2}\right]\delta\left(\frac{\partial v}{\partial x}\right) dx \\
&= EI(x)\frac{\partial^2 v}{\partial x^2}\delta\left(\frac{\partial v}{\partial x}\right)\Bigg|_0^L - \frac{\partial}{\partial x}\left[EI(x)\frac{\partial^2 v}{\partial x^2}\right]\delta v\Bigg|_0^L + \int_0^L \frac{\partial^2}{\partial x^2}\left[EI(x)\frac{\partial^2 v}{\partial x^2}\right]\delta v dx \\
&= \int_0^L \frac{\partial^2}{\partial x^2}\left[EI(x)\frac{\partial^2 v}{\partial x^2}\right]\delta v dx
\end{aligned}$$

The variation of V_P with respect to v is given as

$$\delta V_P = -\int_0^L p(x, t)\delta v dx$$

The variation of V_m with respect to v is given as

$$\delta V_m = \int_0^L m(x)\frac{\partial^2 v}{\partial t^2}\delta v dx$$

Substituting δU_i, δV_P, and δV_m into Eq. (2.74) yields

$$\int_0^L \left\{\frac{\partial^2}{\partial x^2}\left[EI(x)\frac{\partial^2 v}{\partial x^2}\right] + m(x)\frac{\partial^2 v}{\partial t^2} - p(x, t)\right\}\delta v dx = 0$$

Considering $\delta v \neq 0$, the equation of motion of the beam can be obtained as follows:

$$\frac{\partial^2}{\partial x^2}\left[EI(x)\frac{\partial^2 v}{\partial x^2}\right] + m(x)\frac{\partial^2 v}{\partial t^2} = p(x, t)$$

The above equation in this example is exactly the same as that derived from Hamilton's principle in Example 2.8.

2.11 The "set-in-right-position" rule for assembling system matrices and method of computer implementation in Matlab

Based on the principle of total potential energy with a stationary value in elastic system dynamics, Prof. Qingyuan Zeng, the fourth author of this book, proposed the "set-in-right-position" rule for assembling system matrices to formulate the equations of motion in the matrix form directly [7,8]. For the convenience of application of the rule in computer programming, the method of computer implementation in Matlab was developed by Dr. Zhihui Zhou, the first author of this book.

2.11.1 The "set-in-right-position" rule for assembling system matrices

The system is assumed to have n independent displacement coordinates q_i, $i = 1, 2, \cdots, n$. The system is subjected to a set of virtual displacements with the time t fixed transiently. The external forces acting on the system could be considered as constant. Then, the total potential energy of elastic system dynamics Π_d is a function of the system's displacement coordinates q_i, $i = 1, 2, \cdots, n$. According to the principle of total potential energy with a stationary value in elastic system dynamics, $\delta_\varepsilon \Pi_d = 0$, one obtains

$$\delta_\varepsilon \Pi_d = \sum_{i=1}^{n} \frac{\partial \Pi_d}{\partial q_i} \delta q_i = 0 \tag{2.76}$$

Considering $\delta q_i \neq 0$, $i = 1, 2, \cdots, n$, leads to

$$\frac{\partial \Pi_d}{\partial q_i} = 0, i = 1, 2, \cdots, n \tag{2.77}$$

Eq. (2.77) represents n equilibrium equations of the system. It is obvious that the ith equilibrium equation of the system can be obtained through the variation of Π_d with respect to q_i. However, this is not clearly shown from the equation $\partial \Pi_d / \partial q_i = 0$ since other displacement coordinates may be present in this equation, as well as q_i. Thus Eq. (2.76) should be rewritten as follows:

$$\delta_\varepsilon \Pi_d = \delta q_1 \frac{\partial \Pi_d}{\partial q_1} + \delta q_2 \frac{\partial \Pi_d}{\partial q_2} + \cdots + \delta q_n \frac{\partial \Pi_d}{\partial q_n} = 0 \tag{2.78}$$

From Eq. (2.78), one leads to

$$\begin{cases} \delta q_1 \dfrac{\partial \Pi_d}{\partial q_1} = 0 \\ \delta q_2 \dfrac{\partial \Pi_d}{\partial q_2} = 0 \\ \vdots \\ \delta q_n \dfrac{\partial \Pi_d}{\partial q_n} = 0 \end{cases} \tag{2.79}$$

Eq. (2.79) still represents n equilibrium equations due to $\delta q_i \neq 0$, $i = 1, 2, \cdots, n$. As distinct from Eq. (2.77), δq_i is present in Eq. (2.79), which means that $\partial \Pi_d / \partial q_i = 0$ is the ith equilibrium equation of the system. Therefore δq_i represents the ith row in the process of assembling stiffness matrix, damping matrix, mass matrix, and load vector. In addition, $\partial \Pi_d / \partial q_i$ may contain terms related to the displacement coordinate q_j, velocity $\dot{q}_j$ or acceleration $\ddot{q}_j$, $j = 1, 2, \cdots, n$. The subscript j of the displacement coordinate q_j represents the jth column of the stiffness matrix. The coefficients of $\delta q_i \cdot q_j$ in Eq. (2.79) should be added to the original expression (position) of the ith row and the jth column of the stiffness matrix. The subscript j of velocity $\dot{q}_j$ represents the jth column of the damping matrix. The coefficients of $\delta q_i \cdot \dot{q}_j$ in Eq. (2.79) should be added to the original expression (position) of the ith row and the jth column of the damping matrix. The subscript j of acceleration $\ddot{q}_j$ represents the jth column of the mass matrix. The coefficient of $\delta q_i \cdot \ddot{q}_j$ in Eq. (2.79) should be added to the original expression (position) of the ith row and the jth column of the mass matrix. Some terms in Eq. (2.79) are not related to the displacement coordinate q_j, velocity $\dot{q}_j$, or acceleration $\ddot{q}_j$, of which negative value of the coefficients of δq_i should be added to the original expression (position) of the ith row of the load vector. Because the load vector has been moved to the right-hand side of equilibrium equations, the coefficients need to be reversed. This is the "set-in-right-position" rule for assembling system matrices.

According to the "set-in-right-position" rule for assembling system matrices, the system's stiffness matrix $\boldsymbol{K}$ is derived from the first-order variation of the system's strain energy with respect to displacement coordinates, δU_i; the system's damping matrix $\boldsymbol{C}$ is derived from the first-order variation of the system's potential energy of damping forces with respect to displacement coordinates, δV_c; the system's mass matrix $\boldsymbol{M}$ is derived from the system's first-order variation of the system's potential energy of inertial forces with respect to displacement coordinates, δV_m; and the system's load vector $\boldsymbol{Q}$ is derived from the negative value of the first-order variation of the potential energy of external forces with respect to displacement coordinates, $-\delta V_P$. The element and global matrices can be obtained by applying Eq. (2.78) or Eq. (2.79) to the element and whole structure, respectively. Some parts of a structure, such as the portal frame and lateral bracing in a truss bridge, etc., cannot be considered as an element, and some loads act on nodes rather than elements. If the displacement pattern has been provided for these cases, their influence on the stiffness matrix, mass matrix, damping matrix, and load vector of the system can be considered by the variation of the corresponding strain energy, potential energy of the inertial force, damping force, and applied load by using the "set-in-right-position" rule. These procedures make the job of assembling the matrix equations of motion of complex systems easy and clear.

Obviously, the "set-in-right-position" rule described above is essentially different from the computer programming methods which are generally used in the finite element analysis, because this rule is directly derived from $\delta_\varepsilon \Pi_d = 0$ (the principle of total potential energy with a stationary value in elastic system dynamics). Some examples are given below to illustrate the application of this rule.

Example 2.12: The parameters of the system are shown in Fig. 2.16. Use the principle of total potential energy with a stationary value in elastic system dynamics and the "set-in-right-position" rule for assembling the system matrices to formulate the equations of motion in matrix form.

Solution: Two generalized coordinates of the system are selected as x_1 and x_2, as shown in Fig. 2.16.

The elastic strain energy is given by $U_i = (1/2)k_1x_1^2 + (1/2)k_2(x_2 - x_1)^2$

The potential energy of inertial forces is given by $V_m = -(-m_1\ddot{x}_1)x_1 - (-m_2\ddot{x}_2)x_2$.

The potential energy of damping forces is given by

$$V_c = -(-c_1\dot{x}_1x_1) - [-c_2(\dot{x}_2 - \dot{x}_1)(x_2 - x_1)]$$

The potential energy of external forces is given by $V_P = -P_1x_1 - P_2x_2$

The total potential energy of the system is given by $\Pi_d = U_i + V_m + V_c + V_P$

The variation of the total potential energy with respect to displacement coordinates is as follows:

$$\begin{aligned}\delta_\varepsilon \Pi_d &= k_1x_1\delta x_1 + k_2(x_2 - x_1)(\delta x_2 - \delta x_1) - (-m_1\ddot{x}_1)\delta x_1 - (-m_2\ddot{x}_2)\delta x_2 \\ &\quad - (-c_1\dot{x}_1\delta x_1) - [-c_2(\dot{x}_2 - \dot{x}_1)(\delta x_2 - \delta x_1)] - P_1\delta x_1 - P_2\delta x_2 \\ &= [m_1\ddot{x}_1 + k_1x_1 - k_2(x_2 - x_1) + c_1\dot{x}_1 - c_2(\dot{x}_2 - \dot{x}_1) - P_1]\delta x_1 \\ &\quad + [m_2\ddot{x}_2 + k_2(x_2 - x_1) + c_2(\dot{x}_2 - \dot{x}_1) - P_2]\delta x_2\end{aligned}$$

Considering that $\delta x_1 \neq 0$ and $\delta x_2 \neq 0$, the equations of motion of the system can be obtained in accordance with $\delta_\varepsilon \Pi_d = 0$

$$\begin{cases} m_1\ddot{x}_1 + k_1x_1 - k_2(x_2 - x_1) + c_1\dot{x}_1 - c_2(\dot{x}_2 - \dot{x}_1) = P_1 \\ m_2\ddot{x}_2 + k_2(x_2 - x_1) + c_2(\dot{x}_2 - \dot{x}_1) = P_2 \end{cases}$$

The above equations can be rearranged in matrix form as follows:

$$\begin{bmatrix} m_1 & 0 \\ 0 & m_2 \end{bmatrix}\begin{Bmatrix} \ddot{x}_1 \\ \ddot{x}_2 \end{Bmatrix} + \begin{bmatrix} c_1 + c_2 & -c_2 \\ -c_2 & c_2 \end{bmatrix}\begin{Bmatrix} \dot{x}_1 \\ \dot{x}_2 \end{Bmatrix} + \begin{bmatrix} k_1 + k_2 & -k_2 \\ -k_2 & k_2 \end{bmatrix}\begin{Bmatrix} x_1 \\ x_2 \end{Bmatrix} = \begin{Bmatrix} P_1 \\ P_2 \end{Bmatrix} \tag{2.80}$$

The equations of motion of the system are formulated by the principle of total potential energy with a stationary value in elastic system dynamics first. The obtained equations can then be rewritten in matrix form. The assembly of system matrices in such way is inconvenient for complex structures. The procedure of using the "set-in-right-position" rule for assembling system matrices is introduced below.

First, the variation of the total potential energy of the system with respect to displacement coordinates is written as follows:

$$\begin{aligned}\delta_\varepsilon \Pi_d &= k_1x_1\delta x_1 + k_2(x_2 - x_1)(\delta x_2 - \delta x_1) - (-m_1\ddot{x}_1)\delta x_1 - (-m_2\ddot{x}_2)\delta x_2 \\ &\quad - (-c_1\dot{x}_1\delta x_1) - [-c_2(\dot{x}_2 - \dot{x}_1)(\delta x_2 - \delta x_1)] - P_1\delta x_1 - P_2\delta x_2 \\ &= (\delta x_1k_1x_1 + \delta x_2k_2x_2 - \delta x_2k_2x_1 - \delta x_1k_2x_2 + \delta x_1k_2x_1) \\ &\quad + (\delta x_1c_1\dot{x}_1 + \delta x_2c_2\dot{x}_2 - \delta x_2c_2\dot{x}_1 - \delta x_1c_2\dot{x}_2 + \delta x_1c_2\dot{x}_1) \\ &\quad + (\delta x_1m_1\ddot{x}_1 + \delta x_2m_2\ddot{x}_2) \\ &\quad - (\delta x_1P_1 + \delta x_2P_2)\end{aligned}$$

The variational expression above consists of the following four types of terms. The rules for assembling the system matrices corresponding to each type of terms are as follows:

1. $\delta x_i k x_j$: the stiffness coefficient k is added to the original expression of the ith row and the jth column of stiffness matrix $\boldsymbol{K}$.
2. $\delta x_i c \dot{x}_j$: the damping coefficient c is added to the original expression of the ith row and the jth column of damping matrix $\boldsymbol{C}$.
3. $\delta x_i m \ddot{x}_j$: the mass coefficient m is added to the original expression of the ith row and the jth column of mass matrix $\boldsymbol{M}$.
4. $\delta x_i P$: the load coefficient P is added to the original expression of the ith row of load vector $\boldsymbol{Q}$.

Suppose that the positions of x_1 and x_2 in global displacement are set to be the first and second row, respectively. Then, different terms of the total potential energy of the system are assembled into the corresponding positions of the system matrices according to the above rules. Finally, the same equations of motion as Eq.(2.80) could be obtained.

Example 2.13: A planar continuous beam is shown in Fig. 2.25. Discretize the structure with FEM and formulate its equations of motion in matrix form.

Solution:

The FEM is commonly used to analyze structural dynamic problems. The purpose of this example is to illustrate the basic procedures of formulating the equations of motion of a finite element model. Based on the equations, the dynamic properties and responses of a structure can be evaluated using the methods introduced in the following chapters. These contents

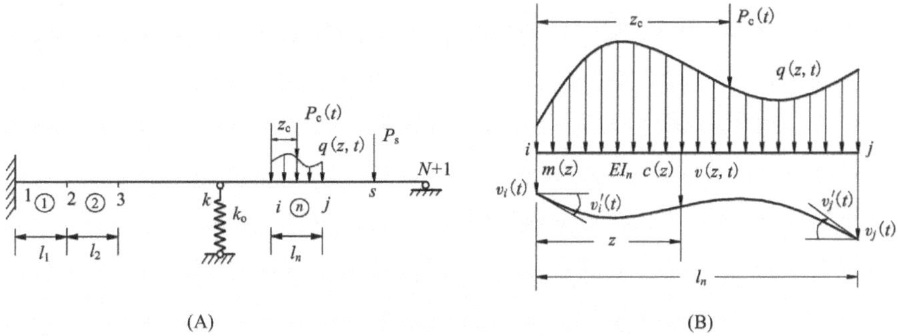

Figure 2.25 Schematic diagram of a planar continuous beam: (A) discretization of a continuous beam; (B) element displacement pattern.

constitute the prototype of the analysis of structural dynamics by using the FEM.

As shown in Fig. 2.25A, the continuous beam can only produce bending deflection in the vertical plane without axial expansion and contraction. The beam is divided into N elements with $(N + 1)$ nodes. The displacement of the nth element is shown in Fig. 2.25B. The vertical translations of nodes i and j are v_i and v_j, respectively, and the downward translation is set to be positive. The rotations of nodes i and j are v_i' and v_j', respectively, and the clockwise direction is set to be positive for rotations. Here, $m(z)$ and $c(z)$ are the mass and the viscous damping coefficient per unit length, respectively.

Solution:

1. Displacement pattern of elements

As shown in Fig. 2.25B, the vector of displacement variables of an element is given by

$$\boldsymbol{q}_e = \begin{Bmatrix} \boldsymbol{q}_i \\ \boldsymbol{q}_j \end{Bmatrix} \tag{2.81}$$

where $\boldsymbol{q}_i = \begin{Bmatrix} v_i \\ v_i' \end{Bmatrix}$, $\boldsymbol{q}_j = \begin{Bmatrix} v_j \\ v_j' \end{Bmatrix}$.

The element vertical deflection function $v(z, t)$ is assumed as

$$v(z, t) = a_0 + a_1 z + a_2 z^2 + a_3 z^3 \tag{2.82}$$

The boundary conditions of the element are given as

$$z = 0, v(0) = v_i, v'(0) = v_i'$$

$$z = l_n, v(l_n) = v_j, v'(l_n) = v_j'$$

Substituting these equations into Eq. (2.82) or its derivative with respect to z, the coefficients, a_0, a_1, a_2, and a_3 can be obtained. Thus one obtains

$$v(z, t) = \boldsymbol{N}\boldsymbol{q}_e \tag{2.83}$$

$$\boldsymbol{N} = \{ N_1 \quad N_2 \quad N_3 \quad N_4 \} \tag{2.84}$$

where

$$N_1 = 1 - 3\left(\frac{z}{l_n}\right)^2 + 2\left(\frac{z}{l_n}\right)^3$$

$$N_2 = z - 2\frac{z^2}{l_n} + \frac{z^3}{l_n^2}$$

$$N_3 = 3\left(\frac{z}{l_n}\right)^2 - 2\left(\frac{z}{l_n}\right)^3$$

$$N_4 = -\frac{z^2}{l_n} + \frac{z^3}{l_n^2}$$

2. Element stiffness matrix

The bending strain energy of the element is given as

$$U_i = \int_0^{l_n} \frac{Md\theta}{2} = \int_0^{l_n} \frac{EI_n v''}{2} \frac{d(v')}{dz} dz = \frac{EI_n}{2} \int_0^{l_n} (v'')^2 dz$$

The variation of U_i with respect to displacement variables is given as follows:

$$\delta_\varepsilon U_i = EI_n \int_0^{l_n} v'' \delta v'' dz$$

Substituting Eq. (2.83) into the above equation, and considering $v'' = \mathbf{N}'' \boldsymbol{q}_e$ and $\delta v'' = \mathbf{N}'' \delta \boldsymbol{q}_e$ yields

$$\delta_\varepsilon U_i = \delta \boldsymbol{q}_e^T \left(EI_n \int_0^{l_n} \mathbf{N}''^T \mathbf{N}'' dz \right) \boldsymbol{q}_e \tag{2.85}$$

Then, the element stiffness matrix could be expressed as

$$\boldsymbol{K}^e = EI_n \int_0^{l_n} \mathbf{N}''^T \mathbf{N}'' dz \tag{2.86}$$

Substituting Eq. (2.84) into Eq. (2.86), and considering $\int_0^{l_n} \mathbf{N}''^T \mathbf{N}'' dz = \int_0^{l_n} \{ N_1'' \;\; z_2'' \;\; N_3'' \;\; N_4'' \}^T \{ N_1'' \;\; N_2'' \;\; N_3'' \;\; N_4'' \} dz$ yields

$$\boldsymbol{K}^e = \begin{matrix} \delta v_i \\ \delta v_i' \\ \delta v_j \\ \delta v_j' \end{matrix} \quad \frac{EI_n}{l_n^3} \begin{matrix} \begin{matrix} v_i & \quad v_i' & \quad v_j & \quad v_j' \end{matrix} \\ \begin{bmatrix} 12 & 6l_n & -12 & 6l_n \\ 6l_n & 4l_n^2 & -6l_n & 2l_n^2 \\ -12 & -6l_n & 12 & -6l_n \\ 6l_n & 2l_n^2 & -6l_n & 4l_n^2 \end{bmatrix} \end{matrix} \tag{2.87}$$

3. Element damping matrix

In this example, external viscous damping is considered. The potential energy of the element damping forces is given as

$$V_c = \int_0^{l_n} c(z) \dot{v}(z,t) v(z,t) dz \tag{2.88}$$

where

$$\dot{v}(z,t) = \frac{dv(z,t)}{dt} = \frac{d}{dt}\left(\mathbf{N}\boldsymbol{q}_e\right) = \mathbf{N}\dot{\boldsymbol{q}}_e \tag{2.89}$$

The variation of V_c with respect to displacement variables is given as follows:

$$\delta_\varepsilon V_c = \int_0^{l_n} c(z)\dot{v}(z,t)\delta v(z,t)dz = \delta \boldsymbol{q}_e^T \left[\int_0^{l_n} c(z)\mathbf{N}^T\mathbf{N}dz\right]\dot{\boldsymbol{q}}_e \tag{2.90}$$

Then, the element damping matrix could be expressed

$$\mathbf{C}^e = \int_0^{l_n} c(z)\mathbf{N}^T\mathbf{N}dz \tag{2.91}$$

When the distributed damping coefficient $c(z)$ is a constant, that is, $c(z) = c$, Eq. (2.91) can be rewritten as

$$\mathbf{C}^e = c\int_0^{l_n} \mathbf{N}^T\mathbf{N}dz = c\begin{bmatrix} \int_0^{l_n} N_1^2 dz & \int_0^{l_n} N_1N_2 dz & \int_0^{l_n} N_1N_3 dz & \int_0^{l_n} N_1N_4 dz \\ \int_0^{l_n} N_2N_1 dz & \int_0^{l_n} N_2^2 dz & \int_0^{l_n} N_2N_3 dz & \int_0^{l_n} N_2N_4 dz \\ \int_0^{l_n} N_3N_1 dz & \int_0^{l_n} N_3N_2 dz & \int_0^{l_n} N_3^2 dz & \int_0^{l_n} N_3N_4 dz \\ \int_0^{l_n} N_4N_1 dz & \int_0^{l_n} N_4N_2 dz & \int_0^{l_n} N_4N_3 dz & \int_0^{l_n} N_4^2 dz \end{bmatrix}$$

$$= \begin{matrix} \delta v_i \\ \delta v_i' \\ \delta v_j \\ \delta v_j' \end{matrix} \quad \frac{cl_n}{420} \begin{matrix} \dot{v}_i & \dot{v}_i' & \dot{v}_j & \dot{v}_j' \end{matrix} \begin{bmatrix} 156 & 22l_n & 54 & -13l_n \\ 22l_n & 4l_n^2 & 13l_n & -3l_n^2 \\ 54 & 13l_n & 156 & -22l_n \\ -13l_n & -3l_n^2 & -22l_n & 4l_n^2 \end{bmatrix} \tag{2.92}$$

4. Element mass matrix

The potential energy of element inertial forces is given as

$$V_m = \int_0^{l_n} m(z)\ddot{v}(z,t)v(z,t)dz$$

where

$$\ddot{v}(z,t) = \frac{d^2v(z,t)}{dt^2} = \frac{d^2}{dt^2}\left(\mathbf{N}\boldsymbol{q}_e\right) = \mathbf{N}\ddot{\boldsymbol{q}}_e \tag{2.93}$$

The variation of V_m with respect to displacement variables is given as

$$\delta_\varepsilon V_m = \int_0^{l_n} m(z)\ddot{v}(z,t)\delta v(z,t)dz = \delta \boldsymbol{q}_e^T \left[\int_0^{l_n} m(z)\mathbf{N}^T\mathbf{N}dz\right]\ddot{\boldsymbol{q}}_e \tag{2.94}$$

Then, the element mass matrix can be expressed as

$$\boldsymbol{M}^e = \int_0^{l_n} m(z)\boldsymbol{N}^T\boldsymbol{N}dz \tag{2.95}$$

When the distributed mass $m(z)$ is constant, that is, $m(z) = m$, Eq. (2.95) can be written as

$$\boldsymbol{M}^e = m\int_0^{l_n} \boldsymbol{N}^T\boldsymbol{N}dz = m\begin{bmatrix} \int_0^{l_n} N_1^2 dz & \int_0^{l_n} N_1N_2 dz & \int_0^{l_n} N_1N_3 dz & \int_0^{l_n} N_1N_4 dz \\ \int_0^{l_n} N_2N_1 dz & \int_0^{l_n} N_2^2 dz & \int_0^{l_n} N_2N_3 dz & \int_0^{l_n} N_2N_4 dz \\ \int_0^{l_n} N_3N_1 dz & \int_0^{l_n} N_3N_2 dz & \int_0^{l_n} N_3^2 dz & \int_0^{l_n} N_3N_4 dz \\ \int_0^{l_n} N_4N_1 dz & \int_0^{l_n} N_4N_2 dz & \int_0^{l_n} N_4N_3 dz & \int_0^{l_n} N_4^2 dz \end{bmatrix}$$

$$= \begin{matrix} \delta v_i \\ \delta v_i' \\ \delta v_j \\ \delta v_j' \end{matrix} \quad \frac{ml_n}{420} \begin{matrix} \ddot{v}_i & \ddot{v}_i' & \ddot{v}_j & \ddot{v}_j' \end{matrix} \begin{bmatrix} 156 & 22l_n & 54 & -13l_n \\ 22l_n & 4l_n^2 & 13l_n & -3l_n^2 \\ 54 & 13l_n & 156 & -22l_n \\ -13l_n & -3l_n^2 & -22l_n & 4l_n^2 \end{bmatrix} \tag{2.96}$$

5. Element load vector

The potential energy of element external forces is given as

$$V_P = -P_c v(c) - \int_0^{l_n} qv(z,t)dz = -P_c v(c) - \int_0^{l_n} q\boldsymbol{N}\boldsymbol{q}_e dz$$

The variation of V_P with respect to displacement variables is given as

$$\delta_\varepsilon V_P = \delta\boldsymbol{q}_e^T\left(-P_c\boldsymbol{N}_{z=z_c}^T - \int_0^{l_n} q\boldsymbol{N}^T dz\right)$$

Then, the element load vector can be expressed as

$$\boldsymbol{Q}^e = \int_0^{l_n} q\boldsymbol{N}^T dz + P_c\boldsymbol{N}_{z=z_c}^T \tag{2.97}$$

Substituting Eq. (2.84) into Eq. (2.97) yields

$$\boldsymbol{Q}^e = \begin{matrix} \delta v_i \\ \\ \delta v_i' \\ \\ \delta v_j \\ \\ \delta v_j' \end{matrix} \begin{Bmatrix} \dfrac{ql_n}{2} + (N_1)_{z=z_c}P_c \\ \\ \dfrac{ql_n^2}{12} + (N_2)_{z=z_c}P_c \\ \\ \dfrac{ql_n}{2} + (N_3)_{z=z_c}P_c \\ \\ -\dfrac{ql_n^2}{12} + (N_4)_{z=z_c}P_c \end{Bmatrix}$$

6. Global matrices of the system

The total potential energy of the beam at any instant of time t includes not only the sum of the total potential energy of all the elements (the element strain energy, the potential energy of the element inertial, damping, and external forces), but also the potential energies of the spring k_0 and the external load P_s acting on node s. The total potential energy of the beam can be given as

$$\Pi_d = \sum_{n=1}^{N} U_i + \sum_{n=1}^{N} V_m + \sum_{n=1}^{N} V_c + \sum_{n=1}^{N} V_P + \frac{1}{2}k_0 v_k^2 - P_s v_s$$

where k_0 is the stiffness coefficient of the supporting spring, v_k is the vertical displacement of node k, P_s is the vertical concentrated load acting on the node s, and v_s is the vertical displacement of node s.

According to the principle of total potential energy with a stationary value in elastic system dynamics, one obtains

$$\delta_\varepsilon \Pi_d = \sum_{n=1}^{N} \delta_\varepsilon U_i + \sum_{n=1}^{N} \delta_\varepsilon V_m + \sum_{n=1}^{N} \delta_\varepsilon V_c + \sum_{n=1}^{N} \delta_\varepsilon V_P + k_0 v_k \delta v_k - P_s \delta v_s = 0$$

which may be expressed alternatively as

$$\sum_{n=1}^{N} \delta \boldsymbol{q}_e^T \boldsymbol{K}^e \boldsymbol{q}_e + \sum_{n=1}^{N} \delta \boldsymbol{q}_e^T \boldsymbol{M}^e \ddot{\boldsymbol{q}}_e + \sum_{n=1}^{N} \delta \boldsymbol{q}_e^T \boldsymbol{C}^e \dot{\boldsymbol{q}}_e - \sum_{n=1}^{N} \delta \boldsymbol{q}_e^T \boldsymbol{Q}^e + k_0 v_k \delta v_k - P_s \delta v_s = 0 \tag{2.98}$$

Factoring out the displacement variation $\delta \boldsymbol{q}^T$, and rearranging Eq. (2.98) in accordance with the "set-in-right-position" rule, leads to

$$\delta \boldsymbol{q}^T (\boldsymbol{M}\ddot{\boldsymbol{q}} + \boldsymbol{C}\dot{\boldsymbol{q}} + \boldsymbol{K}\boldsymbol{q}) = \delta \boldsymbol{q}^T \boldsymbol{Q} \tag{2.99}$$

where

$$\begin{cases} \boldsymbol{M} = \sum_{n=1}^{N} \boldsymbol{M}^e \\ \boldsymbol{C} = \sum_{n=1}^{N} \boldsymbol{C}^e \\ \boldsymbol{K} = \sum_{n=1}^{N} \boldsymbol{K}^e + k_0(\delta v_k \cdot v_k) \\ \boldsymbol{Q} = \sum_{n=1}^{N} \boldsymbol{Q}^e + P_s(\delta v_s) \end{cases} \tag{2.100}$$

where $\boldsymbol{M}$, $\boldsymbol{C}$, $\boldsymbol{K}$, and $\boldsymbol{Q}$ are the system's mass matrix, damping matrix, stiffness matrix, and load vector, respectively; $\boldsymbol{q}$, $\dot{\boldsymbol{q}}$, and $\ddot{\boldsymbol{q}}$ are the displacement, velocity, and acceleration vectors, respectively; and $\delta\boldsymbol{q}$ is the first-order variation of displacement vector.

Eq. (2.100) indicates again that the "set-in-right-position" rule is not mathematical addition for matrices. The term of $k_0(\delta v_k \cdot v_k)$ indicates that the stiffness coefficient k_0 should be added to the entry corresponding to $\delta v_k \cdot v_k$ in the global stiffness matrix. The term of $P_s(\delta v_s)$ indicates that the load P_s should be added to the entry corresponding to δv_s in the global load vector. Considering $\delta\boldsymbol{q}^T \neq \boldsymbol{0}$, the structural dynamic equilibrium equation in matrix form can be expressed as

$$\boldsymbol{M}\ddot{\boldsymbol{q}} + \boldsymbol{C}\dot{\boldsymbol{q}} + \boldsymbol{K}\boldsymbol{q} = \boldsymbol{Q} \tag{2.101}$$

Note that the local coordinate system of the element is completely consistent with the global coordinate system in this example. Therefore no transformation of coordinates is required before assembling element matrices to the global matrices. Otherwise, it is necessary to make the transformation. Detailed information can be found in references related to FEM.

Boundary conditions have not been introduced in Eq. (2.101). For the dynamic finite element problem, boundary conditions are imposed by setting large values for relevant entries or deleting rows and columns in global matrices (vector). As shown in Fig. 2.25A, the nodes 1 and $(N+1)$ are constrained to a certain extent. The vertical translation and the rotation of node 1, as well as the vertical translation of node $(N+1)$, are equal to zero. In the technique of setting large values, a large value k_∞ could be added to the diagonal entry of the system's stiffness matrix corresponding to the constrained displacement variable. The physical meaning can be explained as follows: a spring with the stiffness coefficient k_∞ is imposed to the displacement variable v which needs to be constrained, and the resulting value of v inevitably equals zero due to the large value of the stiffness coefficient. The corresponding strain energy of the spring is $1/2k_\infty v^2$, and its variation with respect to v is $k_\infty \delta v \cdot v$. According to the "set-in-right-position" rule, k_∞ should be added to the entry of the system's stiffness matrix corresponding to $\delta v \cdot v$.

The technique of deleting relevant rows and columns introduces the given supporting conditions into the equations of motion of the structure by means of some necessary revisions. The specific procedure is to remove the rows and columns of the system's matrices which correspond to the constrained displacement variables. The structural displacement vector $\boldsymbol{q}$ is divided into two parts, namely, $\boldsymbol{q}_0$ including all known-displacement entries, and $\boldsymbol{q}_1$ including the remaining unknown-displacement entries, giving

$$\boldsymbol{q} = \begin{Bmatrix} \boldsymbol{q}_0 \\ \boldsymbol{q}_1 \end{Bmatrix} \tag{2.102}$$

Accordingly, the total load vector of the system could be rewritten as follows:

$$\mathbf{Q} = \begin{Bmatrix} \mathbf{Q}_0 \\ \mathbf{Q}_1 \end{Bmatrix} \tag{2.103}$$

$\mathbf{Q}_0$ in Eq. (2.103), corresponding to the reactive forces of the constraints on the structure, is associated with the known-displacement vector $\boldsymbol{q}_0$. Substituting Eqs. (2.102) and (2.103) into the equations of motion, and rearranging the equation in partitioned matrix form, leads to

$$\begin{bmatrix} \boldsymbol{M}_{00} & \boldsymbol{M}_{01} \\ \boldsymbol{M}_{10} & \boldsymbol{M}_{11} \end{bmatrix} \begin{Bmatrix} \ddot{\boldsymbol{q}}_0 \\ \ddot{\boldsymbol{q}}_1 \end{Bmatrix} + \begin{bmatrix} \boldsymbol{C}_{00} & \boldsymbol{C}_{01} \\ \boldsymbol{C}_{10} & \boldsymbol{C}_{11} \end{bmatrix} \begin{Bmatrix} \dot{\boldsymbol{q}}_0 \\ \dot{\boldsymbol{q}}_1 \end{Bmatrix} + \begin{bmatrix} \boldsymbol{K}_{00} & \boldsymbol{K}_{01} \\ \boldsymbol{K}_{10} & \boldsymbol{K}_{11} \end{bmatrix} \begin{Bmatrix} \boldsymbol{q}_0 \\ \boldsymbol{q}_1 \end{Bmatrix} = \begin{Bmatrix} \mathbf{Q}_0 \\ \mathbf{Q}_1 \end{Bmatrix} \tag{2.104}$$

Once the known-displacement vector $\boldsymbol{q}_0 = \boldsymbol{0}$ is given, Eq. (2.104) can be written as two separate equations

$$\boldsymbol{M}_{11}\ddot{\boldsymbol{q}}_1 + \boldsymbol{C}_{11}\dot{\boldsymbol{q}}_1 + \boldsymbol{K}_{11}\boldsymbol{q}_1 = \mathbf{Q}_1 \tag{2.105}$$

$$\boldsymbol{M}_{01}\ddot{\boldsymbol{q}}_1 + \boldsymbol{C}_{01}\dot{\boldsymbol{q}}_1 + \boldsymbol{K}_{01}\boldsymbol{q}_1 = \mathbf{Q}_0 \tag{2.106}$$

Eq. (2.105) is the equations of motion with boundary conditions embedded. The unknown displacement vector $\boldsymbol{q}_1$ can be solved from Eq. (2.105). The dynamic reactive force $\mathbf{Q}_0$ can be obtained by substituting the obtained responses $\boldsymbol{q}_1$ into Eq. (2.106). Therefore Eq. (2.106) is also called the equations of dynamic reactive forces.

The above treatment of boundary conditions is referred to as the technique of deleting rows and columns in program design. That is, the rows and columns corresponding to the constrained displacement variables are deleted from the stiffness, mass, damping matrices, and load vector. Similarly, the equations of motion of a nonsingular system can be formulated as Eq. (2.105).

2.11.2 Method of computer implementation in Matlab for assembling system matrices

Based on the principle of total potential energy with a stationary value in elastic system dynamics and the "set-in-right-position" rule for formulating system matrices, the system matrices (element or global matrix) can be obtained conveniently through the symbolical computation in Matlab. The planar beam element is used as an example here to illustrate the idea and procedures. The source code of the program is attached at the end of this section.

In order to formulate element matrices by means of the Matlab program, the preparations include the following:

1. The node displacement variables of the element are selected. Then, the corresponding variables of velocity and acceleration can be determined. The node displacement variables of the planar beam element are v_i, v_i', v_j, and v_j'. The corresponding velocity variables are $\dot{v}_i$, $\dot{v}_i'$, $\dot{v}_j$, and $\dot{v}_j'$, and the corresponding acceleration variables are $\ddot{v}_i$, $\ddot{v}_i'$, $\ddot{v}_j$, and $\ddot{v}_j'$.
2. The displacement, velocity, and acceleration of the element are determined as follows:

$$v(z,t) = \mathbf{N}\boldsymbol{q}_e, \dot{v}(z,t) = \mathbf{N}\dot{\boldsymbol{q}}_e, \ddot{v}(z,t) = \mathbf{N}\ddot{\boldsymbol{q}}_e \tag{2.107}$$

where

$$\mathbf{N} = \{ N_1 \quad N_2 \quad N_3 \quad N_4 \}, \boldsymbol{q}_e = \{ v_i \quad v_i' \quad v_j \quad v_j' \}^T$$

$$\dot{\boldsymbol{q}}_e = \{ \dot{v}_i \quad \dot{v}_i' \quad \dot{v}_j \quad \dot{v}_j' \}^T, \ddot{\boldsymbol{q}}_e = \{ \ddot{v}_i \quad \ddot{v}_i' \quad \ddot{v}_j \quad \ddot{v}_j' \}^T$$

3. The total potential energy of the element is expressed as the function of node variables of displacement, velocity, and acceleration.
The bending strain energy of the element is

$$U_i = \frac{EI_n}{2}\int_0^{l_n} (v'')^2 dz \tag{2.108}$$

The potential energy of inertial forces of the element is

$$V_m = \int_0^{l_n} m\ddot{v}\,(z,t)v(z,t)dz \tag{2.109}$$

The potential energy of damping forces of the element is

$$V_c = \int_0^{l_n} c\dot{v}(z,t)v(z,t)dz \tag{2.110}$$

The total potential energy of the element can be obtained as follows:

$$\Pi_d = U_i + V_c + V_m \tag{2.111}$$

It should be noted that the specific differentiations and integrals will be done in Matlab. Based on the above preparations, the detailed procedures to formulate the element matrix by means of Matlab program are presented as follows:

- The node displacement variables and their corresponding vector $\boldsymbol{q}_e$, the node velocity variables and their corresponding vector $\dot{\boldsymbol{q}}_e$, and the node acceleration variables and their corresponding vector $\ddot{\boldsymbol{q}}_e$ of the element are symbolically expressed. The sequence of the node displacement variables in the vector $\boldsymbol{q}_e$ determines the position of the terms in the element stiffness (damping and mass) matrix. The sequences of variables in the vectors $\boldsymbol{q}_e$, $\dot{\boldsymbol{q}}_e$, and $\ddot{\boldsymbol{q}}_e$ should be consistent.
- The displacement, velocity, and acceleration variables of nodes are used to express the displacement, velocity, and acceleration of the element, respectively. That is, the element's displacement, velocity, and acceleration are described by symbolical expressions, respectively. For example, the symbolical expressions of $v(z, t)$, $\dot{v}(z, t)$, and $\ddot{v}(z, t)$ of the planar beam element should be written.
- Based on the symbolical expressions of the displacement, velocity, and acceleration of the element, the symbolical expression of the total potential energy Π_d of the element (including the elastic strain energy, potential energy of damping forces, and the potential energy of inertial forces) can be obtained.
- The variation of the total potential energy Π_d is taken with respect to displacement variables, and all the terms in element matrices (stiffness coefficient k_{ij}^e, damping coefficient c_{ij}^e, and mass coefficient m_{ij}^e) can be obtained in the form of Matlab symbol. The specific steps are as follows: (1) the symbolical expression of Π_d is differentiated with respect to the ith variable in $\boldsymbol{q}_e$, and the symbolical expression of the first-order partial derivative of Π_d is obtained, denoted by $\Pi_{d,q_{ei}}$; (2) the symbolical expression of $\Pi_{d,q_{ei}}$ is differentiated with respect to the jth variable in $\boldsymbol{q}_e$, and the resulting derivative is the symbolical expression of k_{ij}^e; (3) the symbolical expression of $\Pi_{d,q_{ei}}$ is differentiated with respect to the jth variable in $\dot{\boldsymbol{q}}_e$, and the resulting derivative is the symbolical expression of c_{ij}^e; (4) the symbolical expression of $\Pi_{d,q_{ei}}$ is differentiated with respect to the jth variable in $\ddot{\boldsymbol{q}}_e$, and the resulting derivative is the symbolical expression of m_{ij}^e; (5) the above process is repeated, and all symbolical expressions of the elements of stiffness, damping, and mass matrices can be obtained. Finally, the element stiffness, damping, and mass matrix can be fully expressed by Matlab symbols.

```
[ (12*E*In)/ln^3,     (6*E*In)/ln^2,    -(12*E*In)/ln^3,     (6*E*In)/ln^2]
[   (6*E*In)/ln^2,     (4*E*In)/ln,      -(6*E*In)/ln^2,      (2*E*In)/ln]
[-(12*E*In)/ln^3,    -(6*E*In)/ln^2,     (12*E*In)/ln^3,  -(6*E*In)/ln^2]
[  (6*E*In)/ln^2,     (2*E*In)/ln,       -(6*E*In)/ln^2,      (4*E*In)/ln]
```

(A)

```
[   (13*c*ln)/35, (11*c*ln^2)/210,    (9*c*ln)/70,       -(13* c*ln^2)/420]
[ (11*c*ln^2)/210,     (c*ln^3)/105,   (13*c*ln^2)/420,     -(c*ln^3)/140]
[    (9*c*ln)/70, (13*c*ln^2)/420,    (13*c*ln)/35,      - (11*c*ln^2)/210]
[ -(13*c*ln^2)/420, - (c*ln^3)/140,   -(11*c*ln^2)/210,   (c*ln^3)/105]
```

(B)

```
[   (13*ln*m)/35, (11*ln^2*m)/210,     (9*ln*m)/70,   - (13*ln^2*m)/420]
[ (11*ln^2*m)/210,   (ln^3*m)/105, (13*ln^2*m)/420,       - (ln^3*m)/140]
[    (9*ln*m)/70, (13*ln^2*m)/420,    (13*ln*m)/35,   - (11*ln^2*m)/210]
[ - (13*ln^2*m)/420,  - (ln^3*m)/140, - (11*ln^2*m)/210,   (ln^3*m)/105]
```

(C)

Figure 2.26 Automatically generated element matrices: (A) element stiffness matrix; (B) element damping matrix; (C) element mass matrix.

The element stiffness, damping, and mass matrices of the planar beam in the form of Matlab symbol, are shown in Fig. 2.26A, B, and C, respectively. The results are exactly the same as the manual derivation. These generated matrices can be directly embedded into the relevant Matlab program, which will provide modules for dynamic analysis program of FEM.

The planar beam element has only four DOFs and its element matrices consist of four rows and four columns. The effort of manual derivation is not too significant. However, a large number of DOFs for elements, a complex element displacement pattern, and the appearance of a higher-order derivative and multiple integrals in the total potential energy expression would lead to cumbersome effort in manual derivation. According to the above procedure, only the displacement pattern and the expression of the total potential energy of the element are required, and the element matrices could be formulated conveniently and accurately. When the expression of the total potential energy includes all the elastic strain energy, potential energy of damping forces, and potential energy of inertial forces of a structural system, the corresponding system matrices could be derived by the same method.

The program for formulating system matrices of the planar beam element is listed as follows:

```
%*************************************************************
% Program for formulating automatically element stiffness, mass
% and damping matrices of planar beam
%*************************************************************

function[K,C,M]=KCM(~)
syms vi vi_z vj vj_z;
syms vi_t vi_zt vj_t vj_zt;
syms vi_tt vi_ztt vj_tt vj_ztt;
syms z ln E In c m;
%-----------------------------------------------------------------
N1=1-3*(z/ln)^2+2*z^3/ln^3;
N2=z-2*z^2/ln+z^3/(ln^2);
N3=3*(z/ln)^2-2*(z/ln)^3;
N4=-z^2/ln+z^3/(ln^2);
N=[N1,N2,N3,N4]; %shape function
qe=[vi,vi_z,vj,vj_z]; %vector of node displacement variables
qe_t=[vi_t,vi_zt,vj_t,vj_zt]; %vector of node velocity variables
qe_tt=[vi_tt,vi_ztt,vj_tt,vj_ztt]; %vector of node acceleration
variables
vz=N*qe'; %vertical displacement function of element
vz_zz=diff(vz,z,2); %second-order derivative of vz with respect to z
Nqe=size(qe,2); %number of DOFs of element
%-----------------------------------------------------------------
%bending strain energy Ui is obtained by integral from 0 to ln with
respect to z
Ui=int(1/2*(E*In*vz_zz^2),z,0, ln);
%potential energy of element damping force Vc is obtained by integral
from 0 to ln with respect to z
Vc=int(c*N*qe_t'*vz,z,0, ln);
%potential energy of element inertial force Vm is obtained by integral
from 0 to ln with respect to z
Vm=int(m*N*qe_tt'*vz,z,0, ln);
%total potential energy of element at any instant of time
Ptotal=Ui+Vc+Vm;
%-----------------------------------------------------------------
Ke=sym((zeros(Nqe,Nqe)));
Ce=sym((zeros(Nqe,Nqe)));
Me=sym((zeros(Nqe,Nqe)));
for i=1:Nqe
Ptotal_qei=diff(Ptotal,qe(i));
for j=1:Nqe
Ke(i,j)=diff(Ptotal_qei,qe(j)); % element stiffness matrix
Ce(i,j)=diff(Ptotal_qei,qe_t(j)); % element damping matrix
```

```
Me(i,j)=diff(Ptotal_qei,qe_tt(j)); % element mass matrix
end
end
Ke; %output element stiffness matrix
Ce; % output element damping matrix
Me; % output element mass matrix
%-----------------------------------------------------------
```

Note that the above program is also applicable to the formulation of the matrices of other types of elements or systems with proper revision.

References

[1] Qiu B. Analytical mechanics. Beijing: China Railway Press; 1998.
[2] Clough RW, Penzien J. Dynamics of structures. 3rd ed. CA: Computers & Structures, Inc; 2003.
[3] Liu J, Du X. Dynamics of structures. Beijing: China Machine Press; 2007.
[4] Zeng Q. Principle of total potential energy with a stationary value in elastic system dynamics. J Huazhong Univ Sci Technol 2000;1:1−3.
[5] Przemienieck JS. Theory of matrix structural analysis. Beijing: National Defense Industry Press; 1974.
[6] Bleich F. Buckling strength of metal structures. Beijing: Science Press; 1965.
[7] Zeng Q, Yang P. The "set-in-right-position" rule for assembling system matrices and finite element method for space analysis of truss bridge. J China Railw Soc 1986;2:48−59.
[8] Zhou Z, Wen Y, Zeng Q. Lectures on dynamics of structures. 2nd ed. Beijing: China Communications Press Co., Ltd; 2017.

Problems

2.1 Explain the difference and relationship between the principle of total potential energy with a stationary value in elastic system dynamics and principle of virtual displacements in dynamics.

2.2 Compare the "set-in-right-position" rule for assembling the system matrices in this book with the "set-in-right-position" rule introduced for general structural mechanics.

2.3 Explain the difference and relationship between the principles of virtual displacements in dynamics and statics.

2.4 A uniform rigid rod with total mass m_1 and length L swings under the action of gravity. A mass m_2 slides along the rod axis and a mass-free spring with stiffness k_2 is connected to the pendulum axis with free length b. It is assumed that the system is frictionless and a large swing angle is considered. The generalized coordinates q_1 and q_2 are shown in Fig. P2.1. Determine the generalized forces corresponding to all the forces acting on the system.

2.5 A vehicle is simplified as a system shown in Fig. P2.2. Vertical vibration occurs when it travels on an uneven track. The irregularity is expressed as $x_s = a \sin \omega t$. When ω is 0.707 times that of the fundamental frequency of the vehicle, evaluate the proportional relationship between the amplitude of the vehicle body and that of unevenness, a. It is known that $M_1 = 4500$ kg, $M_2 = 4500$ kg, $k_1 = 1.683 \times 10^7$ N/m, $k_2 = 3.136 \times 10^8$ N/m.

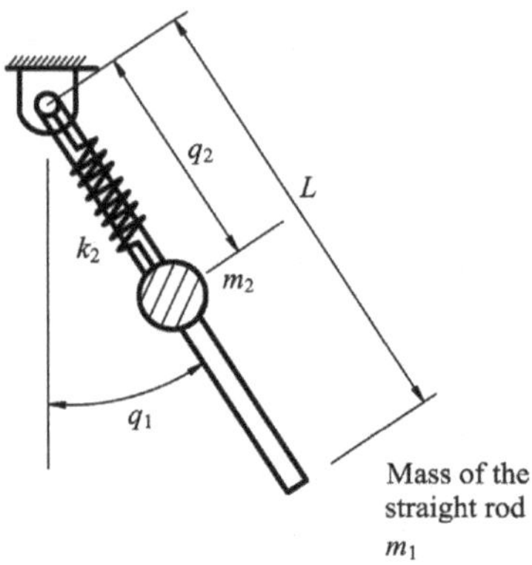

Figure P2.1 Figure of problem 2.4.

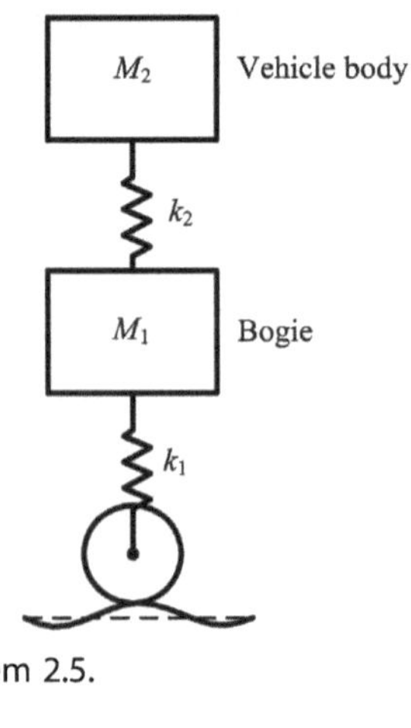

Figure P2.2 Figure of problem 2.5.

2.6 A Winkler beam on an elastic foundation is shown in Fig. P2.3A. The bending deflection in the vertical plane is analyzed by FEM. The beam on the elastic foundation is divided into several planar beam elements with continuously elastic support, as shown in Fig. P2.3B. The length of each element is l, the flexural stiffness is EI, the mass per unit length is $\overline{m}$, and the stiffness coefficient per unit length of the elastic foundation is k_0. The displacement variables of an element are shown in Fig. P2.3C. Derive the stiffness and mass matrices of the element using the principle of total potential energy with a stationary value in elastic system dynamics and the "set-in-right-position" rule for assembling the system matrices.

2.7 A simply supported beam of straddle monorail transit is shown in Fig. P2.4, and only the vertical vibration is investigated. Parameters of

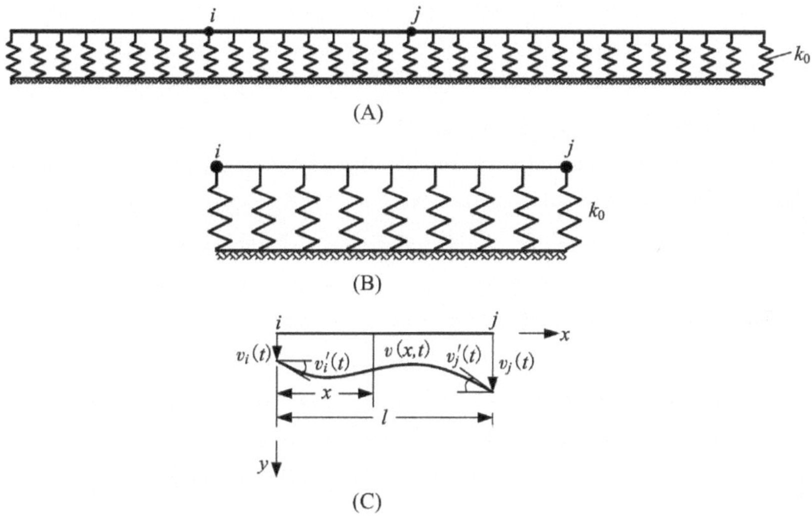

Figure P2.3 Figure of problem 2.6: (A) beam on the elastic foundation; (B) planar beam element with continuously elastic support; (C) element displacement pattern.

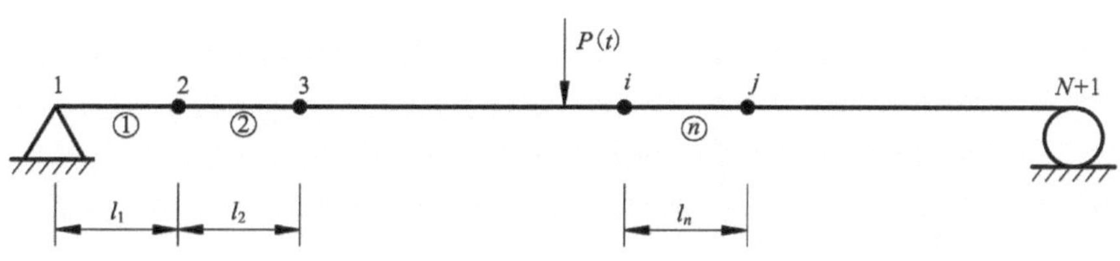

Figure P2.4 Figure of problem 2.7.

the beam are given as follows: the span length $L = 22$ m, the elastic modulus $E = 3.96 \times 10^{10}$ Pa, the moment of inertia $I = 1.586 \times 10^{5}$ m^4, the cross section area $A = 1.0735$ m^2, and the density $\rho = 2551$ kg/m^3. A vertical force $P(t) = P_0 \sin \omega t$ acts at the middle of the beam. Divide the simply supported beam into several planar beam elements and formulate its equations of motion in matrix form.

CHAPTER 3

Analysis of dynamic response of SDOF systems

The general concepts of structural vibration analysis were obtained from the research on the vibration of single-degree-of-freedom (SDOF) systems. These physical concepts are the basis of vibration analysis, which are important for understanding the basic theory of vibration and the dynamic performance of structures. First, the free vibrations of undamped and damped SDOF systems are discussed. Second, the vibration responses of SDOF systems to various external dynamic loads, including harmonic loads, impulsive loads, periodic loads, and base motion, are analyzed. Finally, the methods of time and frequency domains analysis of dynamic response to arbitrary dynamic loads are introduced, including the Duhamel integral method and the Fourier integral method. In addition, typical damping theories and the evaluation of viscous-damping ratio are also introduced in the present chapter.

3.1 Analysis of free vibrations

3.1.1 Undamped free vibrations

A spring–mass system is shown in Fig. 3.1, where the mass of the spring is neglected. The free-vibration equation can be obtained by formulating

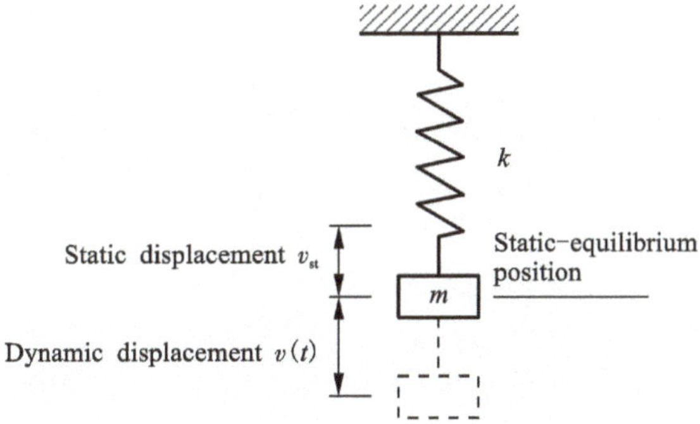

Figure 3.1 Free vibration of a SDOF system. *SDOF*, Single-degree-of-freedom.

Fundamentals of Structural Dynamics
DOI: https://doi.org/10.1016/B978-0-12-823704-5.00003-3

the dynamic equilibrium equation of the mass block. The mass block moves up and down in the vertical plane, and the elastic force of the spring, the inertial force, and the gravity of the mass block should be in equilibrium at any instant of time, that is,

$$m\ddot{v} + k(v + v_{st}) - mg = 0$$

where m is the mass of the mass block, k is the stiffness coefficient of the spring, g is the acceleration of gravity, v_{st} represents the static displacement, and v represents the dynamic displacement at time t measured from static-equilibrium position.

Considering the initial static equilibrium of the mass block, $kv_{st} - mg = 0$, the above equation can be simplified as

$$m\ddot{v} + kv = 0 \tag{3.1}$$

Since Eq. (3.1) is a second-order homogeneous linear differential equation with constant coefficients, its general solution can be expressed as

$$v(t) = C_1 \cos \omega t + C_2 \sin \omega t \tag{3.2}$$

where $\omega = \sqrt{k/m}$, C_1 and C_2 are constants yet undetermined. The values of these two constants can be determined from the initial conditions, that is, the displacement $v(0)$ and velocity $\dot{v}(0)$ at time $t = 0$. Obviously, $v(0) = C_1$ and $\dot{v}(0) = C_2\omega$; one obtains $C_1 = v(0)$ and $C_2 = \dot{v}(0)/\omega$. Eq. (3.2) becomes

$$v = v(0)\cos \omega t + \frac{\dot{v}(0)}{\omega} \sin \omega t \tag{3.3}$$

Eq. (3.3) can be expressed as

$$v = \rho \cos (\omega t - \theta) \tag{3.4}$$

where

$$\begin{cases} \rho = \sqrt{[v(0)]^2 + \left[\frac{\dot{v}(0)}{\omega}\right]^2} \\ \theta = \tan^{-1} \dfrac{\dot{v}(0)}{\omega v(0)} \end{cases} \tag{3.5}$$

The vibration response as described by Eq. (3.4) is plotted in Fig. 3.2. ρ and θ are the amplitude and phase angle of the motion, respectively.

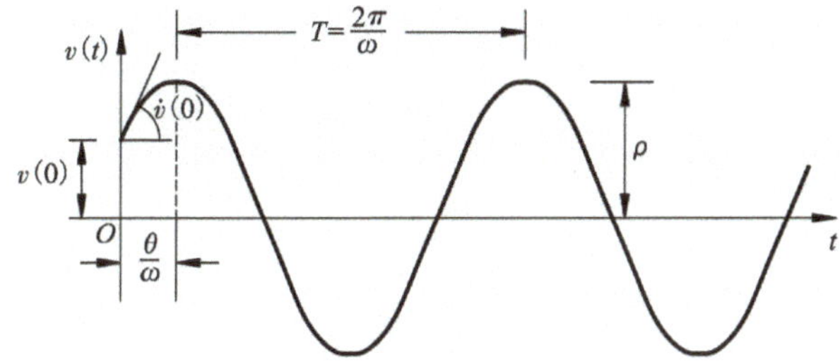

Figure 3.2 Undamped free-vibration response.

ω is the circular frequency of the motion, which is measured in rad/s. For a given dynamic system, k and m are determined, so $\omega = \sqrt{k/m}$ is a constant which is called natural circular frequency. The associated natural period is $T = 2\pi/\omega$, and natural cyclic frequency is $f = 1/T$ measured in Hz.

The natural circular frequency, natural cyclic frequency, and natural period can be expressed in the alternative form

$$f = \frac{1}{2\pi}\sqrt{\frac{g}{v_{st}}}, \quad \omega = \sqrt{\frac{g}{v_{st}}}, \quad T = 2\pi\sqrt{\frac{v_{st}}{g}}$$

where $v_{st} = mg/k$, and g is the acceleration due to gravity. These equations further illustrate that the free-vibration characteristics of the system are completely determined from the parameters of the system, which is independent of the initial conditions.

3.1.2 Damped free vibrations

As shown in Fig. 3.3, a mass–spring–damper system is obtained by introducing a viscous damper into the system in Fig. 3.1. As the viscous damper exerts a viscous-damping force on the moving mass block (see Section 3.4 for damping theory), the direction of the damping force is opposite to that of the velocity of the mass block. The free-vibration equation for the damped system is

$$m\ddot{v} + c\dot{v} + kv = 0$$

Considering $\omega^2 = k/m$, one obtains

$$\ddot{v} + \frac{c}{m}\dot{v} + \omega^2 v = 0 \tag{3.6}$$

where c is the viscous-damping coefficient.

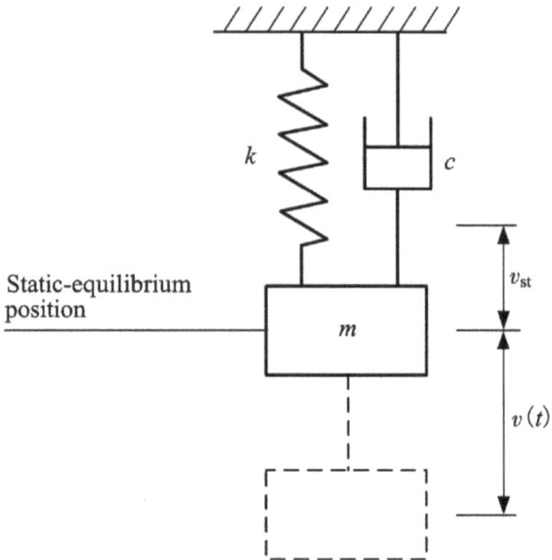

Figure 3.3 Free vibration of a damped SDOF system. *SDOF*, Single-degree-of-freedom.

The solution of Eq. (3.6) has the form

$$v = Ge^{St} \tag{3.7}$$

where G and S are complex constants yet undetermined, and e^{St} denotes the exponential function. Substituting Eq. (3.7) into Eq. (3.6), one gets

$$S^2 + \frac{c}{m}S + \omega^2 = 0 \tag{3.8}$$

Two roots of Eq. (3.8) can be obtained as

$$S_{1,2} = -\frac{c}{2m} \pm \sqrt{\left(\frac{c}{2m}\right)^2 - \omega^2}$$

When the stiffness and mass of the structural system are determined, S is completely dependent on the damping coefficient. When the damping coefficient c is large enough that the quantity under the radical sign is positive, S_1 and S_2 are real numbers, and the system will not oscillate. When the damping coefficient c is small enough so that the quantity under the radical sign is negative, S_1 and S_2 are conjugate complex numbers, and the system will be in oscillatory motion. When the damping coefficient c is equal to a critical value c_c, the quantity under the radical sign is zero, that is, $c = 2m\omega \equiv c_c$. S_1 and S_2 are two identical real numbers. The system motion in this case is the boundary between the two distinct motions mentioned above. The parameter $\xi = c/c_c$ is introduced, which is called the damping ratio. Because the damping coefficient of practical systems is usually difficult to determine directly, the damping ratio ξ is often used to represent the damping characteristic of the system. The characteristics of the solution of Eq. (3.6) in three cases are discussed as follows.

Case 1: Undercritically damped systems ($\xi < 1$)

In this case, $S_{1,2} = -\xi\omega \pm \omega_D i$, where $\omega_D = \omega\sqrt{1-\xi^2}$ is called free-vibration frequency of the damped system. The general solution of Eq. (3.6) can be expressed as (detailed derivation see Refs. [1,2])

$$v = e^{-\xi\omega t}(C_1 \cos \omega_D t + C_2 \sin \omega_D t) \tag{3.9}$$

where C_1 and C_2 are real constants yet undetermined. The initial conditions $v(0)$ and $\dot{v}(0)$ are substituted into Eq. (3.9) and its derivative, respectively, to determine C_1 and C_2. Finally, one can obtain

$$v = e^{-\xi\omega t}\left[v(0) \cos \omega_D t + \frac{\dot{v}(0) + \xi\omega v(0)}{\omega_D} \sin \omega_D t\right] \tag{3.10}$$

Eq. (3.10) can also be rewritten as

$$v = \rho e^{-\xi\omega t} \cos(\omega_D t - \theta_D) \tag{3.11}$$

where

$$\rho = \sqrt{[v(0)]^2 + \left[\frac{\dot{v}(0) + \xi\omega v(0)}{\omega_D}\right]^2} \tag{3.12}$$

$$\theta_D = \tan^{-1} \frac{\dot{v}(0) + \xi\omega v(0)}{\omega_D v(0)} \tag{3.13}$$

The time history of system response given by Eq. (3.10) is plotted in Fig. 3.4. The length of time required for the mass m to complete one cycle of free vibration is the natural period of vibration of the system, which is denoted as T_D. The amplitude of vibration decays with time. Due to $v(t + T_D) \neq v(t)$ and $\dot{v}(t + T_D) \neq \dot{v}(t)$, the damped free vibration can be called isochronous vibration, rather than periodic vibration. However, $T_D = 2\pi/\omega_D = 2\pi/\left(\omega\sqrt{1-\xi^2}\right) \approx 2\pi/\omega$ is still referred to as the period of the damped free vibration.

As shown in Eq. (3.9) and Fig. 3.4, the ratio of two successive positive peaks is

$$\frac{v_m}{v_{m+1}} = e^{\xi\omega T_D} = e^{\xi\omega\frac{2\pi}{\omega_D}} \tag{3.14}$$

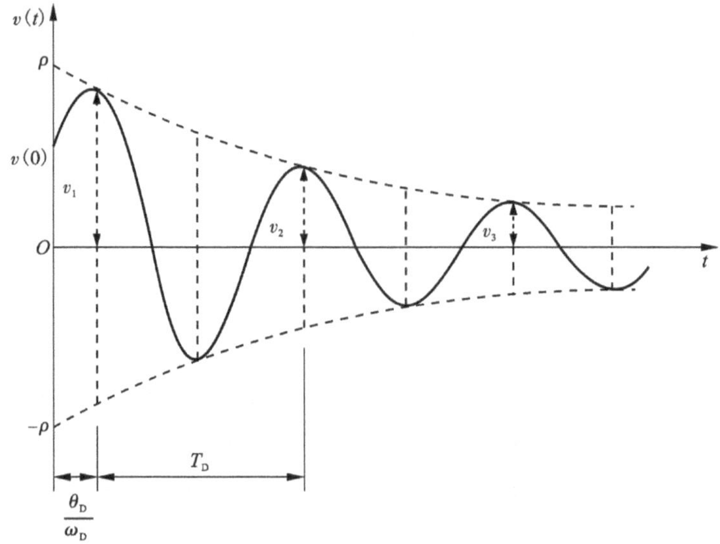

Figure 3.4 Free-vibration response of undercritically-damped system.

Taking the natural logarithm of Eq. (3.14) on both sides, one obtains the logarithmic decrement of damping, δ, defined by

$$\delta = \ln\frac{v_m}{v_{m+1}} = 2\pi\xi\frac{\omega}{\omega_D} = \frac{2\pi\xi}{\sqrt{1-\xi^2}}$$

For most structures, $\xi < 20\%$ and $\xi^2 \ll 1$, δ can be approximated by

$$\delta \approx 2\pi\xi$$

then

$$\frac{v_m}{v_{m+1}} \approx e^{2\pi\xi} = 1 + 2\pi\xi + \frac{1}{2!}(2\pi\xi)^2 + \cdots$$

By retaining the first two terms of the Taylor's series on the right-hand side, the damping ratio ξ is given by

$$\xi \approx \frac{v_m - v_{m+1}}{2\pi v_{m+1}} \tag{3.15}$$

For lightly damped systems, greater accuracy in evaluating the damping ratio ξ can be obtained by considering response peaks which are several cycles apart, say s cycles, then

$$\frac{v_m}{v_{m+s}} = e^{2\pi s\xi\frac{\omega}{\omega_D}} \approx e^{2\pi s\xi} \tag{3.16}$$

By further expanding the exponential function with Taylor's series, one obtains

$$\frac{v_m}{v_{m+s}} \approx e^{2\pi s\xi} = 1 + 2\pi s\xi + \frac{1}{2!}(2\pi s\xi)^2 + \cdots$$

Retaining the first two terms of the series leads to

$$\xi \approx \frac{v_m - v_{m+s}}{2\pi s v_{m+s}} \tag{3.17}$$

When damped free vibrations are observed experimentally, the response peaks v_m and v_{m+s} are easy to be obtained, by which the damping ratio ξ of the structure can be evaluated.

From Eq. (3.16), one obtains

$$s = \frac{1}{2\pi\xi}\ln\left(\frac{v_m}{v_{m+s}}\right)$$

When the peak amplitude decays to 50% (i.e., $v_{m+s} = 50\% v_m$), the number of required cycles, $s_{50\%}$, is as follows:

$$s_{50\%} = \frac{1}{2\pi\xi}\ln\left(\frac{v_m}{0.5v_m}\right) \approx \frac{0.11}{\xi}$$

The relationship between $s_{50\%}$ and ξ is shown in Fig. 3.5. Note that a convenient method for estimating the damping ratio ξ is to count the

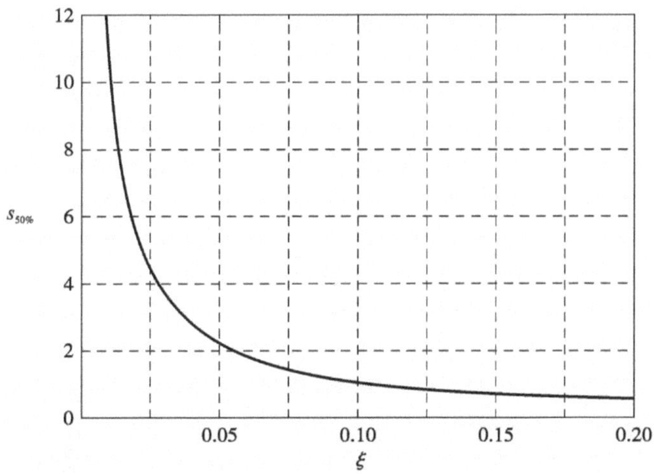

Figure 3.5 Relationship between the number of cycles required to reduce the peak amplitude by 50%, $s_{50\%}$, and the damping ratio ξ.

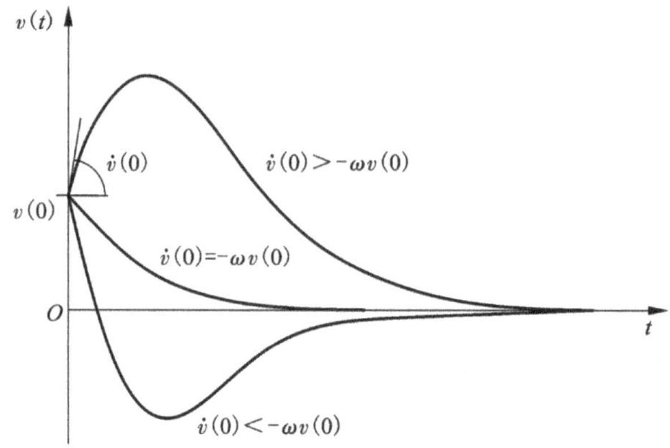

Figure 3.6 Response curves for free vibration of a critically-damped system.

number of cycles required to give a 50% reduction in amplitude, with which the damping ratio ξ can be determined from Fig. 3.5.

Case 2: Critically damping systems ($\xi = 1$)

In this case, Eq. (3.8) has two repeated and negative real roots, that is, $S_{1,2} = -\omega$. The general solution of Eq. (3.6) can be obtained as

$$v = e^{-\omega t}(C_1 t + C_2)$$

where C_1 and C_2 are real constants yet undetermined. Substituting the initial condition $v(0)$ and $\dot{v}(0)$ into the above general solution and its derivative respectively leads to

$$v = e^{-\omega t}[v(0)(1 + \omega t) + \dot{v}(0)t] \tag{3.18}$$

The motion represented by Eq. (3.18) is a nonoscillatory decaying motion. The critical damping is defined as the minimum value of damping that is required to avoid the oscillatory free vibration to occur. For different initial conditions, the vibrational behavior of the system can be represented by the curves shown in Fig. 3.6.

Case 3: Overcritically damped systems ($\xi > 1$)

In this case, Eq. (3.8) has two negative real roots, that is, $S_{1,2} = -\xi\omega \pm k_2$, where $k_2 = \omega\sqrt{\xi^2 - 1}$. The general solution of Eq. (3.6) can be obtained as

$$v = C_1 e^{(-\xi\omega + k_2)t} + C_2 e^{(-\xi\omega - k_2)t}$$

where C_1 and C_2 are real constants yet undetermined.

The above equation can be rewritten as

$$v = e^{-\xi\omega t}(C_1 e^{k_2 t} + C_2 e^{-k_2 t}) \tag{3.19}$$

Considering $\cosh x = (e^x + e^{-x})/2$ and $\sinh x = (e^x - e^{-x})/2$, $e^x = \cosh x + \sinh x$ and $e^{-x} = \cosh x - \sinh x$ can be obtained. Eq. (3.19) can therefore be written as

$$v = e^{-\xi\omega t}[(C_1 + C_2)\cosh(k_2 t) + (C_1 - C_2)\sinh(k_2 t)]$$

Let $C_1 + C_2$ and $C_1 - C_2$ be replaced by $\overline{C}_1$ and $\overline{C}_2$, respectively, one obtains

$$v = e^{-\xi\omega t}\left[\overline{C}_1\cosh(k_2 t) + \overline{C}_2\sinh(k_2 t)\right]$$

By substituting the initial conditions $v(0)$ and $\dot{v}(0)$ into the above equation and its derivative, respectively, $\overline{C}_1$ and $\overline{C}_2$ can be determined as

$$\overline{C}_1 = v(0),\ \overline{C}_2 = \frac{\dot{v}(0) + \xi\omega v(0)}{k_2}$$

Then, Eq. (3.19) becomes

$$v = e^{-\xi\omega t}\left[v(0)\cosh(k_2 t) + \frac{\dot{v}(0) + \xi\omega v(0)}{k_2}\sinh(k_2 t)\right] \tag{3.20}$$

It is easily shown from the form of Eq. (3.20) that the free-vibration response of an overcritically damped system is similar to the motion of a critically damped system as shown in Fig. 3.6. However, the asymptotic return to the zero-displacement position is slower depending on the amount of damping. The damping of civil engineering structures is generally undercritical (see Table 4.1), and overcritically damped systems often occur in mechanical systems.

Example 3.1: A one-story building is idealized as a rigid girder supported by weightless columns, as shown in Fig. 3.7. In order to evaluate the dynamic properties of this structure, a free-vibration test is made, in which the rigid girder is displaced laterally by a hydraulic jack and then suddenly released. During the jacking operation, it is observed that a force of 88.90 kN is required to displace the girder 5.08×10^{-3} m. After the instantaneous release of this initial displacement, the maximum displacement on the first return swing is only 4.064×10^{-3} m, and the period of this displacement cycle is $T_D = 1.4$ seconds. From the experimental data, the following dynamic properties are determined.

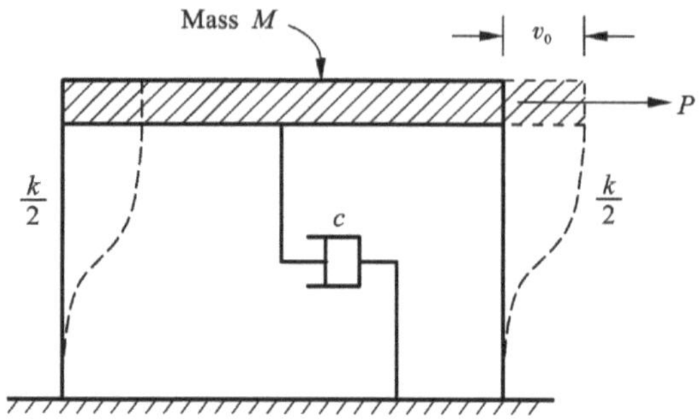

Figure 3.7 Vibration test of a one-story building.

Solution:

1. Effective mass M of the girder:

$$T_D \approx \frac{2\pi}{\omega} = 2\pi\sqrt{\frac{M}{k}}$$

Hence

$$M \approx \left(\frac{T_D}{2\pi}\right)^2 k = \left(\frac{1.40}{2\pi}\right)^2 \times \frac{88.90 \times 10^3}{5.08 \times 10^{-3}} = 8.697 \times 10^5 \text{ kg}$$

2. Damped frequency:

$$\omega_D = \frac{2\pi}{T_D} = \frac{2\pi}{1.40} = 4.48\text{rad/s}$$

3. Damping properties:
   Logarithmic decrement: $\delta = \ln\left[5.08 \times 10^{-3}/\left(4.064 \times 10^{-3}\right)\right] = 0.223$
   Damping ratio: $\xi \approx \delta/(2\pi) = 3.55\%$
   Damping coefficient:

$$c = \xi c_c \approx \xi \cdot 2M\omega_D = 0.0355 \times 2 \times 8.697 \times 10^5 \times 4.48$$
$$= 2.766 \times 10^5 \text{ N/(m} \cdot \text{s)}$$

4. Amplitude after six cycles:

$$v_6 = \left(\frac{v_1}{v_0}\right)^6 v_0 = \left(\frac{4.064 \times 10^{-3}}{5.08 \times 10^{-3}}\right)^6 \times 5.08 \times 10^{-3} = 1.33 \times 10^{-3} \text{ m}$$

3.1.3 Stability of motion

The solutions of free-vibration equation for linear SDOF systems (undamped and viscously damped) have been studied, where the mass m and stiffness k are both positive, and the damping coefficient c satisfies $c \geq 0$. In practice, there is another kind of vibration problem of SDOF systems, such as flutter in bridges, whose equation of motion can be written in the form [2]

$$\ddot{v} + a\dot{v} + bv = 0 \tag{3.21}$$

where the coefficients a and b are not necessarily positive. Since this is a linear differential equation with constant coefficients, the solution of Eq. (3.21) has the form

$$v = Ge^{St} \tag{3.22}$$

where the coefficients G and S are undetermined complex constants. Substituting Eq. (3.22) into Eq. (3.21) gives the eigenvalue equation

$$S^2 + aS + b = 0 \tag{3.23}$$

The two roots of this equation are

$$S_{1,2} = -\frac{a}{2} \pm \sqrt{\left(\frac{a}{2}\right)^2 - b} \tag{3.24}$$

The general solution, then, has the form

$$v = G_1 e^{S_1 t} + G_2 e^{S_2 t} \tag{3.25}$$

The motion of a system governed by Eq. (3.21) is classified according to the following stability categories: (1) asymptotically stable, (2) stable, or (3) unstable (Fig. 3.8). The character of the motion depends on the roots S_1 and S_2, which can be real, purely imaginary, or complex. Let the roots have the general form

$$S = R_e(S) + iI_m(S) \tag{3.26}$$

Correspondingly $S_1 = \alpha_1 + i\beta_1$, $S_2 = \alpha_2 + i\beta_2$.

Since $v(t)$ is real, the roots that are pure imaginary or complex must occur in complex conjugate pairs. That is, when β_1 and β_2 are not equal to zero, $\beta_1 = -\beta_2$ holds.

1. Asymptotically stable motion. If both roots of the eigenvalue equation lie in the left half-plane (i.e., $\alpha_1 < 0$ and $\alpha_2 < 0$), the motion of the

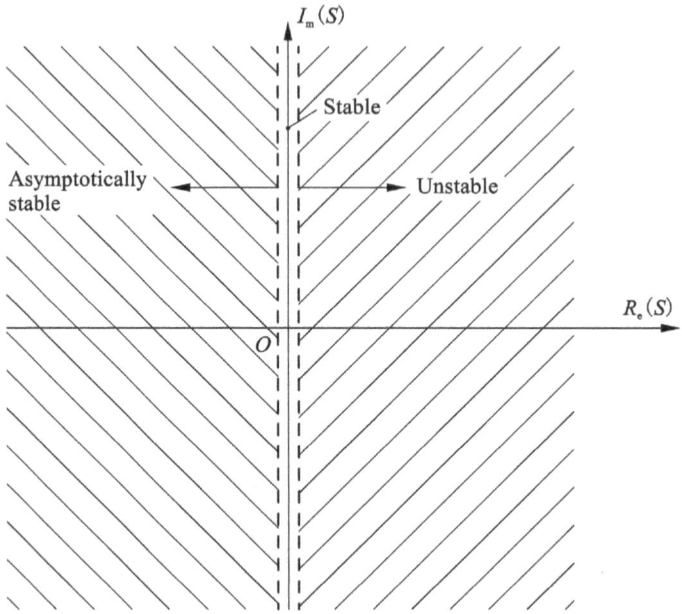

Figure 3.8 Stability relationships in the complex *S* plane.

system is said to be asymptotically stable. That is, the motion will die out with time. Included are the behavior of undercritically damped, critically damped, and overcritically damped systems, as discussed in Section 3.1.2. Response of an undercritically damped system is illustrated in Fig. 3.9A.

2. Stable motion. If the two roots of the eigenvalue equation are purely imaginary complex conjugates (i.e., $\alpha_1 = \alpha_2 = 0$), the motion is said to be stable. The simple harmonic motion of an undamped SDOF system is illustrated in Fig. 3.9B.
3. Unstable motion. If either of the two roots of the eigenvalue equation has a positive real part (i.e., $\alpha_1 > 0$, or $\alpha_2 > 0$, or both), the motion is said to be unstable. There are two types of unstable motion.
   a. Flutter. If the two roots are complex conjugates that lie in the right half-plane, the motion will be a diverging oscillation, as illustrated in Fig. 3.9C. Flutter avoidance is an essential design consideration in the design of airplanes and long-span suspension bridges.
   b. Divergence. If both roots lie on the real axis and at least one of them has a positive real part, nonoscillatory divergent motion will occur, as illustrated in Fig. 3.9D.

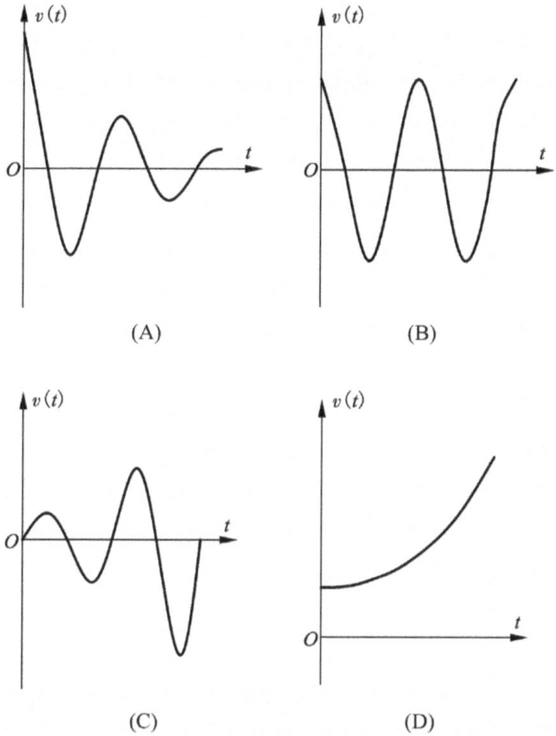

Figure 3.9 Response of four SDOF systems: (A) undercritically damped (decaying) oscillation (asymptotically stable); (B) undamped (harmonic) oscillation (stable); (C) flutter (diverging, unstable); (D) nonoscillatory divergence (unstable). *SDOF*, Single-degree-of-freedom.

Example 3.2: Fig. 3.10 shows an inverted simple pendulum that consists of a mass m at the upper end of a rigid and massless rod whose lower end is connected to a pin support at A. The mass m is also supported laterally by two linear springs of spring constant k. (1) Determine the linearized equation of motion of this system for small-angle oscillation, that is, for $\theta \ll 1$. (2) Solve for the free vibration of the pendulum with initial conditions $\theta(0) = \theta_0$ and $\dot{\theta}(0) = 0$.

Solution:

1. Derive the equation of motion for small-angle oscillation. First the free-body diagram of the pendulum in a slightly displaced configuration is drawn as Fig. 3.11. Next the equation of motion for fixed-axis rotation about the pin at A is formulated as follows:

$$mg(L\sin\theta) - 2f_s(L\cos\theta) = mL^2\ddot{\theta} \tag{3.27}$$

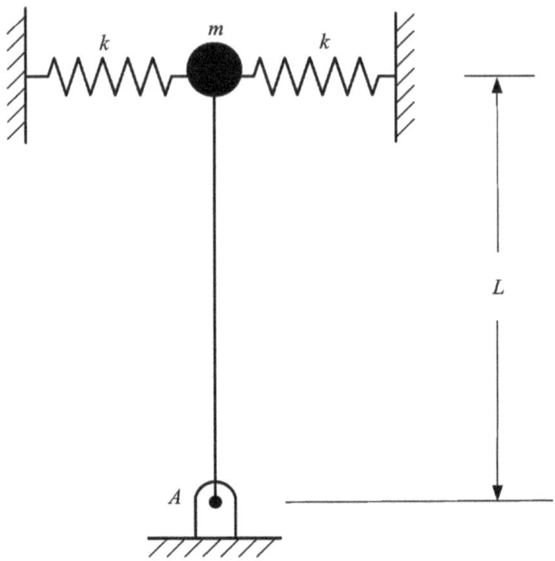

Figure 3.10 Inverted-pendulum SDOF system. *SDOF*, Single-degree-of-freedom.

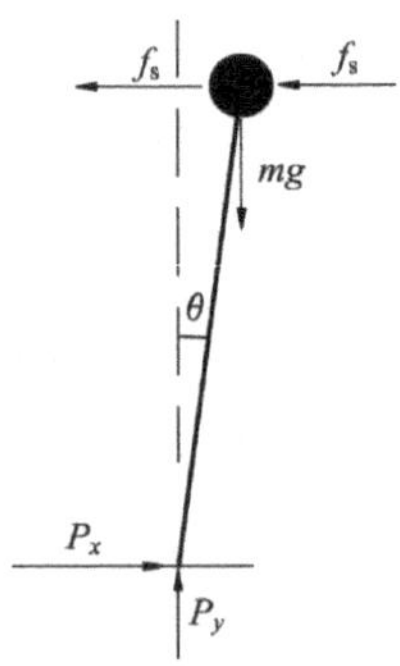

Figure 3.11 Free-body diagram.

The spring forces are given by

$$f_s = k(L\ \sin\theta) \tag{3.28}$$

Considering $\sin\theta \approx \theta$ and $\cos\theta \approx 1$ due to the small value of θ, and substituting Eq. (3.28) into Eq. (3.27), leads to linearized equation of motion of this system

$$\ddot{\theta} + \left(\frac{2k}{m} - \frac{g}{L}\right)\theta = 0 \tag{3.29}$$

which has the form

$$\ddot{\theta} + b\theta = 0 \tag{3.30}$$

2. Obtain the free-vibration solution of Eq. (3.29). As before, the solution of this equation is assumed to be of the form

$$v = Ge^{St} \tag{3.31}$$

Substituting this into Eq. (3.29) leads to

$$S^2 + \left(\frac{2k}{m} - \frac{g}{L}\right) = 0 \tag{3.32}$$

Clearly, the solutions that satisfy Eq. (3.32) will depend on the sign of the term in parentheses, that is, on the sign of the effective stiffness

$$b = \frac{2k}{m} - \frac{g}{L} \tag{3.33}$$

If $b > 0$, the solution of Eq. (3.29) will be oscillatory at the natural frequency of $\sqrt{2k/m - g/L}$. However, if $b < 0$, the solution of Eq. (3.29) will have the form

$$\theta = G_1 e^{\sqrt{(g/L - 2k/m)}t} + G_2 e^{-\sqrt{(g/L - 2k/m)}t} \tag{3.34}$$

Finally, the solution that corresponds to the initial conditions $\theta(0) = \theta_0$, $\dot{\theta}(0) = 0$ is

$$\theta = \frac{\theta_0}{2}\left[e^{\sqrt{(g/L - 2k/m)}t} + e^{-\sqrt{(g/L - 2k/m)}t}\right] \tag{3.35}$$

Clearly, the second term in Eq. (3.35) dies out with time, but the first term grows with time in a nonoscillatory fashion. This type of behavior, which is called divergence, is illustrated in Fig. 3.9D.

3.2 Response of SDOF systems to harmonic loads

The exciting force that varies with time t by sine or cosine law is called a simple harmonic load. For example, a rotating machine has an eccentric mass m_1, and the distance from the mass m_1 to the rotation axis is e, as shown in Fig. 3.12. The angular velocity of the machine rotating in the clockwise direction is $\overline{\omega}$, and associated centrifugal force is $P_0 = m_1 e\overline{\omega}^2$. Its

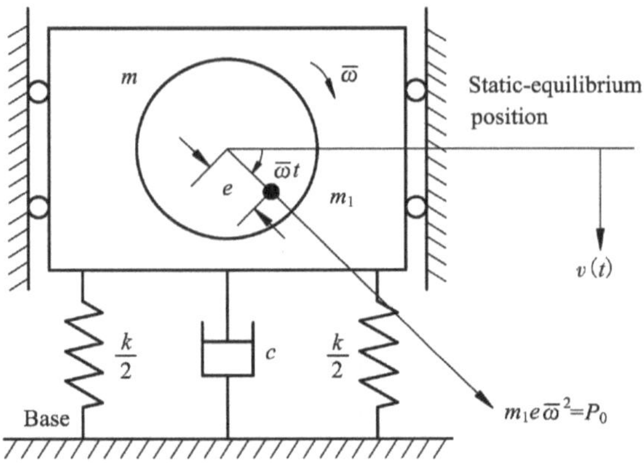

Figure 3.12 Vertical vibration induced by a rotating machine.

horizontal and vertical components are equal to $P_0 \cos \overline{\omega}t$ and $P_0 \sin \overline{\omega}t$, respectively, which are both harmonic loads.

Any periodic load can be represented by Fourier series, which is the sum of several harmonic loads. The response of linear systems to periodic loads can be obtained by superimposing the responses produced by the harmonic components. Therefore it is valuable to analyze the responses of the system to harmonic loads, since they not only show the basic rules of motion but also exhibit the general characteristics of the system to periodic loads.

In Fig. 3.12, the rotating machine is completely fixed in the horizontal direction and only vertical vibration occurs. In the vertical direction, the machine is supported by the foundation represented by a spring and a viscous damper. The elastic stiffness coefficient of the foundation is k, the damping coefficient is c, and the total mass of the machine is m. The vertical displacement at any instant of time t, $v(t)$, is measured from the static-equilibrium position. The equation of motion for the system in the vertical direction is formulated as follows:

$$m\ddot{v} + c\dot{v} + kv = P_0 \sin \overline{\omega}t \tag{3.36}$$

According to the theory of ordinary differential equation, the general solution of Eq. (3.36) consists of a complementary free-vibration solution v_c and a particular solution v_p.

Assuming damping ratio $\xi < 1$, the complementary solution v_c is given by Eq. (3.9). The particular solution v_p of Eq. (3.36) can be expressed as

$$v_p = D_1 \cos \overline{\omega}t + D_2 \sin \overline{\omega}t \tag{3.37}$$

where D_1 and D_2 are constants yet undetermined. Substituting Eq. (3.37) into Eq. (3.36), two algebraic equations can be obtained by considering the coefficients of the sine and cosine terms on both sides of Eq. (3.36) to be identical, which are then used to determine D_1 and D_2. This method is physically straightforward, but is cumbersome to solve for D_1 and D_2. In the following discussion, complex number method is used to obtain v_p.

Set a complex equation of motion [3]

$$m\ddot{Z} + c\dot{Z} + kZ = P_0 e^{i\overline{\omega}t}$$

where the response Z and applied load $P_0 e^{i\overline{\omega}t}$ are both complex numbers.

Setting $Z = a + ib$ (a and b are real functions of time t) and substituting it into the complex equation leads to

$$(m\ddot{a} + c\dot{a} + ka) + i(m\ddot{b} + c\dot{b} + kb) = P_0 \cos \overline{\omega}t + iP_0 \sin \overline{\omega}t$$

Considering the imaginary parts on both sides of the equation to be identical, gives

$$m\ddot{b} + c\dot{b} + kb = P_0 \sin \overline{\omega}t$$

By comparing the above equation with Eq. (3.36), it can be seen that the imaginary part of the complex solution Z of the complex equation is the particular solution v_p of Eq. (3.36).

Substituting the general solution of the form $Z = Ae^{i\overline{\omega}t}$ into the complex equation leads to

$$A = \frac{P_0}{k - m\overline{\omega}^2 + ic\overline{\omega}} = \frac{P_0}{\sqrt{\left(k - m\overline{\omega}^2\right)^2 + (c\overline{\omega})^2} e^{i\theta}}$$

Hence, the complex response is expressed as

$$Z = Ae^{i\overline{\omega}t} = \frac{P_0 e^{i(\overline{\omega}t - \theta)}}{\sqrt{\left(k - m\overline{\omega}^2\right)^2 + (c\overline{\omega})^2}}$$

By taking the imaginary part of Z, one obtains

$$v_p = \frac{P_0 \sin(\overline{\omega}t - \theta)}{\sqrt{\left(k - m\overline{\omega}^2\right)^2 + (c\overline{\omega})^2}} \tag{3.38}$$

Eq. (3.38) is rewritten as

$$v_p = \rho \sin(\overline{\omega}t - \theta) \tag{3.39}$$

where $\rho = P_0D/k$ is the amplitude of the steady-state response, and the coefficient D is given by

$$D = \frac{\rho}{P_0/k} = \left[\left(1-\beta^2\right)^2 + (2\xi\beta)^2\right]^{-\frac{1}{2}} \tag{3.40}$$

where $\beta = \overline{\omega}/\omega$ denotes the frequency ratio. Since P_0/k is the displacement of the system to static force P_0, D is the ratio of the maximum steady-state response to static displacement of the system, which is called the dynamic magnification factor for system displacement.

The phase angle θ in Eq. (3.39) is given by

$$\theta = \tan^{-1}\frac{c\overline{\omega}}{k - m\overline{\omega}^2} = \tan^{-1}\frac{2\xi\beta}{1-\beta^2} \tag{3.41}$$

The general solution of Eq. (3.36) can be obtained as follows:

$$v = v_c + v_p = e^{-\xi\omega t}(C_1 \cos\omega_D t + C_2 \sin\omega_D t) + \rho\sin(\overline{\omega}t - \theta) \tag{3.42}$$

Substituting the initial conditions $v(0)$ and $\dot{v}(0)$ into the Eq. (3.42) leads to

$$\begin{aligned} v = {} & e^{-\xi\omega t}\left[v(0)\cos\omega_D t + \frac{\dot{v}(0) + \xi\omega v(0)}{\omega_D}\sin\omega_D t\right] + \\ & \rho e^{-\xi\omega t}\left[\sin\theta\cos\omega_D t + \frac{\xi\omega\sin\theta - \overline{\omega}\cos\theta}{\omega_D}\sin\omega_D t\right] + \rho\sin(\overline{\omega}t - \theta) \end{aligned} \tag{3.43}$$

The first term in Eq. (3.43) represents the free vibration at natural frequency ω_D, which is determined from the initial conditions of the system. This vibration does not occur under zero initial conditions (i.e., $v(0) = 0$ and $\dot{v}(0) = 0$). The second term in Eq. (3.43) represents a harmonic motion at the natural frequency of ω_D, of which the amplitude is related to the excitation. This vibration, which is independent of the initial conditions, is purely generated as a companion of the forced vibration. So it is called the associated free vibration. Due to the presence of damping, the two terms mentioned above decay rapidly, so it is called the transient response, as shown in Fig. 3.13A. The third term of Eq. (3.43) represents the forced vibration, as shown in Fig. 3.13B, which is an oscillatory motion at the exciting frequency $\overline{\omega}$ regardless of the initial conditions. The amplitude ρ does not vary with time, so it is called the steady-state response. It lags behind the applied load by phase angle θ.

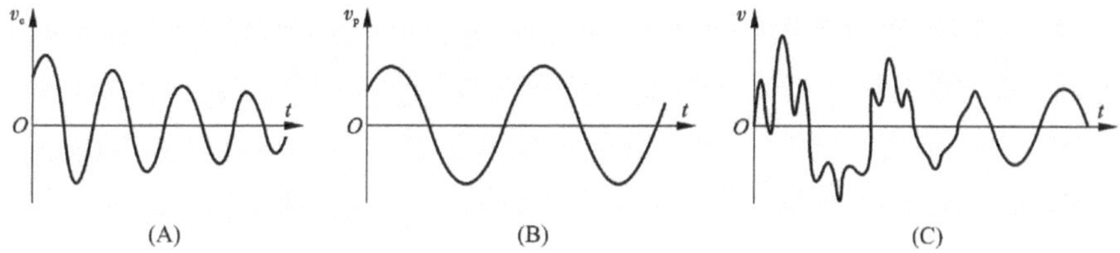

Figure 3.13 Response of SDOF system in forced vibration: (A) transient vibration; (B) steady-state vibration; (C) resultant motion. *SDOF*, Single-degree-of-freedom.

The resultant motion of the transient and steady-state vibrations is shown in Fig. 3.13C. The dynamic response of the system is completely controlled by the steady-state vibration after the transient vibration vanishes over time due to the damping effect.

Some important features of the response to harmonic loads are provided as follows:

1. Variations of D and θ with β and ξ

The variation of D with β and ξ is shown in Fig. 3.14, which is called the amplitude–frequency characteristic curve of the vibration system. When $\beta \rightarrow 0$, $D \rightarrow 1$. This can be explained as follows: since the applied force changes slowly, it is almost a constant force in a short time, which is similar to the static force. When $\beta \gg 1$, D approaches zero. This observation can be explained as follows: when $\overline{\omega}$ is very large, the direction of the applied force changes rapidly and the vibrating system cannot respond timely to the dynamic load of high frequency due to the inertial effect. Thus the system is almost stationary. Fig. 3.14 shows that the influence of the damping on dynamic magnification factor can almost be ignored when the two extremes of $\beta \rightarrow 0$ and $\beta \gg 1$ occur.

When $\beta = 1$, Eq. (3.40) gives

$$\begin{cases} D_{\beta=1} = \dfrac{1}{2\xi} \\[2ex] \rho_{\beta=1} = \dfrac{P_0}{2k\xi} \end{cases} \tag{3.44}$$

In this situation, the amplitude and the dynamic magnification factor are both very large, but strictly speaking, they are not the maximum values of D and ρ. Taking the derivative of Eq. (3.40) with

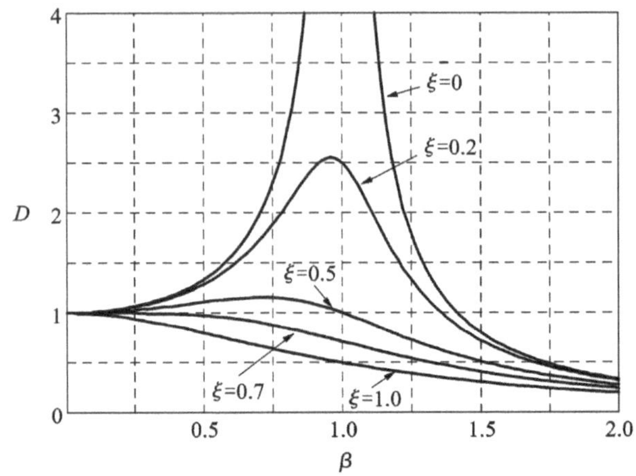

Figure 3.14 Variation of dynamic magnification coefficient with the damping ratio and frequency ratio.

respect to β and letting it be zero, one obtains the frequency ratio for obtaining $\rho_{\max}$ and $D_{\max}$

$$\beta_m = \sqrt{1 - 2\xi^2} \tag{3.45}$$

This equation applies to the system with $\xi \leq 1/\sqrt{2}$. When the damping is small, the maximum value of the dynamic magnification factor appears around $\beta = 1$, and the first expression of Eq. (3.44) can be taken as the maximum value of the dynamic magnification factor approximately. When $\xi > 1/\sqrt{2}$, the system does not produce an amplified response, that is, $D < 1$.

In Fig. 3.12, the applied force varies according to $\sin \overline{\omega} t$. The steady-state response, as given by Eq. (3.39), varies according to $\sin(\overline{\omega} t - \theta)$. Therefore the steady-state response lags behind the applied force by a phase angle θ. The phase angle θ is determined from Eq. (3.41) and its variation with damping ratio ξ and frequency ratio β is shown in Fig. 3.15, which depicts the phase–frequency characteristic curve.

For the case of zero damping ($\xi = 0$), in the range of $\beta < 1$, $\theta = 0$, indicating that the forced-vibration response is in phase with the applied force. In the range of $\beta > 1$, $\theta = \pi$, indicating that the response is out of phase with the applied force. When $\beta = 1$, it can be seen from Eq. (3.41) that the angle θ is indeterminate. In the case of damped systems, the phase angle θ varies continuously as the

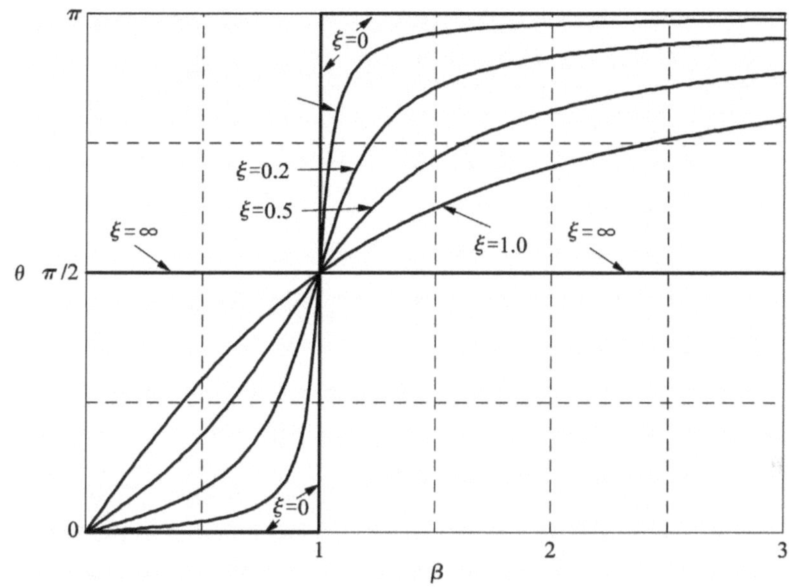

Figure 3.15 Variation of the phase angle with damping ratio and frequency ratio.

frequency ratio increases. When $\beta = 1$, $\theta = \pi/2$ as long as the damping is present regardless of its magnitude. This means that the resonant response lags behind applied force by a quarter of a cycle, and the natural frequency can be measured by means of this characteristic, which is called phase resonance method. If the frequency ratio β is far away from resonant region, the small damping ratio would only have a minor effect on the phase angle. When $\beta \ll 1$, $\theta \approx 0$, indicating that the response is in phase with the applied force when exciting frequency is very low. When $\beta \gg 1$, $\theta \approx \pi$, indicating that they are out of phase each other. Thus in these cases, the influence of damping on the phase angle may be ignored.

2. Dynamic equilibrium in the steady-state vibration [3]

In accordance with the d'Alembert principle, the system is in dynamic equilibrium at any instant of time. Eq. (3.36) can be rewritten as

$$-m\ddot{v} - c\dot{v} - kv + P_0 \sin \overline{\omega} t = 0$$

which represents the equilibrium of the forces acting on the system at any instant of time t. These forces include the inertial force $-m\ddot{v}$, damping force $-c\dot{v}$, elastic restoring force $-kv$, and applied force

$P_0 \sin \omega t$. Considering the steady-state response expressed by Eq. (3.39), $c = 2m\xi\omega$, $\beta = \overline{\omega}/\omega$, and $\rho = P_0 D/k$, leads to

Inertial force: $-m\ddot{v} = m\overline{\omega}^2 \rho \sin(\overline{\omega}t - \theta) = P_0 D\beta^2 \sin(\overline{\omega}t - \theta)$

Damping force:

$$-c\dot{v} = -\rho c\overline{\omega} \cos(\overline{\omega}t - \theta) = -\rho c\overline{\omega} \sin\left(\overline{\omega}t - \theta + \frac{\pi}{2}\right)$$
$$= P_0 D(2\xi\beta) \sin\left(\overline{\omega}t - \theta - \frac{\pi}{2}\right)$$

Elastic restoring force: $-kv = -k\rho \sin(\overline{\omega}t - \theta) = P_0 D \sin(\overline{\omega}t - \theta - \pi)$

Applied force: $P_0 \sin \overline{\omega}t$

The inertial, damping, elastic, and applied forces are harmonic loads with common frequency of $\overline{\omega}$ but with different amplitudes and phase angles. The inertial force lags behind the applied force by phase angle θ, the damping force lags behind the applied force by phase angle $\theta + \pi/2$, and the elastic force lags behind the applied force by phase angle $\theta + \pi$. To ensure consistency with the direction of the applied force in Fig. 3.12, the above forces are represented by the projection of associated force amplitude vectors (i.e., the length of vector is the amplitude of time-varying force) on the imaginary axis of the complex plane, as shown in Fig. 3.16. The phase relationships among these forces can be seen intuitively from Fig. 3.16. The amplitude and phase angle of the inertial, damping, and elastic forces are functions of the frequency ratio β. When β takes different values, there are three situations as follows:

a. When $\beta \ll 1$, the exciting frequency is very low, and the system vibrates slowly. In this case, the inertial and damping forces are both small, and the phase angle θ is also small. The steady-state displacement is almost in phase with the applied force and the applied force is almost equilibrated by the elastic resistance. In this situation, the dynamic magnification factor $D \approx 1$, and the damping effect is minor.

b. When $\beta = 1$, $\theta = \pi/2$. The steady-state displacement lags behind the applied force by phase angle $\pi/2$ and the steady-state velocity is almost in phase with the applied force. The inertial and applied forces are equilibrated by elastic resistance and damping force, respectively. The damping has a strong influence in this case. Since the dynamic magnification factor D approaches its maximum

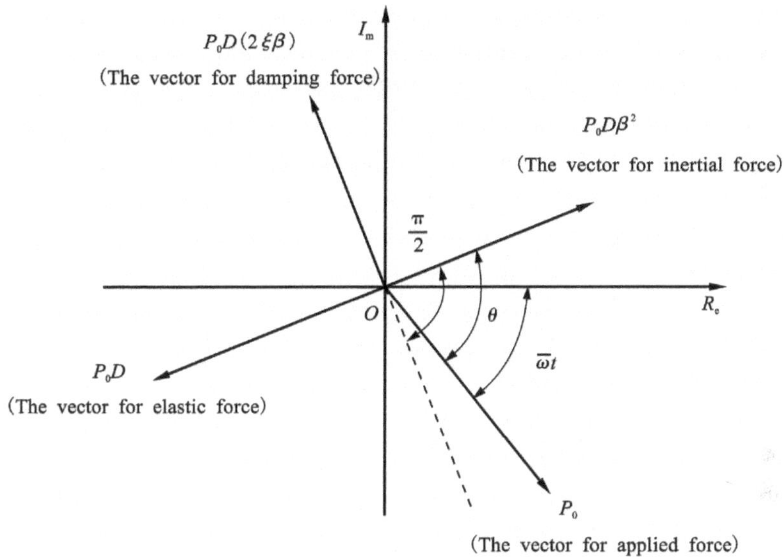

Figure 3.16 Equilibrium of forces in steady-state vibration.

value, the response amplitude also reaches its maximum value and the system is in the most unfavorable state.

c. When $\beta \gg 1$, $\theta \approx \pi$. Note from Fig. 3.14 that $D \approx 0$ in this case. Since the exciting frequency is far higher than the natural frequency of the system, the applied force varies very quickly, the direction of the system motion also changes frequently. The acceleration response is relatively large, and the displacement and velocity are small. Therefore the elastic resistance and the damping force are small, and the applied force is almost used to equilibrate the inertial force.

3. Resonant response

When the exciting frequency $\bar{\omega}$ is equal to or close to the natural frequency ω, a large amplitude vibration occurs, which is called resonance. Substituting Eq. (3.45) into Eq. (3.40), the dynamic magnification factor and steady-state response amplitude at resonance are obtained as

$$\begin{cases} D_{\max} = \dfrac{1}{2\xi\sqrt{1-\xi^2}} \\ \rho_{\max} = \dfrac{P_0 D_{\max}}{k} = \dfrac{P_0}{2k\xi\sqrt{1-\xi^2}} \end{cases} \tag{3.46}$$

Since the structural damping ratio $\xi \ll 1$, the differences between Eqs. (3.46) and (3.44) are very small. Thus it is generally considered that the resonance occurs when $\beta = 1$. In the above analysis, the time history of the system reaching maximum value at resonance is not given. For a more complete understanding of resonance response, considering $\beta = 1$, assuming the zero initial condition (that is, the system starts to vibrate from rest), and recalling that $\theta = \pi/2$, $\rho = P_0/(2k\xi)$, and $\omega_D = \omega\sqrt{1-\xi^2}$ in this case, one obtains from Eq. (3.43)

$$v_{\beta=1} = \frac{1}{2\xi}\frac{P_0}{k}\left[e^{-\xi\omega t}\left(\frac{\xi}{\sqrt{1-\xi^2}}\sin \omega_D t + \cos \omega_D t\right) - \cos \overline{\omega} t\right] \tag{3.47}$$

The ratio of $v_{\beta=1}$ to the displacement $v_{st} = P_0/k$ generated by the static load P_0, is called the resonance response ratio, that is,

$$R(t) = \frac{v_{\beta=1}}{v_{st}} = \frac{1}{2\xi}\left[e^{-\xi\omega t}\left(\frac{\xi}{\sqrt{1-\xi^2}}\sin \omega_D t + \cos \omega_D t\right) - \cos \overline{\omega} t\right] \tag{3.48}$$

Considering the general structural damping ratio $\xi < 0.2$, ignoring the effect of $\xi/\sqrt{1-\xi^2}\sin \omega_D t$ on the response amplitude, and noting that $\beta = 1$ and $\overline{\omega} = \omega_D \approx \omega$, leads to

$$R(t) = \frac{1}{2\xi}\left(e^{-\xi\omega t} - 1\right)\cos \omega t \tag{3.49}$$

For zero damping, Eq. (3.48) is indeterminate. By applying the L'Hospital's rule, and considering $\overline{\omega} = \omega_D = \omega$ and $\sqrt{1-\xi^2} \to 1$, one obtains resonance response ratio for the undamped system as follows:

$$R(t) = \lim_{\xi \to 0} \frac{\frac{d}{d\xi}\left[e^{-\xi\omega t}\left(\frac{\xi}{\sqrt{1-\xi^2}}\sin \omega_D t + \cos \omega_D t\right) - \cos \overline{\omega} t\right]}{\frac{d}{d\xi}(2\xi)} = \frac{1}{2}(\sin \omega t - \omega t\cos \omega t) \tag{3.50}$$

The response ratios described by Eqs. (3.49) and (3.50) are plotted in Fig. 3.17, which shows the increasing responses of undamped and damped systems in the case of resonance. For the undamped system, the response

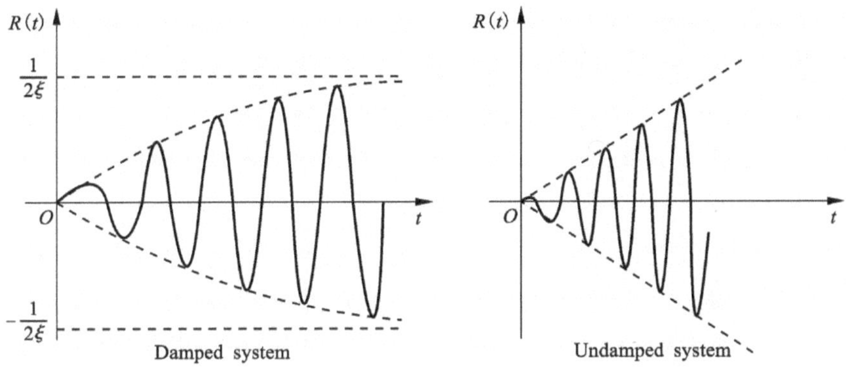

Figure 3.17 Response to resonant load ($\beta = 1$) for at-rest initial conditions.

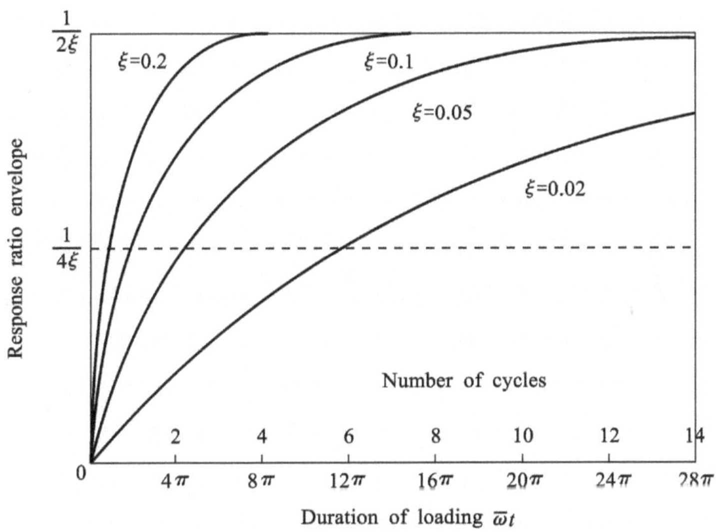

Figure 3.18 Rate of buildup of resonant response from rest.

increases continuously. Unless the exciting frequency varies, the system eventually breaks down due to the ever-increasing amplitude of vibration. The effects of damping will restrict the development of resonance amplitude of $R(t)$ to $1/(2\xi)$ in the damped system. The number of required cycles to reach the maximum amplitude approximately depends on the damping ratio ξ. The envelope function of resonance response ratio is plotted against frequency in Fig. 3.18 for discrete damping ratios. The smaller the damping ratio, the more cycles are required to reach the maximum amplitude. For example, when $\xi = 0.1$, about six cycles are needed; when $\xi = 0.05$, about 14 cycles are required.

Note that the analysis of resonance response ratio is based on the linear elastic theory. The actual system is not linear elastic when it enters the phase of large amplitude vibration. For example, the dynamic characteristics of the locally yielded system are significantly different from those of linear elastic systems. Whether the resonance response ratio $R(t)$ of the damped system can reach $1/(2\xi)$ or not depends on the variation of the system properties. Nonetheless, the dynamic response will become quite large when the system is at or close to resonance, which inevitably influences the normal operation or may even damage the structure. Hence, resonance should be avoided completely for structural design. The structure should generally be kept away from the range of $0.75 < \beta < 1.25$, which is often referred to as the resonance region.

Example 3.3: A portable harmonic-load machine provided an effective means for evaluating the dynamic properties of structures in the field. By operating the machine at two different frequencies and measuring the resulting structural-response amplitude and phase relationship in each case, the mass, damping, and stiffness of a SDOF structure can be determined by using experimental data. In a test of this type on a single-story building, the shaking machine was operated at frequencies of $\overline{\omega}_1 = 16$ rad/s and $\overline{\omega}_2 = 25$ rad/s with a force amplitude of 2 222.64 N in each case. The response amplitudes and phase relationships measured in the two cases were $\rho_1 = 1.83 \times 10^{-4}$ m, $\theta_1 = 15$ degrees, $\rho_2 = 3.68 \times 10^{-4}$ m, and $\theta_2 = 55$ degrees. Determine the dynamic properties of this structure.

Solution:

For the convenience of evaluation, the steady-state response amplitude ρ from Eq. (3.40) is rewritten as

$$\begin{aligned}
\rho &= \frac{P_0 D}{k} = \frac{P_0}{k}\left[\left(1-\beta^2\right)^2 + (2\xi\beta)^2\right]^{-\frac{1}{2}} \\
&= \frac{P_0}{k}\frac{1}{1-\beta^2}\left[1+\left(\frac{2\xi\beta}{1-\beta^2}\right)^2\right]^{-\frac{1}{2}} = \frac{P_0}{k}\frac{1}{1-\beta^2}\left(1+\tan^2\theta\right)^{-\frac{1}{2}} \qquad (3.51) \\
&= \frac{P_0}{k}\frac{1}{1-\beta^2}\left(\sec^2\theta\right)^{-\frac{1}{2}} = \frac{P_0 \cos\theta}{k(1-\beta^2)}
\end{aligned}$$

By substituting $k(1-\beta^2) = k - \overline{\omega}^2 m$ into Eq. (3.51), one gets

$$k - \overline{\omega}^2 m = \frac{P_0 \cos\theta}{\rho}$$

Then, introducing two sets of test data to the above equation leads to the following matrix equation

$$\begin{bmatrix} 1 & -16^2 \\ 1 & -25^2 \end{bmatrix} \begin{Bmatrix} k \\ m \end{Bmatrix} = \begin{Bmatrix} 2222.64 \times 0.966/(1.83 \times 10^{-4}) \\ 2222.64 \times 0.574/(3.68 \times 10^{-4}) \end{Bmatrix}$$

which can be solved to give

$$k = 1.75 \times 10^7 \text{ N/m} = 1.75 \times 10^4 \text{ kN/m} \qquad m = 22.39 \times 10^3 \text{ kg} = 22.39 \text{ t}$$

So, the natural frequency of the building is given by

$$\omega = \sqrt{\frac{k}{m}} = \sqrt{\frac{1.75 \times 10^7}{22.39 \times 10^3}} = 27.9 \text{ rad/s}$$

From Eq. (3.41), one obtains

$$\xi = \frac{1-\beta^2}{2\beta} \tan\theta \tag{3.52}$$

Substituting any group of the test data, say the first group, into Eq. (3.52) leads to

$$\xi = \frac{1-(16/27.9)^2}{2 \times (16/27.9)} \tan(15°) = 0.157$$

Therefore the damping coefficient is given by

$$\begin{aligned} c &= 2m\omega\xi = 2 \times 22.39 \times 10^3 \times 27.9 \times 0.157 \\ &= 1.961 \times 10^5 \text{ N} \cdot \text{s/m} = 196.1 \text{ kN} \cdot \text{s/m} \end{aligned}$$

3.3 Vibration caused by base motion and vibration isolation

3.3.1 Vibration caused by base motion

The base motion, which is regarded as an external excitation, also causes vibration of the system. For example, the ground motion causes the vibration of the building and the wave undulation causes the ship to jump

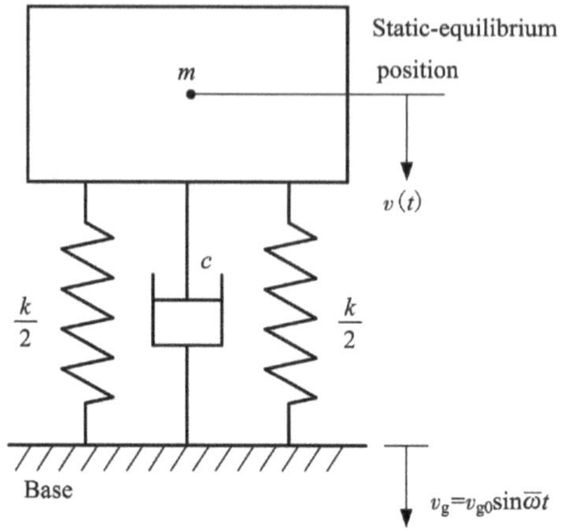

Figure 3.19 Vertical vibration of the object caused by the base motion.

upward and downward, etc. Assume that the mass block shown in Fig. 3.19 is limited to move in the vertical direction. The vertical vibration v of the mass block m is caused by the harmonic vibration $v_g = v_{g0}\sin\overline{\omega}t$ of the foundation. The equation of motion for the mass block is given by ignoring the mass of the spring and damper

$$m\ddot{v} + c(\dot{v} - \dot{v}_g) + k(v - v_g) = 0$$

Thus

$$m\ddot{v} + c\dot{v} + kv = c\dot{v}_g + kv_g \tag{3.53}$$

The particular solution is obtained by the complex number method (only the steady-state response solution is discussed here), and the complex equation is set as

$$m\ddot{Z} + c\dot{Z} + kZ = c\dot{v}_{gc} + kv_{gc} \tag{3.54}$$

where $v_{gc} = v_{g0}e^{i\overline{\omega}t}$.

According to the discussion in the previous section, the imaginary part of Z represents the particular solution v_p of Eq. (3.53). Substituting the general solution of the form, $Z = Ae^{i\overline{\omega}t}$, into Eq. (3.54) yields

$$A = \frac{v_{g0}(k + ic\overline{\omega})}{k - m\overline{\omega}^2 + ic\overline{\omega}} = \frac{v_{g0}[k^2 + (c\overline{\omega})^2]}{\sqrt{[k(k - m\overline{\omega}^2) + (c\overline{\omega})^2]^2 + (mc\overline{\omega}^3)^2}e^{i\theta}} \tag{3.55}$$

where

$$\theta = \tan^{-1}\frac{mc\overline{\omega}^3}{k(k - m\overline{\omega}^2) + (c\overline{\omega})^2} \tag{3.56}$$

Thus

$$Ae^{i\overline{\omega}t} = \frac{v_{g0}[k^2 + (c\overline{\omega})^2]e^{i(\overline{\omega}t-\theta)}}{\sqrt{[k(k-m\overline{\omega}^2)+(c\overline{\omega})^2]^2 + (mc\overline{\omega}^3)^2}}$$

The particular solution v_p of Eq. (3.53) can be expressed as the imaginary part of $Ae^{i\overline{\omega}t}$, that is,

$$v_p = \frac{v_{g0}[k^2 + (c\overline{\omega})^2]\sin(\overline{\omega}t - \theta)}{\sqrt{[k(k-m\overline{\omega}^2)+(c\overline{\omega})^2]^2 + (mc\overline{\omega}^3)^2}} \equiv \rho\sin(\overline{\omega}t - \theta) \tag{3.57}$$

where

$$\rho = \frac{v_{g0}[k^2 + (c\overline{\omega})^2]}{\sqrt{[k(k-m\overline{\omega}^2)+(c\overline{\omega})^2]^2 + (mc\overline{\omega}^3)^2}} \tag{3.58}$$

Considering $\rho = \left|Ae^{i\overline{\omega}t}\right| = |A|$ and Eq. (3.55), one obtains

$$\frac{\rho}{v_{g0}} = \sqrt{\frac{k^2 + (c\overline{\omega})^2}{(k-m\overline{\omega}^2)^2 + (c\overline{\omega})^2}} = \sqrt{\frac{1 + (2\xi\beta)^2}{(1-\beta^2)^2 + (2\xi\beta)^2}} = D\sqrt{1 + (2\xi\beta)^2} \tag{3.59}$$

The variation of steady-state response amplitude with frequency ratio β and damping ratio ξ is analyzed below. Eq. (3.59) is plotted in Fig. 3.20 for discrete damping ratios. Fig. 3.20 shows that when $\beta > \sqrt{2}$, $\rho/v_{g0} < 1$; when $\beta \gg 1$, $\rho/v_{g0} \to 0$, that is, $\rho \to 0$, indicating that the ground motion is not transmitted to the mass m. Additionally, when $\beta > \sqrt{2}$, the larger the damping, the larger the amplitude, indicating that the damping has an adverse effect in this case. Hence, the damping should be kept as small as possible. If $\xi \to 0$, Eq. (3.59) produces

$$\lim_{\xi \to 0}\frac{\rho}{v_{g0}} = \frac{1}{|1 - \beta^2|} \tag{3.60}$$

This equation indicates that as long as a low stiffness spring is adopted, that is, the natural frequency ω of the system is much lower than the ground vibrating frequency $\overline{\omega}$ ($\beta \gg 1$), the mass m is almost stationary

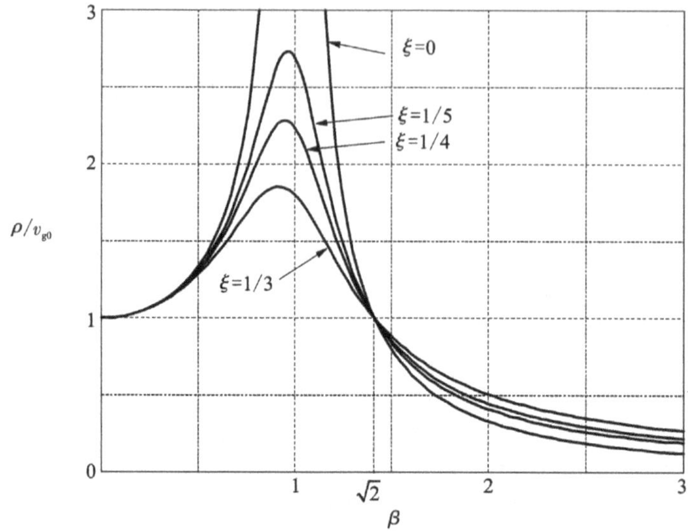

Figure 3.20 Amplitude of vibration due to base motion versus frequency ratio β.

regardless of the vibration of the ground. For example, when $\beta = 5$, the amplitude ρ of mass m is only $1/24$ of the base-motion amplitude v_{g0}. Plastic sheets are usually placed between the instrument and the instrument panel in a car so that the vibration of the car is transmitted to the instrument as little as possible.

Base motion is sometimes measured using the acceleration record. For example, earthquakes are recorded using a three-dimensional (east—west, north—south, and up—down) accelerometer. The damage due to earthquakes is mainly caused by the large deformation of the members resulting from the large motion of the building relative to the ground. Therefore engineering designs often concern the motion of the system relative to the base. Suppose the ground acceleration in Fig. 3.19 is measured as

$$\ddot{v}_g = \ddot{v}_{g0} \sin \overline{\omega} t \tag{3.61}$$

The displacement of mass m relative to the base is expressed as

$$v_r = v - v_g \tag{3.62}$$

Hence,

$$\begin{cases} v = v_r + v_g \\ \dot{v} = \dot{v}_r + \dot{v}_g \\ \ddot{v} = \ddot{v}_r + \ddot{v}_g \end{cases} \tag{3.63}$$

Substituting Eq. (3.63) into Eq. (3.53) yields

$$m\ddot{v}_r + c\dot{v}_r + kv_r = -m\ddot{v}_g = -m\ddot{v}_{g0}\sin\overline{\omega}t \tag{3.64}$$

Because Eq. (3.64) is similar to Eq. (3.36), the solution of Eq. (3.36) can be used as long as P_0 is substituted for $-m\ddot{v}_{g0}$. The steady-state response of mass m relative to the base is

$$v_r = -\frac{m\ddot{v}_{g0}}{k}D\sin(\overline{\omega}t - \theta) \tag{3.65}$$

where

$$\theta = \tan^{-1}\frac{2\xi\beta}{1-\beta^2}$$

By ignoring the damping effect, Eq. (3.65) becomes

$$v_r = -\frac{m\ddot{v}_{g0}}{k|1-\beta^2|}\sin\overline{\omega}t \tag{3.66}$$

Example 3.4: Deflections will develop in concrete bridges due to creep. If the bridge consists of a long series of identical spans, these deformations will cause a harmonic excitation for a vehicle traveling over the bridge at constant speed. The springs and shock absorbers of the vehicle are intended to provide a vibration—isolation system which limits the vertical motions transmitted from road to occupants. Fig. 3.21 shows an idealized model of this type of system in which the vehicle mass is 1814 kg, and its spring stiffness is defined by a test which showed that adding 444.52 N force caused a deflection of 2.032×10^{-3} m. The bridge surface profile is

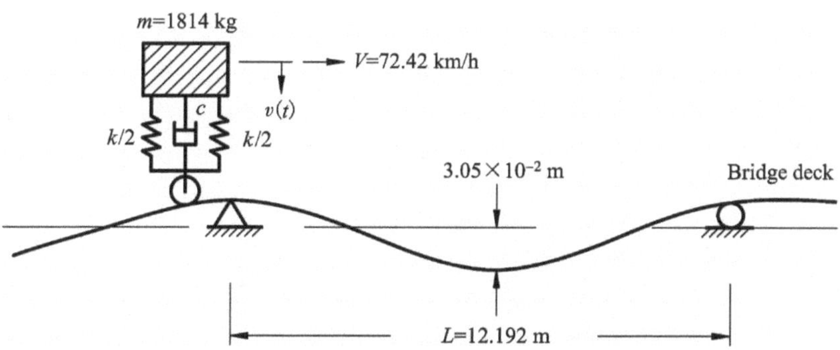

Figure 3.21 Idealized vehicle traveling over an uneven bridge deck.

represented by a sine curve having a wavelength (girder span) of 12.292 m and a (single) amplitude of 3.05×10^{-2} m. Based on these data, it is desired to predict the steady-state vertical motions in the vehicle traveling at a speed of 72.42 km/h. The damping ratio was selected as 40%.

Solution:
The speed of the traveling vehicle is $V = 72.42\ \text{km/h} = 20.12\ \text{m/s}$
The period of the excitation due to unevenness of the bridge deck is

$$T_p = 12.192/20.12 = 0.606\ \text{s}$$

The natural period of the vehicle is

$$T = 2\pi\sqrt{\frac{m}{k}} = 2\pi\sqrt{\frac{1814}{444.52/(2.032 \times 10^{-3})}} = 0.572\text{s}$$

Therefore the frequency ratio is

$$\beta = \frac{\overline{\omega}}{\omega} = \frac{T}{T_p} = \frac{0.572}{0.606} = 0.944$$

By considering $\xi = 0.40$ and $v_{g0} = 3.05 \times 10^{-2}$ m, the steady-state vertical amplitude is calculated from Eq. (3.59) as follows:

$$\rho = \frac{v_{g0}\sqrt{1 + (2\xi\beta)^2}}{\sqrt{\left(1-\beta^2\right)^2 + (2\xi\beta)^2}} = \frac{3.05 \times 10^{-2}\sqrt{1 + (2 \times 0.4 \times 0.944)^2}}{\sqrt{(1-0.944^2)^2 + (2 \times 0.4 \times 0.944)^2}} = 0.05\ \text{m}$$

If no damping is present in the vehicle ($\xi = 0$), the amplitude would be

$$\rho = \frac{3.05 \times 10^{-2}}{\sqrt{(1-0.944^2)^2}} = \frac{3.05 \times 10^{-2}}{0.11} = 0.277\ \text{m}$$

Such a large amplitude is beyond the elastic range of the spring, and thus has little practical significance. But it does demonstrate the important function of shock absorbers in limiting the motions resulting from the waviness of the bridge (or road) surface.

3.3.2 Vibration isolation

The vibration of the structural system conversely exerts a reaction on the base, and the design for the vibration isolation should be considered. The rotating machine with an eccentric mass produces an unbalanced force as shown in Fig. 3.12. If the machine is mounted directly on a rigid foundation, this unbalanced force is transmitted to the base completely, which may cause nearby equipments and buildings to vibrate and generate strong noise. To reduce the transmission of the unbalanced force, the bottom of the machine is usually equipped with springs, rubbers, corks, felts, and other materials, which is equivalent to a spring and a damper linking the bottom of the machine and the base.

When the machine vibrates vertically, as shown in Fig. 3.19, the resultant force transmitted to the base is the sum of spring force kv and the damping force $c\dot{v}$, which is called base force. The base force is given by

$$kv + c\dot{v} = k\rho \sin(\overline{\omega}t - \theta) + c\overline{\omega}\rho \cos(\overline{\omega}t - \theta) = F_T \sin(\overline{\omega}t - \theta + \alpha) \quad (3.67)$$

where

$$F_T = \sqrt{(k\rho)^2 + (c\overline{\omega}\rho)^2} = \rho\sqrt{k^2 + (c\overline{\omega})^2} \quad (3.68)$$

$$\alpha = \tan^{-1}\frac{c\overline{\omega}}{k} \quad (3.69)$$

where F_T is the amplitude of the base force, and α is the phase angle by which the displacement v lags behind the base force.

The ratio of the maximum base force to the amplitude of applied force, which is known as the transmissibility (TR) of the supporting system, can be expressed as

$$TR = \frac{F_T}{P_0} = \frac{\rho\sqrt{k^2 + (c\overline{\omega})^2}}{k\rho/D} = D\sqrt{1 + \left(\frac{2m\xi\omega\overline{\omega}}{k}\right)^2} = D\sqrt{1 + (2\xi\beta)^2} \quad (3.70)$$

In addition, Eq. (3.59) also gives

$$TR = \frac{\rho}{v_{g0}} = D\sqrt{1 + (2\xi\beta)^2} \quad (3.71)$$

Therefore it should be noted that the requirement of the unbalanced force not being transmitted to the base, that is, $F_T \ll P_0$, is identical to that of the base vibration not being transmitted to the structural system, that is, $\rho \ll v_{g0}$.

To facilitate the design of the vibration–isolation system, the isolation efficiency $IE = 1 - TR$ should be adopted to represent the vibration isolation effect. According to Fig. 3.20, $\beta \gg \sqrt{2}$ and very small damping should be ensured to obtain a good isolation effect. When $\beta < \sqrt{2}$, $TR \geq 1$ holds regardless of the damping, indicating that a practical vibration–isolation system is efficient in the range of $\beta > \sqrt{2}$.

The expressions of the transmissibility with zero damping and the corresponding IE are as follows:

$$TR = \frac{1}{\beta^2 - 1}, \quad IE = 1 - TR = \frac{\beta^2 - 2}{\beta^2 - 1} \tag{3.72}$$

where $\beta \geq \sqrt{2}$. When $\beta \rightarrow \infty$, $IE = 1$, indicating that the vibration is isolated completely. When $\beta = \sqrt{2}$, $IE = 0$, indicating that the isolation effect vanishes.

Noting that $\beta^2 = \overline{\omega}^2/\omega^2 = \overline{\omega}^2 m/k = \overline{\omega}^2 W/(kg) = \overline{\omega}^2 v_{st}/g$ (g is the acceleration of gravity, $v_{st} = W/k$, and v_{st} is the static displacement of the isolated object due to its self-weight W) and $\overline{\omega} = 2\pi\overline{f}$ ($\overline{f}$ is the exciting cyclic frequency), the relationship between $\overline{f}$ and IE is obtained from the second expression of Eq. (3.72) as follows:

$$\overline{f} = \frac{\overline{\omega}}{2\pi} = \frac{1}{2\pi}\sqrt{\frac{g(2 - IE)}{v_{st}(1 - IE)}} \tag{3.73}$$

The IE can be calculated if $\overline{f}$ and v_{st} are known. Conversely, v_{st} can be calculated with known $\overline{f}$ and IE, and then the stiffness coefficient k of the support pad can also be determined. Ref. [1] provided the vibration–isolation design chart based on Eq. (3.73), with which it is convenient to directly find the required data.

Example 3.5: A reciprocating machine with the mass of 9.071×10^3 kg is known to develop a vertical harmonic force of amplitude 2.22 kN at its operating frequency of 40 Hz. To limit the vibrations excited in the building in which this machine is to be installed, the machine is supported by a spring at each corner of its rectangular base. The support stiffness required for each spring to limit the total harmonic force transmitted from the machine to the building to 0.355 6 kN needs to be designed.

Solution:

The transmissibility in this case is obtained as

$$TR = \frac{F_T}{P_0} = \frac{0.3556}{2.22} = 0.16$$

From the first expression of Eq. (3.72), one obtains

$$\beta^2 = \frac{1}{TR} + 1 = 7.25$$

By considering $\beta^2 = \overline{\omega}^2/\omega^2 = \overline{\omega}^2 m/k$, total spring stiffness is given by

$$k = \frac{\overline{\omega}^2 m}{\beta^2} = \frac{(2\pi\overline{f})^2 m}{\beta^2} = \frac{(2\pi \times 40)^2 \times 9.071 \times 10^3}{7.25}$$
$$= 7.90 \times 10^7 \text{ N/m} = 7.90 \times 10^4 \text{ kN/m}$$

Therefore the required maximum stiffness of each spring is

$$\frac{k}{4} = \frac{7.90 \times 10^4}{4} = 1.98 \times 10^4 \text{ kN/m}$$

3.4 Introduction to damping theory

Engineering practice shows that the free vibration of a system gradually decays and eventually stops. External forces must be continuously applied to maintain the steady-state vibration of the system. These indicate the energy dissipation of the system in the process of vibration. Microscopically, the thermal effects produced by the relative motion of the material molecules during structural vibration are irreversible. Local inelastic deformation also occurs due to the inhomogeneity of the material, which cause the materials to dissipate energy in the process of structural vibration. The friction at connection points of structures (such as the friction at bolt joints of steel structures), as well as the opening and closing of microcracks in concrete, tends to dissipate energy due to friction resulting from relative motion. The surrounding medium resists the structural vibration (e.g., the aircraft is resisted by the atmosphere, and the submarine is resisted by the sea water) and also dissipates the vibration energy. When the structural vibration energy is transferred to the foundation, it is partly dissipated by the internal friction of soil. The mechanism of the energy dissipation of the system is usually referred to as damping, which is generally represented by damping forces.

In practical engineering, it is difficult to find an accurate damping-force model due to the combined effects of several factors acting on structures. For simplification, a highly idealized damping-force model is generally used in structural vibration analysis, when some type of damping dominates the damping effects. The damping theories corresponding to different energy-dissipation mechanisms have different damping-force models, such as viscous, hysteretic, and frictional damping-force models. Here, three commonly used damping theories and associated damping-force models are introduced.

3.4.1 Viscous-damping theory

3.4.1.1 Viscous-damping-force model

In viscous-damping theory, it is assumed that viscous-damping force is proportional to the velocity, that is,

$$F_{vd} = c\dot{v} \tag{3.74}$$

where F_{vd} is the viscous-damping force, c is the viscous-damping coefficient and $\dot{v}$ is the velocity. The damping force always opposes the velocity $\dot{v}$. The viscous-damping hypothesis leads to a linear differential equation of motion of the system, which is relatively easy to solve. Hence, it is widely used in dynamic analysis.

3.4.1.2 Problems of viscous damping

The experimental results show that the viscous-damping hypothesis is not ideally consistent with the energy-dissipation rule of actual structures. To analyze the problems of viscous damping, the energy-dissipation mechanism of viscous damping is investigated first.

The steady-state displacement response of a SDOF system subjected to applied load $P(t) = P_0\sin\overline{\omega}t$ is $v(t) = \rho\sin(\overline{\omega}t - \theta)$, and the corresponding velocity is

$$\dot{v}(t) = \rho\overline{\omega}\cos(\overline{\omega}t - \theta) = \pm\rho\overline{\omega}\sqrt{1 - \sin^2(\overline{\omega}t - \theta)} = \pm\rho\overline{\omega}\sqrt{1 - \left(\frac{v}{\rho}\right)^2}$$

The damping force (to be consistent with the notations used in other chapters, the viscous-damping force is denoted by F_d) is:

$$F_d = c\dot{v} = \pm c\rho\overline{\omega}\sqrt{1 - \left(\frac{v}{\rho}\right)^2}$$

This equation can also be rewritten as

$$\left(\frac{F_d}{c\overline{\omega}\rho}\right)^2 + \left(\frac{v}{\rho}\right)^2 = 1 \tag{3.75}$$

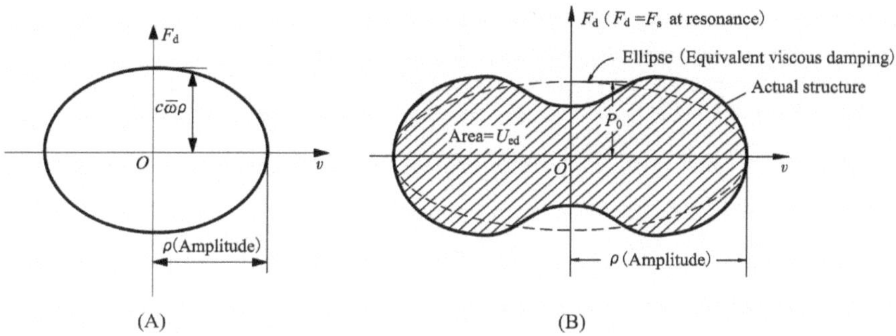

Figure 3.22 Hysteresis curves of (A) viscous damping and (B) equivalent viscous damping.

Eq. (3.75) indicates that the relationship between damping force F_d and displacement v is an ellipse as shown in Fig. 3.22A. This curve is called a hysteresis curve or hysteresis loop, which represents the hysteresis characteristic of a viscous-damping system in steady-state vibrations.

Both the damping force F_d and the displacement v vary with time. The work done by the damping force F_d in one cycle (the work is actually negative, and the following equation gives the magnitude of work) is

$$W_d = \int_0^T F_d \frac{dv}{dt} dt = \int_0^T c\dot{v}\dot{v} dt = c\int_0^T \dot{v}^2 dt = c\rho^2\overline{\omega}^2 \int_0^T \cos^2(\overline{\omega}t - \theta) dt = \pi c\rho^2\overline{\omega}$$

The above work is easily proven to be equal to the area of the ellipse shown in Fig. 3.22A. U_{vd} denotes the energy dissipated in one cycle for a viscously damped system, which is given by

$$U_{vd} = \pi c\rho^2\overline{\omega} \tag{3.76}$$

which indicates that the energy dissipation is proportional to the exciting frequency due to the viscous-damping hypothesis, that is, the higher the vibration frequency, the greater the energy dissipated in a cycle. However, experiments have proven that the energy dissipation in a cycle for many structural systems is independent of the vibration frequency. Therefore the correction of the viscous-damping hypothesis or other damping hypotheses is required.

3.4.1.3 Equivalent viscous damping

Experiments show that the influence of damping on structural vibrations mainly depends on the amount of dissipated energy and is rarely related to the specific process of energy dissipation. Based on this consideration, the concept of equivalent viscous damping was presented.

Although the actual structure is not a viscously damped system, the system can be assumed to have an equivalent viscous damping to take advantage of the simplified analysis. The equivalent viscous-damping model was developed by assuming that the dissipated energy by the assumed system is equal to the energy dissipated by the actual structure in one cycle. It is also assumed that the displacement amplitudes of the assumed and actual systems are identical.

Suppose that the area surrounded by the actual structural hysteretic curve as illustrated by the solid line is equal to the ellipse area shown by the dashed line in Fig. 3.22B. Hence,

$$U_{ed} = \pi c_{eq} \rho^2 \overline{\omega} \tag{3.77}$$

The equivalent viscous-damping coefficient c_{eq} and equivalent viscous-damping ratio ξ_{eq} can be obtained

$$c_{eq} = \frac{U_{ed}}{\pi \rho^2 \overline{\omega}}, \quad \xi_{eq} = \frac{c_{eq}}{2m\omega} \tag{3.78}$$

The energy dissipation in a cycle, U_{ed}, for the actual structure can be measured by the resonance experiment. It is stated in Section 3.2 that damping force F_d is out of phase with applied load, and their magnitudes are identical at resonance ($\overline{\omega} = \omega$). Thus the value of damping force can be obtained as long as the applied load is measured. The corresponding value of the displacement can also be measured so that the hysteretic curve of the actual structure can be obtained. The specific test method is outlined in Section 3.5. The area surrounded by this curve is U_{ed}. Considering the stiffness coefficient $k = m\omega^2$, the equivalent viscous-damping coefficient and equivalent viscous-damping ratio at resonance ($\overline{\omega} = \omega$ and $\rho = \rho_{\beta=1}$) can be written as

$$c_{eq} = \frac{U_{ed}}{\pi \rho_{\beta=1}^2 \omega}, \quad \xi_{eq} = \frac{c_{eq}}{2m\omega} = \frac{U_{ed}}{2\pi m \rho_{\beta=1}^2 \omega^2} = \frac{U_{ed}}{2\pi k \rho_{\beta=1}^2} \tag{3.79}$$

Once the equivalent viscous-damping ratio is obtained, the equation of the viscous-damping system obtained in the past can be applied as long as the damping ratio ξ in the original equation is replaced by the equivalent viscous-damping ratio ξ_{eq}.

3.4.2 Hysteretic-damping theory

Although the viscous-damping model simplifies the equations of motion, the experimental results were rarely consistent with this type of energy loss. The equivalent viscous-damping concept defined by the energy loss per cycle makes it possible that the theory is close to the experimental results in many experiments. However, the viscous-damping mechanism depending on the exciting frequency as mentioned above is inconsistent with a large number of experimental results. Most experimental results showed that the damping forces are almost independent of the exciting frequency.

The mathematical model provided by the hysteretic-damping theory is independent of frequency. The damping force in this case is defined to be proportional to the displacement and in phase with the velocity. The hysteretic-damping mechanism, which can well represent the energy-dissipation mechanism of inter friction of the material, is also known as material damping or structural damping. The relationship between the hysteretic-damping force and displacement can be expressed as

$$F_{hd} = \zeta k|v| \tag{3.80}$$

where F_{hd} is the hysteretic-damping force, k is the elastic stiffness coefficient, $|v|$ is the absolute value of displacement, and ζ is the hysteretic-damping coefficient, which is the ratio of the damping force to the elastic resistance. The hysteretic-damping force is in phase with the velocity $\dot{v}$. The relationship between the hysteretic-damping force and displacement in a complete cycle is shown in Fig. 3.23. When the displacement

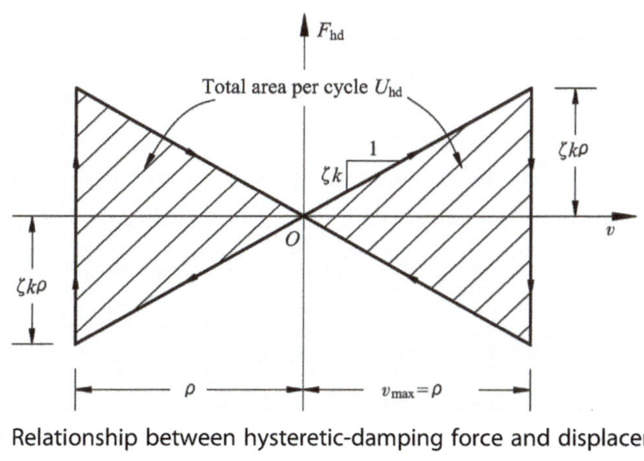

Figure 3.23 Relationship between hysteretic-damping force and displacement.

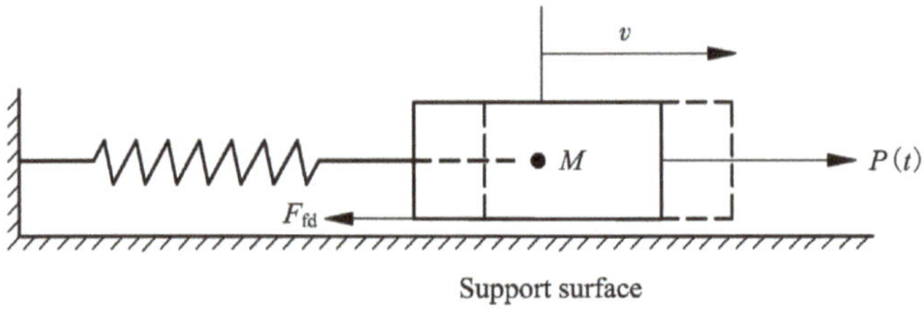

Figure 3.24 Coulomb friction.

increases, the damping force is similar to the linear elastic force. When the displacement decreases, the damping force acts opposite to the sense of the displacement.

The hysteretic energy loss per cycle as given by this mechanism is

$$U_{hd} = 2\zeta k\rho^2 \tag{3.81}$$

The hysteretic energy loss per cycle is independent of the exciting frequency, which is different from U_{ed} in Eq. (3.77).

3.4.3 Frictional damping theory

The friction between components of the system and the friction between the system and the support surface are often called the dry friction or Coulomb friction, such as the friction between the beam and the bearing, and that between the mass M and the support surface in Fig. 3.24. The frictional force is assumed to be

$$F_{fd} = \mu N \tag{3.82}$$

where F_{fd} is the frictional force, μ is the friction coefficient, and N is the magnitude of normal pressure on the support surface. The frictional force always opposes the velocity $\dot{v}$.

The experiment has proven that μ is almost constant (less than the static friction coefficient) in the case of low velocity. When μ is constant, the frictional force is constant regardless of the velocity, and this damping mechanism is called Coulomb damping. The solution of the equations of motion for a system exhibiting Coulomb damping is more complex.

3.5 Evaluation of viscous-damping ratio

The mass, stiffness, damping, and other physical parameters of the system must be determined to analyze the vibration response of a SDOF system.

In most cases, the parameters of mass and stiffness are easily evaluated using simple physical methods. It is usually not feasible to determine the damping coefficient by similar means because the damping mechanism in most actual structures is seldom fully understood. In fact, the energy-dissipation mechanism of actual structures is more complicated than the viscously damped model. However, it is possible to determine an appropriate equivalent viscous-damping coefficient by experimental methods. A brief treatment of the methods commonly used for this purpose is introduced in this section [1].

3.5.1 Free-vibration decay method

According to Eq. (3.16) in Section 3.1, the damping ratio is

$$\xi = \frac{\delta_s}{2\pi s \frac{\omega}{\omega_D}} \approx \frac{\delta_s}{2\pi s} \tag{3.83}$$

where $\delta_s = \ln(v_m / v_{m+s})$ represents the logarithmic decrement over s cycles. After the free vibrations of the system are activated by any means, the amplitudes of the mth and $(m+s)$th cycles can be measured, and the damping ratio ξ can be calculated from Eq. (3.83). Since the first natural mode of the system dominates the system vibration, the damping ratio ξ obtained is actually associated with the first principal vibration. This method requires the least number of instruments and the free vibrations of the system can be easily initiated. So it is the simplest and most commonly used method.

3.5.2 Resonant amplification method

The following relation can be obtained from the second equation of Eq. (3.44)

$$\xi = \frac{P_0}{2k\rho_{\beta=1}} = \frac{v_{st}}{2\rho_{\beta=1}} \tag{3.84}$$

where $v_{st} = P_0/k$ is the static displacement caused by the static load P_0, and $\rho_{\beta=1}$ is the resonance amplitude of the steady-state response of the system. It is difficult to exert a precise resonant load in practice, but the maximum response amplitude $\rho_{\max}$ can be determined conveniently according to the frequency–response curve of the system, shown in Fig. 3.25. It occurs when the exciting frequency $\overline{\omega}$ is slightly smaller than

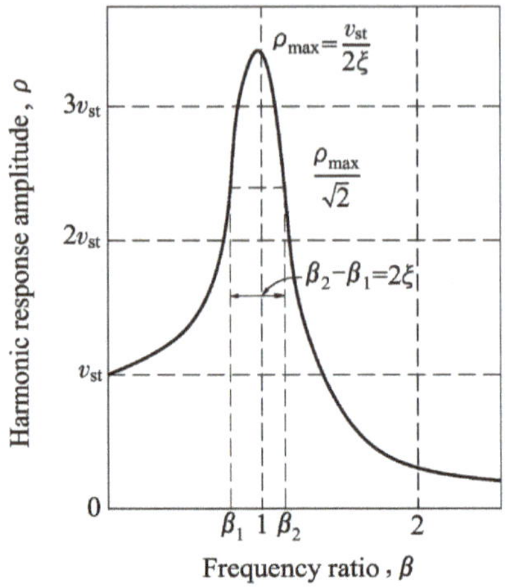

Figure 3.25 Frequency–response curve.

the natural frequency ω (i.e., β is slightly less than unity). The damping ratio ξ can be evaluated from the experimental data using

$$\xi = \frac{P_0}{2k\rho_{max}\sqrt{1-\xi^2}} \approx \frac{v_{st}}{2\rho_{max}} \tag{3.85}$$

This method of determining the damping ratio requires only simple instrumentation to measure the dynamic response amplitudes at discrete values of frequency and fairly simple dynamic-load equipment. However, obtaining the static displacement v_{st} may present a problem because the typical harmonic-load system cannot produce a load at zero frequency.

3.5.3 Half-power (band-width) method

According to Eq. (3.40), the frequency–response curve is controlled by the amount of damping of the system. Therefore the damping ratio can be obtained from many different properties of the curve, such as resonant amplification method. Another convenient method is the half-power method. In this method, the damping ratio is determined according to two specified exciting frequencies at which the amplitude is equal to $1/\sqrt{2}\rho_{max}$. Since the input power at these frequencies is approximately half of the resonance power, it is called the half-power method.

From Eq. (3.40), the amplitude is $\rho = P_0D/k = v_{st}\left[\left(1-\beta^2\right)^2+(2\xi\beta)^2\right]^{-\frac{1}{2}}$. Introducing Eq. (3.45) leads to $\rho_{max} = P_0/\left(2k\xi\sqrt{1-\xi^2}\right) = v_{st}/\left(2\xi\sqrt{1-\xi^2}\right)$.

The condition that the amplitude ρ is equal to $1/\sqrt{2}\rho_{\max}$ can be satisfied if

$$\frac{1}{\sqrt{2}}\frac{v_{st}}{2\xi\sqrt{1-\xi^2}}=\frac{v_{st}}{\sqrt{(1-\beta^2)^2+(2\xi\beta)^2}}$$

The square of the frequency ratio is solved as

$$\beta_{1,2}^2=1-2\xi^2\pm 2\xi\sqrt{1-\xi^2}$$

Since ξ is very small for general engineering structures, the term ξ^2 in the above expression can be ignored, thus

$$\beta_{1,2}=\sqrt{1\pm 2\xi}$$

Two half-power frequency ratios are obtained by retaining only the first two terms in the Taylor series as follows:

$$\beta_1=\sqrt{1-2\xi}\approx 1-\xi,\ \beta_2=\sqrt{1+2\xi}\approx 1+\xi \tag{3.86}$$

Subtracting β_1 from β_2, one obtains

$$\beta_2-\beta_1=2\xi$$

and then

$$\xi=\frac{1}{2}(\beta_2-\beta_1)=\frac{1}{2}\left(\frac{\overline{\omega}_2}{\omega}-\frac{\overline{\omega}_1}{\omega}\right)=\frac{1}{2}\left(\frac{\overline{f}_2-\overline{f}_1}{f}\right) \tag{3.87}$$

Adding β_1 into β_2 yields

$$\beta_2+\beta_1=2$$

and then

$$f=\frac{1}{2}(\overline{f}_2+\overline{f}_1) \tag{3.88}$$

Eq. (3.87) indicates that the damping ratio is equal to half of the difference between the two half-power frequency ratios, which can be determined from the frequency–response curve. As shown in Fig. 3.25, a horizontal line is drawn across the curve at $1/\sqrt{2}$ times its peak value, and the difference mentioned above can be easily obtained. It is evident that this method of obtaining the damping ratio avoids the need for determining the static displacement v_{st}, however, it does require that the frequency–response curve be obtained accurately at its peak and at the level of $1/\sqrt{2}\rho_{\max}$.

Why the above method is commonly referred to as the half-power method is clarified below. The steady-state response of a SDOF system is $v(t) = \rho\sin(\overline{\omega}t - \theta)$ under the action of the applied load $P(t) = P_0\sin\overline{\omega}t$. The energy input to the system by force $P(t)$ (i.e., the work done by the applied load $P(t)$) is equal to the energy dissipated by the viscous damping in a cycle (i.e., the work done by the damping force F_d). The average rate of input energy (i.e., average power input) of the system can therefore be calculated as follows:

The input energy of the force $P(t) = P_0\sin\overline{\omega}t$ in one cycle of the steady-state response is

$$W_p = \int P(t)\mathrm{d}v = \int_0^{2\pi/\overline{\omega}} P(t)\,\dot{v}\,\mathrm{d}t = P_0\rho\overline{\omega}\int_0^{2\pi/\overline{\omega}} \sin\overline{\omega}t\cos(\overline{\omega}t - \theta)\mathrm{d}t = P_0\rho\pi\sin\theta \tag{3.89}$$

Combining Eqs. (3.40) and (3.41) leads to

$$\sin\theta = \frac{2\xi\beta}{\sqrt{(1-\beta^2)^2 + (2\xi\beta)^2}} = \frac{\rho}{P_0/k}(2\xi\beta)$$

Substituting the above equation into Eq. (3.89), one obtains

$$W_p = 2\pi\xi k\beta\rho^2 \tag{3.90}$$

The corresponding average rate of input energy is

$$P_{p,avg} = \frac{W_p}{T} = \frac{W_p}{2\pi/\overline{\omega}} = \xi m\omega^3(\beta\rho)^2 \tag{3.91}$$

In addition, the energy dissipated by the viscous damping in one cycle is

$$W_d = \int F_d dv = \int_0^{2\pi/\overline{\omega}} c\dot{v}\dot{v}dt = c\rho^2\overline{\omega}^2\int_0^{2\pi/\overline{\omega}} \cos^2(\overline{\omega}t - \theta)dt = \pi c\overline{\omega}\rho^2$$

The corresponding average rate of energy dissipation is

$$P_{d,avg} = \frac{W_d}{T} = \frac{W_d}{2\pi/\overline{\omega}} = \frac{1}{2}c\overline{\omega}^2\rho^2 = \xi m\omega^3(\beta\rho)^2 \tag{3.92}$$

The comparison of Eqs. (3.91) and (3.92) demonstrates that the average rate of input energy of force $P(t)$ in steady-state vibration is equal to the average rate of energy dissipated by the system damping. Eq. (3.91) shows that the average rate of input energy is proportional to $(\beta\rho)^2$.

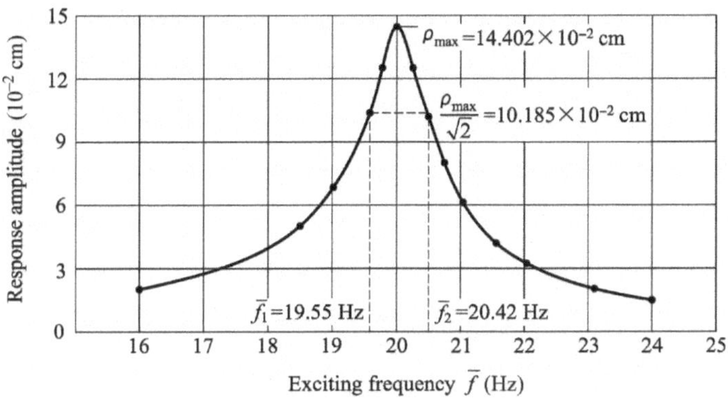

Figure 3.26 Frequency—response experiment to determine damping ratio.

When $\beta = \beta_m$, that is, at resonance, the amplitude ρ reaches its maximum value $\rho_{\max}$. The corresponding average rate of input energy is $P_{p,m}$. When $\rho_1 = \rho_2 = 1/\sqrt{2}\rho_{\max}$, the average rates of input energy at frequency ratios β_1 and β_2 corresponding to ρ_1 and ρ_2 are

$$P_{\beta_1} = \left(\frac{\beta_1\rho_1}{\beta_m\rho_{\max}}\right)^2 P_{p,m} = \left(\frac{\beta_1}{\beta_m}\right)^2 \frac{P_{p,m}}{2}$$

$$P_{\beta_2} = \left(\frac{\beta_2\rho_2}{\beta_m\rho_{\max}}\right)^2 P_{p,m} = \left(\frac{\beta_2}{\beta_m}\right)^2 \frac{P_{p,m}}{2}$$

where β_m is given by Eq. (3.45). While the average rate of input energy P_{β_1} at β_1 is somewhat less than one-half of the rate of input energy at β_m and the average rate of input energy P_{β_2} at β_2 is somewhat greater. The mean value of P_{β_1} and P_{β_2} is very close to one-half of $P_{P,m}$ at β_m. Therefore this approach is called the half-power method.

Example 3.6: The data of the frequency response curve of a SDOF system is shown in Fig. 3.26, from which the damping ratio of the system was evaluated.

Solution:

1. Determine the peak response

$$\rho_{\max} = 14.402 \times 10^{-2} \text{ cm}$$

2. Draw a horizontal line at $1/\sqrt{2}$ times the peak level.

3. Determine the two frequencies at which the horizontal line cuts the frequency—response curve:

$$\bar{f}_1 = 19.55\ \text{Hz},\ \bar{f}_2 = 20.42\ \text{Hz}$$

4. Evaluate the damping ratio by combining Eqs. (3.87) and (3.88)

$$\xi = \frac{1}{2}\left[\frac{\bar{f}_2 - \bar{f}_1}{(\bar{f}_1 + \bar{f}_2)/2}\right] = \frac{\bar{f}_2 - \bar{f}_1}{\bar{f}_2 + \bar{f}_1} = \frac{20.42 - 19.55}{20.42 + 19.55} = \frac{0.87}{39.97} = 2.18\%$$

3.5.4 Resonance energy loss per cycle method

According to the discussion in Section 3.4, the equivalent viscous-damping coefficient and the equivalent viscous-damping ratio at resonance ($\overline{\omega} = \omega,\ \rho = \rho_{\beta=1}$) can be expressed as the function of the dissipated energy per cycle U_{ed}.

$$c_{eq} = \frac{U_{ed}}{\pi\rho_{\beta=1}^2\omega},\ \xi_{eq} = \frac{c_{eq}}{2m\omega} = \frac{U_{ed}}{2\pi m\rho_{\beta=1}^2\omega^2} = \frac{U_{ed}}{2\pi k\rho_{\beta=1}^2} \tag{3.93}$$

If the dissipated energy per cycle U_{ed}, the amplitude $\rho_{\beta=1}$ at resonance, and structural stiffness coefficient k can be measured, the equivalent viscous-damping ratio at resonance can be determined from the above equation.

Due to the equilibrium between the applied load F_s and the damping force F_d at resonance described in Section 3.2, the applied-load/displacement diagram in one cycle can be regarded as the damping-force/displacement diagram. If the structure has linear viscous damping, the diagram is an ellipse in accordance with Eq. (3.75) (its area is $\pi\rho_{\beta=1}P_0$ and P_0 is the amplitude of the excitation at resonance), as illustrated by the dashed line in Fig. 3.27. If the nonlinear viscous damping is present, the above diagram is not an ellipse, but a solid line in Fig. 3.27. The area enclosed by the solid line is denoted by U_{ed}, and the maximum amplitude v_{max} is the same as $\rho_{\beta=1}$ at resonance.

The stiffness k of the structure can be measured by the same instrumentation used to obtain the energy loss per cycle, only by operating the system very slowly at essentially static conditions. The applied force/displacement diagram under static conditions can be measured, from which the maximum elastic strain energy U_s of the structure can be obtained. The static force/displacement diagram obtained in this way will be of the form shown in Fig. 3.28, if the structure is linearly elastic. The stiffness is

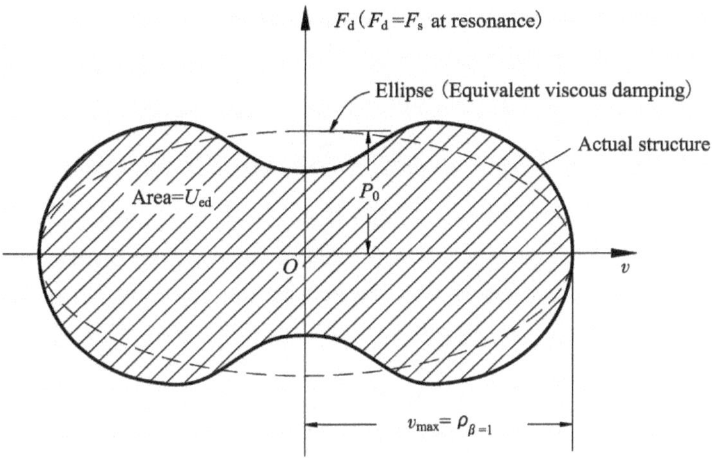

Figure 3.27 Actual and equivalent damping energy dissipation per cycle at resonance.

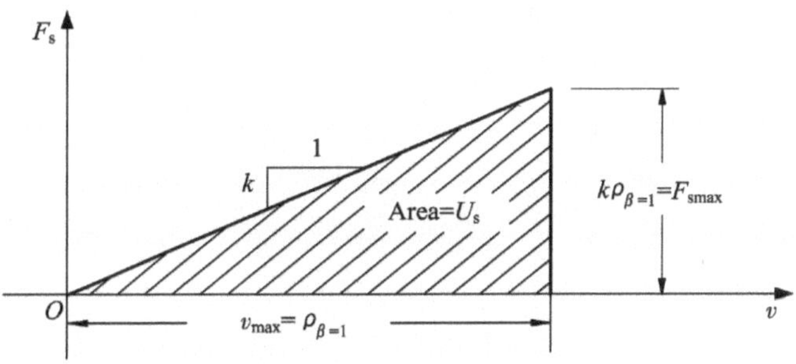

Figure 3.28 Elastic stiffness and elastic strain energy.

obtained as the slope of the straight line. The maximum elastic force of the structure is $k\rho_{\beta=1}$ and the maximum elastic strain energy of the structure is $U_s = 1/2k\rho^2_{\beta=1}$ at resonance. Therefore the structural stiffness coefficient can also be expressed as

$$k = \frac{2U_s}{\rho^2_{\beta=1}} \tag{3.94}$$

Substituting Eq. (3.94) into the second equation of Eq. (3.93) yields

$$\xi_{eq} = \frac{U_{ed}}{4\pi U_s} \tag{3.95}$$

The equivalent viscous-damping ratio determined from a test at $\overline{\omega} = \omega$ would not be correct at any other exciting frequency, but it would be a satisfactory approximation. This equivalent method is widely used in engineering and is equally applicable to MDOF systems.

3.6 Response of SDOF systems to periodic loads

The inertial effect caused by reciprocating machines and the dynamic pressure generated by the stern thruster are both periodic loads. The variation with time for periodic loads can be generally illustrated by the curve shown in Fig. 3.29. Since any periodic load can be expressed as the combination of many harmonic components by means of Fourier series expansion, for linear elastic structures, the response to each simple harmonic load is superposed to obtain the total response due to periodic loads. In addition, the response analysis for linear systems to periodic loads is instructive for evaluating the response to nonperiodic loads.

The Fourier series expansion of periodic load $P(t)$ may be defined as

$$P(t) = \frac{A_0}{2} + \sum_{n=1}^{\infty} (A_n \cos n\overline{\omega}_1 t + B_n \sin n\overline{\omega}_1 t) \tag{3.96}$$

in which

$$\overline{\omega}_1 = \frac{2\pi}{T_P},$$

$$A_0 = \frac{2}{T_P}\int_0^{T_P} P(t)dt,$$

$$A_n = \frac{2}{T_P}\int_0^{T_P} P(t)\cos n\overline{\omega}_1 t dt, \quad n = 1, 2, 3, \cdots$$

$$B_n = \frac{2}{T_P}\int_0^{T_P} P(t)\sin n\overline{\omega}_1 t dt, \quad n = 1, 2, 3, \cdots$$

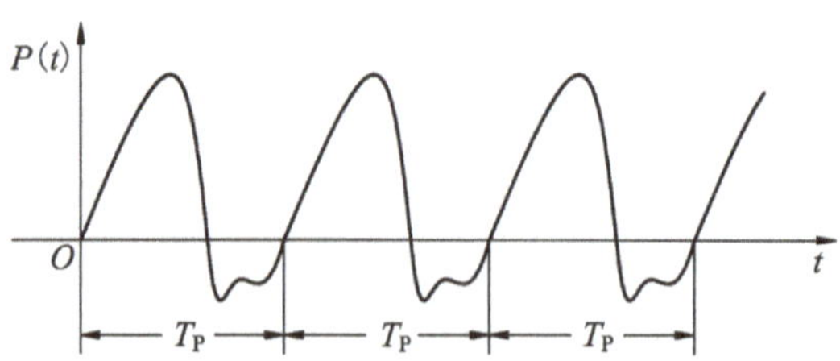

Figure 3.29 Arbitrary periodic load.

where T_P is the fundamental period of the load, as shown in Fig. 3.29, $\overline{\omega}_1$ is the fundamental frequency of $P(t)$, and A_n and B_n are called Fourier coefficients.

The equation of motion of a SDOF system subjected to periodic load is

$$m\ddot{v} + c\dot{v} + kv = \frac{A_0}{2} + \sum_{n=1}^{\infty} (A_n \cos n\overline{\omega}_1 t + B_n \sin n\overline{\omega}_1 t) \quad (3.97)$$

The steady-state response of the system is obtained using Eq. (3.39)

$$v(t) = \frac{A_0}{2k} + \frac{1}{k}\sum_{n=1}^{\infty} \left[A_n D_n \cos(n\overline{\omega}_1 t - \theta_n) + B_n D_n \sin(n\overline{\omega}_1 t - \theta_n)\right] \quad (3.98)$$

where $A_0/(2k)$ is the static displacement caused by constant load $A_0/2$,

$$D_n = \left[\left(1-\beta_n^2\right)^2 + (2\xi\beta_n)^2\right]^{-\frac{1}{2}}, \quad \beta_n = \frac{n\overline{\omega}_1}{\omega}, \quad \theta_n = \tan^{-1}\frac{2\xi\beta_n}{1-\beta_n^2} \quad (3.99)$$

Example 3.7: The undamped SDOF system in Fig. 3.30A is subjected to a periodic load which is shown in Fig. 3.30B. Assume that the fundamental period of the load is 4/3 times the natural period of the system, and then evaluate the steady-state response of the system.

Solution:

The frequency of each harmonic component and the associated Fourier coefficients are calculated as follows:

$$\overline{\omega}_n = n\overline{\omega}_1 = n\frac{2\pi}{T_P}$$

$$\frac{A_0}{2} = \frac{1}{T_P}\int_0^{\frac{T_P}{2}} P_0 \sin\frac{2\pi t}{T_P}\, dt = \frac{P_0}{\pi}$$

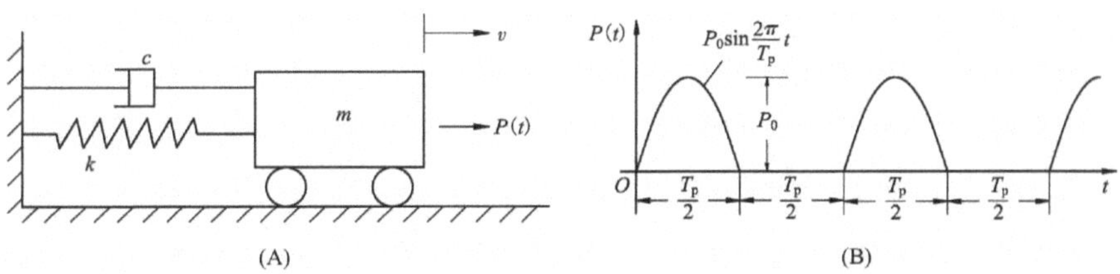

Figure 3.30 Analysis of response to periodic load: (A) SDOF system; (B) periodic load. *SDOF*, Single-degree-of-freedom.

$$A_n = \frac{2}{T_P}\int_0^{\frac{T_P}{2}} P_0 \sin\frac{2\pi t}{T_P}\cos\frac{2n\pi t}{T_P}dt = \begin{cases} 0 & , \quad n = 1, 3, 5, \cdots \\ \dfrac{P_0}{\pi}\dfrac{2}{1-n^2} & , \quad n = 2, 4, 6, \cdots \end{cases}$$

$$B_n = \frac{2}{T_P}\int_0^{\frac{T_P}{2}} P_0 \sin\frac{2\pi t}{T_P}\sin\frac{2n\pi t}{T_P}dt = \begin{cases} \dfrac{P_0}{2} & , \quad n = 1 \\ 0 & , \quad n > 1 \end{cases}$$

Substituting the frequencies and Fourier coefficients into Eq. (3.96) leads to the Fourier series representation of $P(t)$

$$P(t) = \frac{P_0}{\pi}\left(1 + \frac{\pi}{2}\sin \overline{\omega}_1 t - \frac{2}{3}\cos 2\overline{\omega}_1 t - \frac{2}{15}\cos 4\overline{\omega}_1 t - \frac{2}{35}\cos 6\overline{\omega}_1 t + \cdots\right)$$

where $\overline{\omega}_1 = 2\pi/T_P$. Since $T/T_P = 3/4$, $\beta_1 = \overline{\omega}_1/\omega = T/T_P = 3/4$, $\beta_2 = 2\overline{\omega}_1/\omega = 3/2$, $\beta_4 = 4\overline{\omega}_1/\omega = 3$, and $\beta_6 = 6\overline{\omega}_1/\omega = 9/2$.

Assuming the system without damping, one obtains $D_n = 1/\left(1-\beta_n^2\right)$ and $\theta_n = 0$.

Hence,

$$B_1 D_1 = \frac{P_0}{2}\frac{1}{1-\beta_1^2} = \frac{P_0}{2}\frac{1}{1-\left(\frac{3}{4}\right)^2} = \frac{8P_0}{7},$$

$$A_2 D_2 = \frac{P_0}{\pi}\frac{2}{1-2^2}\frac{1}{1-\beta_2^2} = -\frac{2P_0}{3\pi}\frac{1}{1-\left(\frac{3}{2}\right)^2} = \frac{8P_0}{15\pi},$$

$$A_4 D_4 = \frac{P_0}{\pi}\frac{2}{1-4^2}\frac{1}{1-\beta_4^2} = -\frac{2P_0}{15\pi}\frac{1}{1-3^2} = \frac{P_0}{60\pi},$$

$$A_6 D_6 = \frac{P_0}{\pi}\frac{2}{1-6^2}\frac{1}{1-\beta_6^2} = -\frac{2P_0}{35\pi}\frac{1}{1-\left(\frac{9}{2}\right)^2} = \frac{8P_0}{2695\pi} \approx \frac{P_0}{337\pi}.$$

By substituting the above parameters into Eq. (3.98), the steady-state response can be obtained as

$$v(t) = \frac{1}{k}\left(\frac{P_0}{\pi} + B_1 D_1 \sin \overline{\omega}_1 t + A_2 D_2 \cos 2\overline{\omega}_1 t + A_4 D_4 \cos 4\overline{\omega}_1 t + A_6 D_6 \cos 6\overline{\omega}_1 t + \cdots\right)$$

$$= \frac{P_0}{k\pi}\left(1 + \frac{8\pi}{7}\sin \overline{\omega}_1 t + \frac{8}{15}\cos 2\overline{\omega}_1 t + \frac{1}{60}\cos 4\overline{\omega}_1 t + \frac{1}{337}\cos 6\overline{\omega}_1 t + \cdots\right)$$

The analytical result shows an important concept that the higher the frequency of the load component, the smaller the activated dynamic response. This statement is consistent with the conclusions from Fig. 3.14.

3.7 Response of SDOF systems to impulsive loads

The impacts of atomic shock wave and bomb explosion on buildings, as well as those of rail joints and road pits on vehicles, are typical impulsive loads, as illustrated in Fig. 3.31. The structures subject to a particular impulsive load will reach their maximum response in a very short time. The damping force cannot dissipate too much energy in such a short time, and the damping is of little significance in controlling the maximum response of structures to impulsive loads. Therefore damping is generally not considered when the response of structures to impulsive loads is evaluated. To understand the dynamic characteristics of structures to impulsive loads, the dynamic responses of SDOF structures to three typical kinds of impulsive loads are discussed respectively [1].

3.7.1 Sine-wave impulsive load

As shown in Fig. 3.32, structural responses to sine-wave impulsive load can be divided into two phases, the first corresponding to the forced-vibration phase in the interval during which the load acts and the second corresponding to the free-vibration phase which follows.

Phase I: During this phase ($0 \leq t \leq t_1$), the SDOF structure is subjected to a single half-sine-wave load shown in Fig. 3.32. Assume that the system starts from rest, say $v(0) = \dot{v}(0) = 0$, and the damping effect is neglected,

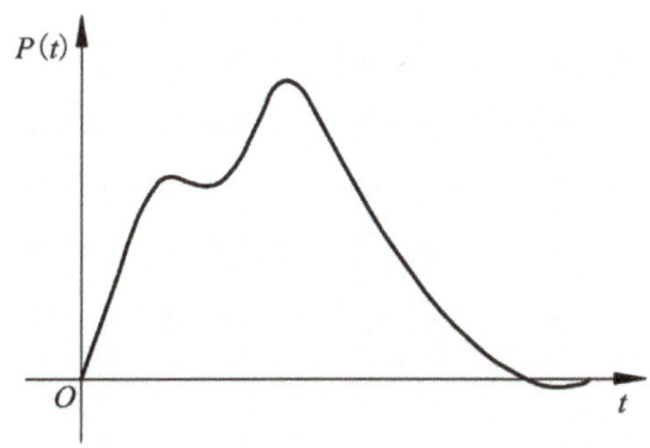

Figure 3.31 Arbitrary impulsive load.

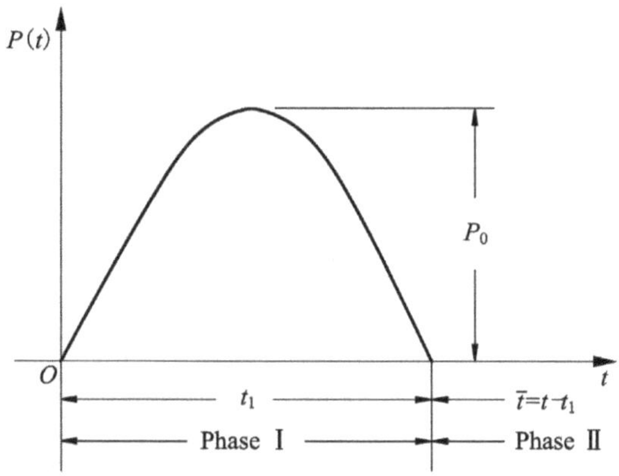

Figure 3.32 Half-sine-wave impulse load.

that is, $\xi = 0$, $\omega_D = \omega$. Considering these conditions, the dynamic response of the system is obtained from Eq. (3.43)

$$v(t) = \frac{P_0}{k(1-\beta^2)}(\sin \overline{\omega} t - \beta \sin \omega t), \ 0 \leq t \leq t_1 \tag{3.100}$$

Introducing a parameter $\alpha = t/t_1$ and considering $\overline{\omega} = 2\pi/(2t_1) = \pi/t_1$ leads to the time history of response ratio during Phase I

$$R(\alpha) = \frac{1}{1-\beta^2}\left[\sin\pi\alpha - \beta\sin\left(\frac{\pi\alpha}{\beta}\right)\right], \quad 0 \leq \alpha \leq 1 \tag{3.101}$$

Since Eq. (3.101) is indeterminate for $\beta = 1$, L'Hospital's rule must be applied to obtain a useable expression for this special case. Taking this action, one obtains

$$R(\alpha) = \frac{1}{2}(\sin\pi\alpha - \pi\alpha\cos\pi\alpha), \quad \beta = 1, \ 0 \leq \alpha \leq 1 \tag{3.102}$$

Phase II: The free vibration which occurs during this phase ($\bar{t} = t - t_1 \geq 0$), depends on the displacement $v(t_1)$ and velocity $\dot{v}(t_1)$ at the end of Phase I. According to Eq. (3.3), the system response during Phase II is given by

$$v(t) = v(t_1)\cos \omega\bar{t} + \frac{\dot{v}(t_1)}{\omega}\sin \omega\bar{t} \tag{3.103}$$

From Eq. (3.100), $v(t_1)$ and $\dot{v}(t_1)$ are obtained as

$$v(t_1) = \frac{P_0}{k(1-\beta^2)}(\sin\overline{\omega}t_1 - \beta\sin\omega t_1) = -\frac{P_0\beta}{k(1-\beta^2)}\sin\frac{\pi}{\beta} \tag{3.104}$$

$$\dot{v}(t_1) = \frac{P_0}{k(1-\beta^2)}(\overline{\omega}\cos\overline{\omega}t_1 - \beta\omega\cos\omega t_1) = -\frac{P_0\overline{\omega}}{k(1-\beta^2)}\left(1+\cos\frac{\pi}{\beta}\right) \tag{3.105}$$

Substituting $v(t_1)$ and $\dot{v}(t_1)$ into Eq. (3.103) leads to

$$v(t) = \frac{-P_0\beta}{k(1-\beta^2)}\left\{\sin\frac{\pi}{\beta}\cos\omega(t-t_1) + \left(1+\cos\frac{\pi}{\beta}\right)\sin\omega(t-t_1)\right\}, \quad t \geq t_1 \tag{3.106}$$

Similarly, introducing $\alpha = t/t_1$ leads to the response ratio in Phase II

$$R(\alpha) = \frac{-\beta}{1-\beta^2}\left\{\sin\frac{\pi}{\beta}\cos\left[\frac{\pi}{\beta}(\alpha-1)\right] + \left(1+\cos\frac{\pi}{\beta}\right)\sin\left[\frac{\pi}{\beta}(\alpha-1)\right]\right\}, \quad \alpha \geq 1 \tag{3.107}$$

where $\pi/\beta(\alpha-1) = \omega(t-t_1)$. Eq. (3.107) is indeterminate for $\beta = 1$. Using L'Hospital's rule once again leads to

$$R(\alpha) = \frac{\pi}{2}\cos[\pi(\alpha-1)], \quad \beta = 1, \ \alpha \geq 1 \tag{3.108}$$

Using Eqs. (3.101) and (3.102) for Phase I and Eqs. (3.107) and (3.108) for Phase II, the time histories of response ratio can be plotted for different values of β, as illustrated by the solid lines in Fig. 3.33. The values of β selected for this figure are 1/10, 1/4, 1/3, 1/2, 1, and 3/2, and the corresponding values of t_1/T are equal to 5, 2, 3/2, 1, 1/2, and 1/3, respectively,

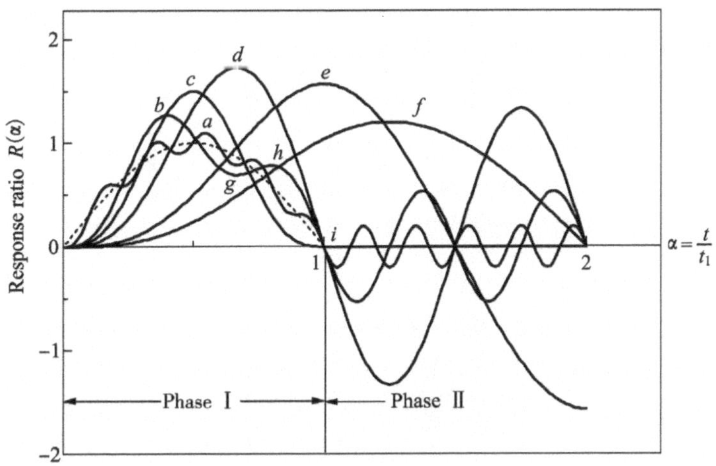

Figure 3.33 Response ratios due to half-sine pulse.

where T is the natural period of the structure. For the purpose of comparison, the quasistatic response ratio $[P(t)/k]/(P_0/k) = P(t)/P_0$, which has a peak value equal to unity, is shown by the dashed line. For $t_1/T = 1/2$ ($\beta = 1$), the maximum response at point e occurs exactly at the end of Phase I. For any value of t_1/T less than 1/2 ($\beta > 1$), the maximum response will occur during Phase II; while for any value of t_1/T greater than 1/2 ($\beta < 1$), the maximum response will occur in Phase I. Obviously, the maximum response depends on the ratio of the load duration to the natural period of the structure, t_1/T.

Although it is important to understand the complete time-history behavior, as shown in Fig. 3.33, engineers are usually only interested in the maximum response as represented by points a, b, c, d, e, and f. If the maximum response occurs during Phase I, the value of α at which it occurs, can be determined by differentiating Eq. (3.101) with respect to α and equating to zero, thus obtaining

$$\frac{dR(\alpha)}{d\alpha} = \left(\frac{\pi}{1-\beta^2}\right)\left(\cos\pi\alpha - \cos\frac{\pi\alpha}{\beta}\right) = 0 \tag{3.109}$$

from which

$$\cos\pi\alpha = \cos\frac{\pi\alpha}{\beta}$$

This equation is satisfied when

$$\pi\alpha = \pm\frac{\pi\alpha}{\beta} + 2\pi n, \quad n = 0, \pm 1, \pm 2, \cdots \tag{3.110}$$

Solving for α gives

$$\alpha = \frac{2\beta n}{\beta \pm 1}, \; n = 0, \pm 1, \pm 2, \cdots \tag{3.111}$$

Eq. (3.111) is valid only when α falls in Phase I (i.e., $0 \leq \alpha \leq 1$). As previously shown, this condition is met only when $0 \leq \beta \leq 1$. To satisfy the two conditions above, it is necessary that the positive and negative values of n should be used along with the plus and minus signs in Eq. (3.111), respectively. Note that the zero value of n can be dropped from consideration as it yields $\alpha = 0$ which simply conforms that the zero-velocity initial condition has been satisfied.

To understand the use of Eq. (3.111) further, the cases shown in Fig. 3.33 and Table 3.1 are considered. For $\beta = 1$, using the plus sign and

Table 3.1 Maximum response ratio to half-sine pulse.

β	1/10	1/4	1/3	1/2	1	3/2
t_1/T	5	2	3/2	1	1/2	1/3
Point corresponding to maximum response	*a*	*b*	*c*	*d*	*e*	*f*
α corresponding to maximum response	6/11	2/5	1/2	2/3	1	–
Value of *n* in Eq. (3.111)	3	1	1	1	1	–
Use of plus and minus signs in Eq. (3.111)	+	+	+	+	+	–
R_{max}	1.0997	1.268	1.50	1.73	1.57	1.20

$n = +1$, one obtains $\alpha = 1$, which is substituted into Eq. (3.102) to yield $R(1) = \pi/2$, as shown by point *e* in Fig. 3.33. When $\beta = 1/2$, Eq. (3.111) has only one valid solution, namely the solution using the plus sign and $n = +1$. The resulting value of α is 2/3, which is substituted into Eq. (3.101) to give $R(2/3) = 1.73$, as shown by point *d*. For $\beta = 1/3$, the use of a plus sign in Eq. (3.111) gives $\alpha = 1/2$ and 1 when $n = +1$ and $+2$, respectively. Substituting the values of α into Eq. (3.101) yields $R(1/2) = 3/2$ and $R(1) = 0$, as shown in Fig. 3.33 by points *c* and *i*, respectively. Because $\dot{R}(1)$ is 0 in this case, there is no free vibration in Phase II. For $\beta = 1/4$, two maxima (points *b* and *h*) and one minima (point *g*) are clearly present in Phase I. Points *b* and *h* correspond to using the plus sign along with $n = +1$ and $+2$, respectively, which leads to $\alpha = 2/5$ and $4/5$, respectively. Point *g* corresponds to using the minus sign along with $n = -1$ giving $\alpha = 2/3$. It is now clear that using the plus sign in Eq. (3.111) along with positive values of *n* yields α-values for the maxima, while using the minus sign along with negative values of *n* yields α-values for the minima. Substituting the above values of α into Eq. (3.101) gives $R(2/5) = 1.268$, $R(4/5) = 0.784$, and $R(2/3) = 0.693$, which correspond to points *b*, *h*, and *g*, respectively. If one examines additional cases by further reducing the value of β, the numbers of maxima and minima will continuously increase in Phase I, such as only one peak for $\beta = 1/3$, three peaks for $\beta = 1/4$, and nine peaks for $\beta = 1/10$. In the limit, as $\beta \to 0$, the response–ratio curve will approach the quasistatic response curve shown by the dashed line in Fig. 3.33, and R_{max} approaches unity.

Finally, consider the case of $\beta = 3/2$, which has its maximum response in Phase II as indicated by point *f*. In this case of free vibration, it is not necessary to determine the value of α corresponding to the maximum response because the desired maximum value is obtained directly by simply taking the vector sum of the two orthogonal components in Eq. (3.106), giving

$$\begin{aligned} D = R_{max} &= \left(\frac{-\beta}{1-\beta^2}\right)\left[\left(1+\cos\frac{\pi}{\beta}\right)^2+\left(\sin\frac{\pi}{\beta}\right)^2\right]^{1/2} \\ &= \left(\frac{-\beta}{1-\beta^2}\right)\left[2\left(1+\cos\frac{\pi}{\beta}\right)\right]^{1/2} \end{aligned} \tag{3.112}$$

Finally, using the trigonometric identity

$$\left[2\left(1+\cos\frac{\pi}{\beta}\right)\right]^{1/2} = 2\cos\frac{\pi}{2\beta}$$

Eq. (3.112) can be simplified to

$$D = \left(\frac{-2\beta}{1-\beta^2}\right)\cos\frac{\pi}{2\beta} \tag{3.113}$$

Therefore for the above case of $\beta = 3/2$, $D = 1.2$.

3.7.2 Rectangular impulsive load

The rectangular impulse load is shown in Fig. 3.34. The response can also be divided into forced-vibration phase (Phase I) and free-vibration phase (Phase II).

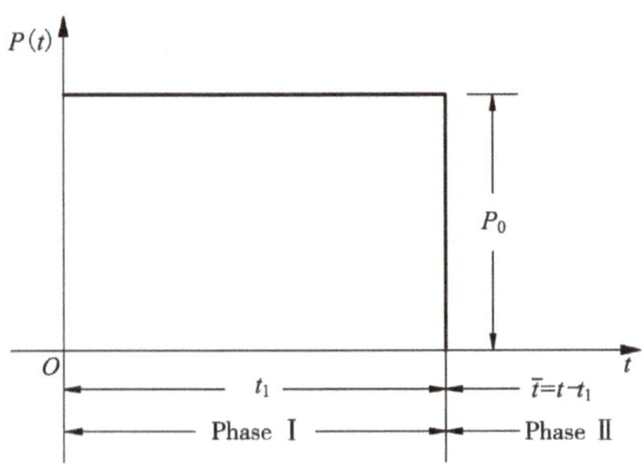

Figure 3.34 Rectangular impulse load.

Phase I: In this phase ($0 \le t \le t_1$), the equation of motion for the SDOF system is $m\ddot{v} + kv = P_0$. Its particular solution is $v_p = P_0/k$, and the complementary free-vibration solution is $v_c = C_1 \cos \omega t + C_2 \sin \omega t$. Thus the general solution is given by

$$v = C_1 \cos \omega t + C_2 \sin \omega t + \frac{P_0}{k}$$

Suppose the system starts from rest, that is, $v(0) = \dot{v}(0) = 0$, one obtains $C_1 = -P_0/k$ and $C_2 = 0$. Therefore

$$v = \frac{P_0}{k}(1 - \cos \omega t), \quad 0 \le t \le t_1 \tag{3.114}$$

Note that the maximum response occurs in Phase I, when $t_1 \ge T/2$ ($t_1/T \ge 1/2$). The time t_m corresponding to the maximum response is

$$t_m = \frac{\pi}{\omega} = \frac{T}{2}$$

and the maximum response $v_{\max}$ is

$$v_{\max} = \frac{P_0}{k}\left[1 - \cos\left(\frac{\pi}{\omega} \cdot \omega\right)\right] = \frac{2P_0}{k}$$

When $t_1 < T/2$ ($t_1/T < 1/2$), the maximum response occurs in Phase II. The maximum amplitude and dynamic magnification factor can be determined according to the following free vibration.

Phase II: In this phase ($t \ge t_1$), the system freely vibrates due to the displacement $v(t_1)$ and velocity $\dot{v}(t_1)$ at $t = t_1$. The amplitude is

$$\rho = \left\{ [v(t_1)]^2 + \left[\frac{\dot{v}(t_1)}{\omega}\right]^2 \right\}^{\frac{1}{2}} \tag{3.115}$$

where $v(t_1) = P_0/k(1 - \cos \omega t_1)$ and $\dot{v}(t_1) = P_0/k\omega \sin \omega t_1$.

Considering the relations $\omega = 2\pi/T$ and $1 - \cos(2\pi t_1/T) = 2\sin^2(\pi t_1/T)$ leads to

$$\rho = \frac{P_0}{k}\left[\left(1 - \cos\frac{2\pi t_1}{T}\right)^2 + \sin^2\frac{2\pi t_1}{T}\right]^{\frac{1}{2}}$$

$$= \frac{P_0}{k}\left[2\left(1 - \cos\frac{2\pi t_1}{T}\right)\right]^{\frac{1}{2}} = \frac{2P_0}{k}\left|\sin\frac{\pi t_1}{T}\right| \tag{3.116}$$

Therefore when $t_1/T < 1/2$, the dynamic magnification factor is

$$D = 2\sin\frac{\pi t_1}{T} \tag{3.117}$$

Note that the dynamic magnification factor D varies with the ratio t_1/T.

3.7.3 Triangular impulsive load

Triangular impulse load is shown in Fig. 3.35. The equation of motion of the undamped SDOF system in Phase I is given by

$$m\ddot{v} + kv = P_0\left(1 - \frac{t}{t_1}\right)$$

In Phase I, the general solution is

$$v = C_1\cos\omega t + C_2\sin\omega t + \frac{P_0}{k}\left(1 - \frac{t}{t_1}\right) \tag{3.118}$$

Suppose the system starts from rest, that is, $v(0) = \dot{v}(0) = 0$, one obtains $C_1 = -P_0/k$ and $C_2 = P_0/(kt_1\omega)$. Substituting them into Eq. (3.118) gives

$$v = \frac{P_0}{k}\left(\frac{1}{\omega t_1}\sin\omega t - \cos\omega t + 1 - \frac{t}{t_1}\right), \quad 0 \le t \le t_1 \tag{3.119}$$

Taking the derivative of Eq. (3.119) with respect to time t leads to

$$\dot{v} = \frac{dv}{dt} = \frac{P_0\omega}{k}\left(\frac{\cos\omega t}{\omega t_1} + \sin\omega t - \frac{1}{\omega t_1}\right), \quad 0 \le t \le t_1 \tag{3.120}$$

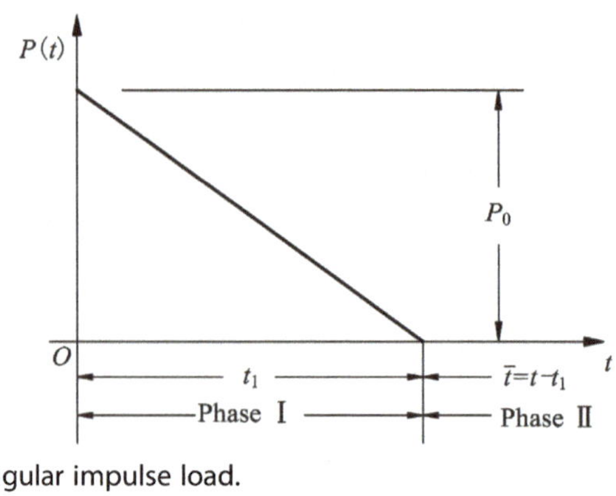

Figure 3.35 Triangular impulse load.

Letting Eq. (3.120) equate zero, and using the relations $\sin 2\alpha = 2\sin\alpha\cos\alpha$ and $\cos 2\alpha = 1 - 2\sin^2\alpha$, leads to time t_m corresponding to the maximum displacement response

$$t_m = \frac{2}{\omega}\tan^{-1}(\omega t_1)$$

Substituting the above equation into Eq. (3.119) gives the maximum displacement response

$$v_{\max} = \frac{2P_0}{k}\left(1 - \frac{1}{\omega t_1}\tan^{-1}\omega t_1\right)$$

and the corresponding dynamic magnification factor is

$$D = \frac{v_{\max}}{P_0/k} = 2\left(1 - \frac{1}{\omega t_1}\tan^{-1}\omega t_1\right), \quad t_1/T \geq 0.371$$

The above two equations are valid only when $t_m \leq t_1$. In this case, the maximum response occurs in Phase I. Substituting $t_m = 2/\omega\tan^{-1}(\omega t_1)$ into $t_m \leq t_1$ leads to $\omega t_1 \leq \tan(1/2\omega t_1)$ or $\omega t_1 \geq 0.742\pi$, that is, when $t_1/T \geq 0.371$, the maximum response occurs in Phase I.

When $t_1/T < 0.371$, the maximum displacement response occurs in the free-vibration phase. Free vibration is produced by the displacement $v(t_1)$ and velocity $\dot{v}(t_1)$ at time $t = t_1$. From Eqs. (3.119) and (3.120), the displacement and velocity at $t = t_1$ are evaluated, respectively, as

$$v(t_1) = \frac{P_0}{k}\left(\frac{\sin\omega t_1}{\omega t_1} - \cos\omega t_1\right)$$

$$\dot{v}(t_1) = \frac{P_0\omega}{k}\left(\frac{\cos\omega t_1}{\omega t_1} + \sin\omega t_1 - \frac{1}{\omega t_1}\right)$$

Therefore the amplitude (the value of maximum response) of free vibration can be obtained from Eq. (3.115) as

$$\rho = \frac{P_0}{k}\left[\left(\frac{\sin\omega t_1}{\omega t_1} - \cos\omega t_1\right)^2 + \left(\frac{\cos\omega t_1}{\omega t_1} + \sin\omega t_1 - \frac{1}{\omega t_1}\right)^2\right]^{\frac{1}{2}}$$

$$= \frac{P_0}{k}\left[1 + \frac{2}{(\omega t_1)^2} - \frac{2}{\omega t_1}\left(\frac{\cos\omega t_1}{\omega t_1} + \sin\omega t_1\right)\right]^{\frac{1}{2}}$$

Table 3.2 Dynamic amplification factor (D) under triangular impulse load.

t_1/T	0.20	0.371	0.40	0.50	0.75	1.00	1.50	2.00
D	0.60	1.00	1.05	1.20	1.42	1.55	1.69	1.76

The corresponding dynamic magnification factor is

$$D = \frac{\rho}{P_0/k} = \left[1 + \frac{2}{(\omega t_1)^2} - \frac{2}{\omega t_1}\left(\frac{\cos \omega t_1}{\omega t_1} + \sin \omega t_1\right)\right]^{\frac{1}{2}}, \quad t_1/T < 0.371 \tag{3.121}$$

The above discussions show that for triangular impulse load with short duration ($t_1/T < 0.371$), the maximum response $v_{\max}$ appears in Phase II; when $t_1/T \geq 0.371$, the maximum response $v_{\max}$ is present in Phase I. The dynamic amplification factors D corresponding to discrete load durations are shown in Table 3.2.

3.7.4 Response ratios to different types of impulsive loads

The response ratios $R(t)$ of a SDOF system to four types of impulse loads are shown in Fig. 3.36A–D. It can be seen from these figures that: (1) the maximum response generally occurs at the first peak; (2) the maximum response occurs during the forced-vibration phase for impulse loads with long duration (i.e., t_1/T is relatively large); and (3) the maximum response is present in the free-vibration phase for impulse loads with short duration (i.e., t_1/T is relatively small).

3.7.5 Response spectra (shock spectra)

The above expressions of the dynamic amplification factors show that the maximum response of the undamped SDOF structure to impulsive loads depends only on the ratio of the duration of the impulse to the natural period of the system. Therefore for various kinds of impulsive loads, the relations between the dynamic amplification factor D and t_1/T are plotted in Fig. 3.37. These plots are called the displacement–response spectra (response spectra) under impulsive loads. These response spectra can be used to estimate the maximum response of simple structures to given impulsive load with adequate accuracy.

Although response spectra described above have been developed for the undamped SDOF system, they can also be used for damped systems

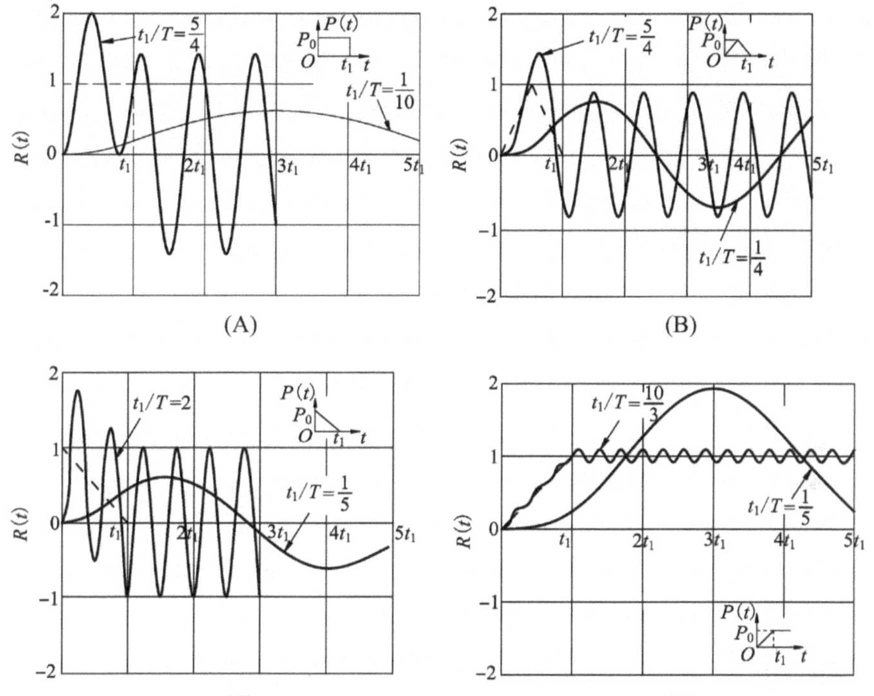

Figure 3.36 Response ratios to different impulsive loads: (A) rectangular impulse; (B) symmetrical triangular impulse; (C) instant triangular impulse; (D) step force with finite rise time.

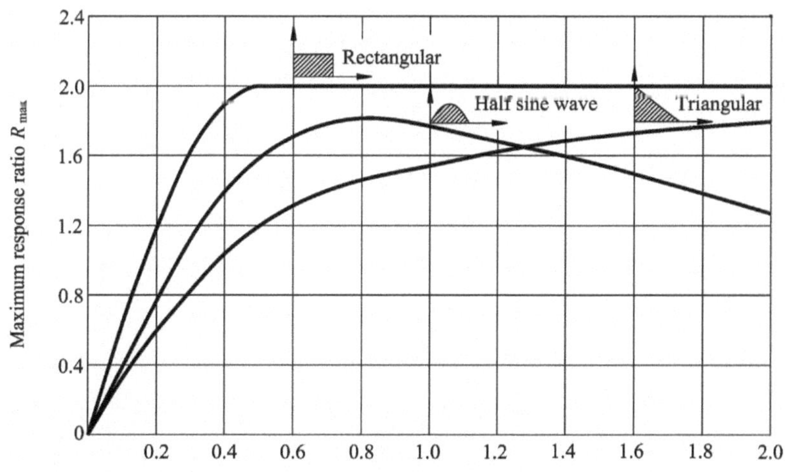

Figure 3.37 Displacement–response spectra (shock spectra) for three types of impulse.

since the damping in the practical range of interest has little effect on the maximum response produced by short-duration impulsive loads.

Example 3.8: Fig. 3.38A shows a SDOF building subjected to the triangular blast load (Fig. 3.38B). Based on the data in the figure, the maximum displacement and elastic force of the structure are estimated using the response spectra in Fig. 3.37.

Solution:

The natural period of the structure is given by

$$T = \frac{2\pi}{\omega} = 2\pi\sqrt{\frac{m}{k}} = 2\pi\sqrt{\frac{2.766 \times 10^5}{1.75 \times 10^9}} = 0.079\text{s}$$

The ratio of impulse duration to natural period is

$$\frac{t_1}{T} = \frac{0.05}{0.079} = 0.63$$

The dynamic magnification factor is obtained as $D = 1.33$ from Fig. 3.37. Thus the maximum displacement is calculated as

$$v_{\max} = \frac{P_0 D}{k} = \frac{4445.28 \times 1.33}{1.75 \times 10^6} = 3.38 \times 10^{-3}\text{m}$$

The maximum elastic force of the structure is

$$F_{smax} = kv_{\max} = 1.75 \times 10^6 \times 3.38 \times 10^{-3} = 5.915 \times 10^3\text{kN}$$

If the blast-pressure impulse had been only one-tenth as long ($t_1 = 0.005$ s), the dynamic magnification factor for the impulse duration ($t_1/T = 0.005/0.079 = 0.063$) would be only 0.2. In this case, the elastic

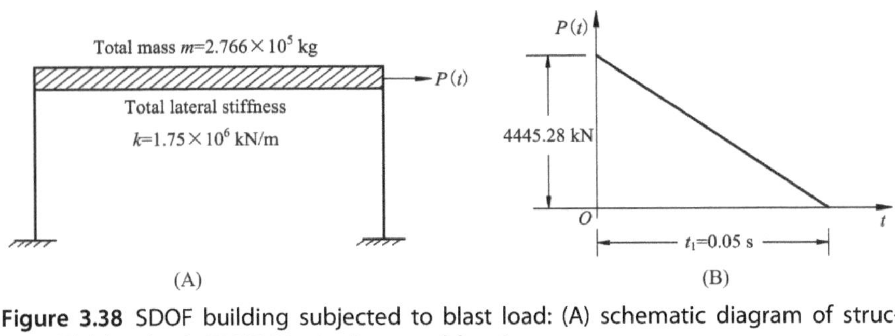

Figure 3.38 SDOF building subjected to blast load: (A) schematic diagram of structure; (B) blast load. *SDOF*, Single-degree-of-freedom.

restoring force becomes 889.47 kN. Note that a large part of the impulse load with very short duration is resisted by the inertial force of the structure, so the stresses produced are much smaller than those produced by the loads of longer duration.

3.7.6 Approximate analysis of response to impulsive loads

The following two conclusions can be drawn by observing Figs. 3.35 and 3.36:

1. For impulse loads of long duration, for example, $t_1/T > 1$, the dynamic magnification factor depends principally on the rate of increase of the load to its maximum value. A step load of sufficient duration produces a dynamic magnification factor of 2; a very gradual increase causes a dynamic magnification factor of 1, as shown in Fig. 3.36D.
2. For impulse loads of short duration, say $t_1/T = 0.2$, the maximum displacement $v_{\max}$ depends principally upon the magnitude of the applied impulse $I = \int_0^{t_1} P(t)dt$ and is not strongly influenced by the form of the impulsive loads. This conclusion can be verified as follows:

For the rectangular impulse, $D_a = 1.176$, $v_{a,\max} = 1.176P_0/k$, and $I_a = P_0t_1$; for the half-sine-wave impulse, $D_b = 0.763$, $v_{b,\max} = 0.763P_0/k$, and $I_b = 0.637P_0t_1$, for the triangular impulse, $D_c = 0.60$, $v_{c,\max} = 0.60P_0/k$, and $I_c = 1/2P_0t_1$; thus

$$v_{a,\max}:v_{b,\max}:v_{c,\max} = 1.176:0.763:0.6 = 1:0.649:0.510$$

$$I_a:I_b:I_c = 1:0.637:0.5 \approx v_{a,\max}:v_{b,\max}:v_{c,\max}$$

The above conclusions provide a way to approximately analyze the impulsive response as outlined below. The equation of motion of the SDOF system is $m\ddot{v} + kv = P(t)$ under the aforementioned impulsive loads. Integrating the differential equation from $t = 0$ to $t = t_1$, one obtains

$$m\int_0^{t_1} \ddot{v}dt = \int_0^{t_1} [P(t) - kv]dt$$

that is,

$$m\Delta\dot{v} = \int_0^{t_1} [P(t) - kv]dt \tag{3.122}$$

where $\Delta\dot{v} = \int_0^{t_1} \ddot{v}dt$ is the velocity increment of the mass m produced by applied load. When t_1 is very small, assuming the average acceleration a leads to

$$\Delta\dot{v} = \dot{v}(t_1) \approx at_1$$

$$\Delta v = v(t_1) \approx \frac{1}{2}at_1^2$$

It may be observed that for a small value of t_1 the displacement increment Δv is of the order of t_1^2, and the velocity increment $\Delta\dot{v}$ is of the order of t_1. Hence, the impulse is also of the order of t_1, and the elastic force term kv will be the order of t_1^2. The elastic force is much smaller than the external load $P(t)$, so it can be ignored in Eq. (3.122). Therefore Eq. (3.122) can be expressed approximately as

$$m\Delta\dot{v} \approx \int_0^{t_1} P(t)dt \quad \text{or} \quad \Delta\dot{v} = \frac{1}{m}\int_0^{t_1} P(t)dt \tag{3.123}$$

After the applied load ends, the mass m vibrates freely due to the displacement $v(t_1)$ and velocity $\dot{v}(t_1)$ at $t = t_1$. Considering the effect of impulsive load, and ignoring damping leads to the free-vibration response

$$v(\bar{t}) = v(t_1)\cos\omega\bar{t} + \frac{\dot{v}(t_1)}{\omega}\sin\omega\bar{t}$$

where $\bar{t} = t - t_1$. Since $v(t_1)$ is small, the first term can be ignored. Considering $\dot{v}(t_1) = \Delta\dot{v}$ and Eq. (3.123), one obtains

$$v(\bar{t}) \approx \frac{1}{m\omega}\left(\int_0^{t_1} P(t)dt\right)\sin\omega\bar{t} \tag{3.124}$$

Example 3.9: Determine the response of the system to the impulsive load, as shown in Fig. 3.39.

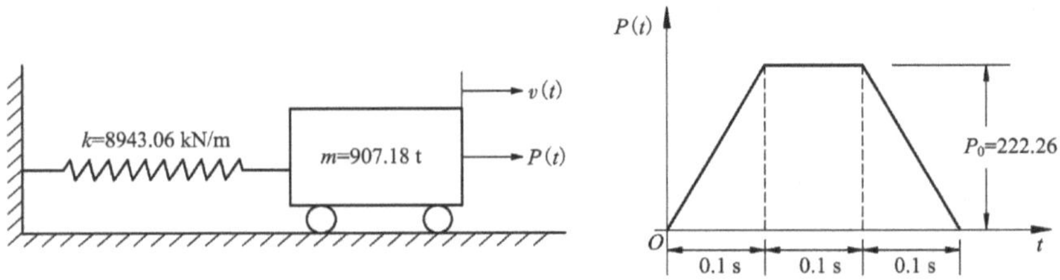

Figure 3.39 Approximate impulse–response analysis.

Solution:

The natural frequency of the SDOF system is

$$\omega = \sqrt{\frac{k}{m}} = \sqrt{\frac{8943.06 \times 10^3}{907.18 \times 10^3}} = 3.14 \text{ rad/s}$$

The duration of the applied load is $t_1 = 0.3$ s.

The total impulse due to applied load is

$$\int_0^{t_1} P(t)dt = 2 \times \frac{1}{2} \times 0.1 \times 222.26 + 0.1 \times 222.26 = 44.45 \text{ kN} \cdot \text{s}$$

The displacement can be obtained by Eq. (3.124) as

$$v(\bar{t}) = \frac{44.45}{907.18 \times 3.14} \sin\omega\bar{t}$$

When $\sin\omega\bar{t} = 1$, the maximum displacement is

$$v_{\max} = \frac{44.45}{907.18 \times 3.14} \approx 1.56 \times 10^{-2}\text{m}$$

and the maximum elastic force of the spring is

$$F_{smax} = kv_{\max} = 8943.06 \times 1.56 \times 10^{-2} = 139.55 \text{ kN}$$

Since the natural period of this system is $T = 2\pi/\omega = 2\pi/3.14 = 2$ s, the ratio of load duration to period is $t_1/T = 0.3/2 = 0.15$. For such a short-duration load, the approximate analysis is quite accurate. In fact, the exact maximum displacement determined from direct integration of the equation of motion is 0.015 3 m, and the relative error in the approximate result is less than 2%.

3.8 Time-domain analysis of dynamic response to arbitrary dynamic loads

In the above analysis of vibration response, the dynamic loads can be expressed by explicit time functions. For the case of simple function of loads, the analytical solution of the response can be obtained by solving the equation of motion of the system. When the dynamic load is a complex time function, the analytical solution of the response cannot be obtained. In addition, many practical loads, such as wind load, seismic load, wave force, and ground- or track-induced force on vehicles, can only be measured through tests and cannot be expressed as explicit

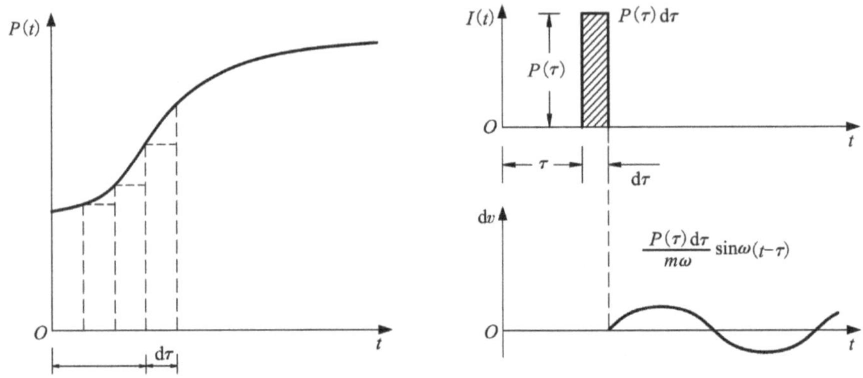

Figure 3.40 Derivation of Duhamel integral (undamped system).

analytical functions. The common methods for analyzing response to these types of loads include the time-domain (e.g., Duhamel integral method) and frequency-domain [e.g., Fourier integral transform and discrete Fourier transform (DFT)] analyses. This section introduces the Duhamel integral method.

As shown in Fig. 3.40, the loading process of $P(t)$ can be regarded as the action of a series of differential-impulses $P(\tau)d\tau$. The velocity increment of the system induced by this differential-impulse $P(\tau)d\tau$ is $P(\tau)d\tau/m$, and the corresponding displacement increment can be ignored, which has been demonstrated in Section 3.7. When damping is absent, the free vibration developed at the subsequent time t ($t \geq \tau$) due to the velocity increase can be given by

$$dv = \frac{1}{m\omega} P(\tau)d\tau \sin\omega(t-\tau), \quad t \geq \tau$$

Then, all the differential displacement responses are superposed to obtain the total displacement response of the undamped system at time t, that is,

$$v(t) = \int_0^t dv = \frac{1}{m\omega}\int_0^t P(\tau)\sin\omega(t-\tau)d\tau \tag{3.125}$$

For the case of viscously damped systems, the displacement response would be

$$v(t) = \frac{1}{m\omega_D}\int_0^t e^{-\xi\omega(t-\tau)} P(\tau)\sin\omega_D(t-\tau)d\tau \tag{3.126}$$

The displacement expressed by the Duhamel integral is purely produced by the applied load, without considering the initial conditions. Therefore if the system vibrates from the initial displacement $v(0)$ and velocity $\dot{v}(0)$, the responses caused by the applied load and initial conditions should be added together to obtain the total dynamic response.

For the undamped systems, the total displacement response is

$$v(t) = v(0)\cos \omega t + \frac{\dot{v}(0)}{\omega}\sin \omega t + \frac{1}{m\omega}\int_0^t P(\tau)\sin \omega(t-\tau)d\tau \tag{3.127}$$

For the viscously damped systems, the total displacement response is

$$\begin{aligned} v(t) &= e^{-\xi\omega t}\left[v(0)\cos \omega_D t + \frac{\dot{v}(0) + \xi\omega v(0)}{\omega_D}\sin \omega_D t\right] \\ &+ \frac{1}{m\omega_D}\int_0^t e^{-\xi\omega(t-\tau)}P(\tau)\sin \omega_D(t-\tau)d\tau \end{aligned} \tag{3.128}$$

If the applied-load function is of simple analytic form, the integrals in Eqs. (3.125) and (3.126) can be evaluated directly. However, this usually is not possible in most practical cases as the load is known only from experimental data. The Duhamel integral must be evaluated numerically, as detailed in Ref. [1]. Additionally, the step-by-step integral method introduced in Chapter 7, Step-by-step integration method, can also be used to solve the response of the SDOF system.

Since Duhamel integral equations are derived from the superposition principle, this method can only be applied to linear systems without time-varying characteristics during vibration. The Duhamel integral method is not applicable to time-varying systems, such as buildings undergoing elastic—plastic vibration due to strong earthquake.

Example 3.10: A SDOF system is subjected to the rectangular impulsive load, as shown in Fig. 3.34. The initial displacement and velocity are zero, the mass of the system is m, the stiffness coefficient is k, and the damping effect is not considered. Evaluate the response of the system during different phases.

Solution:

The load $P(t)$ can be expressed as the following piecewise function

$$P(t) = \begin{cases} P_0, & 0 \le t \le t_1 \\ 0, & t > t_1 \end{cases}$$

In Phase I ($0 \leq t \leq t_1$), the system is subjected to instantaneously applied force P_0. Considering zero initial conditions, Eq. (3.127) can be evaluated as

$$
\begin{aligned}
v(t) &= \frac{1}{m\omega}\int_0^t P_0 \sin \omega(t-\tau)d\tau \\
&= \left[\frac{P_0}{m\omega^2}\cos \omega(t-\tau)\right]_0^t, \qquad 0 \leq t \leq t_1 \\
&= \frac{P_0}{k}(1-\cos \omega t)
\end{aligned}
$$

In Phase II ($t > t_1$), the system is not subjected to any external force. The displacement response can be evaluated by using the initial conditions at the end of Phase I, as introduced in Section 3.7. Here, the response in Phase II is also evaluated by the Duhamel integral as follows:

$$
v(t) = \frac{1}{m\omega}\int_0^{t_1} P_0 \sin \omega(t-\tau)d\tau = \frac{2P_0}{k}\sin\frac{\omega t_1}{2}\sin \omega\left(t-\frac{t_1}{2}\right), \qquad t > t_1
$$

The evaluated responses agree well with the analytical solutions in Section 3.7.

3.9 Frequency-domain analysis of dynamic response to arbitrary dynamic loads

As shown in Section 3.8, the time-domain analysis method can be used to determine the response of any linear SDOF system subjected to arbitrary dynamic loads. However, it is sometimes more convenient to perform the analysis in the frequency domain. Furthermore, the frequency-domain analysis method plays an important role in experimental modal analysis and random vibration analysis.

In Section 3.6, the periodic load is expanded into many simple harmonic components. The response of the structure to each simple harmonic component is evaluated, and the total response of the structure is obtained by superposing each simple harmonic responses. An imaginary periodic load can be obtained when an arbitrary load is supposed to have an infinite period, and in fact, a finite value of time much larger than actual load duration is taken as the period, as shown in Fig. 3.41. This idea can be extended to the frequency-domain analysis of structural

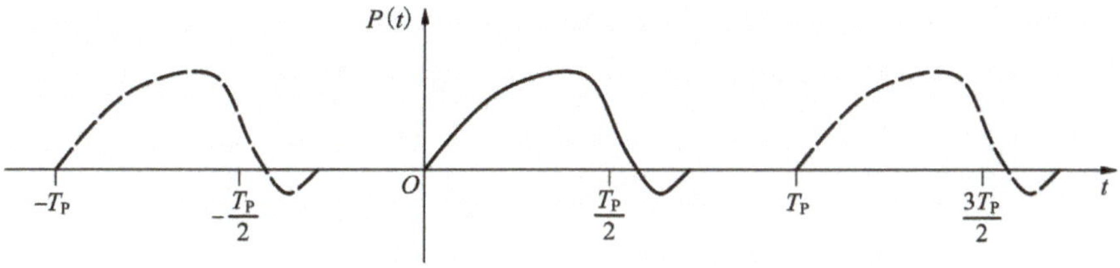

Figure 3.41 Arbitrary load represented by Fourier series.

response to arbitrary loads. In order to facilitate the derivation, it is necessary to express the periodic load and system response in complex form, and introduce the concept of complex Fourier series and complex frequency response function.

3.9.1 Express the system response to periodic loads in complex form

By using Euler's equations

$$\cos\theta = \frac{1}{2}\left(e^{i\theta} + e^{-i\theta}\right), \quad \sin\theta = -\frac{i}{2}\left(e^{i\theta} - e^{-i\theta}\right)$$

Eq. (3.96) can be written, in complex form, as

$$\begin{aligned} P(t) &= \frac{A_0}{2} + \sum_{n=1}^{\infty}(A_n\cos n\overline{\omega}_1 t + B_n\sin n\overline{\omega}_1 t) \\ &= \frac{A_0}{2} + \sum_{n=1}^{\infty}\left[\frac{A_n}{2}\left(e^{in\overline{\omega}_1 t} + e^{-in\overline{\omega}_1 t}\right) - i\frac{B_n}{2}\left(e^{in\overline{\omega}_1 t} - e^{-in\overline{\omega}_1 t}\right)\right] \\ &= \frac{A_0}{2} + \sum_{n=1}^{\infty}\left[\left(\frac{A_n}{2} - i\frac{B_n}{2}\right)e^{in\overline{\omega}_1 t} + \left(\frac{A_n}{2} + i\frac{B_n}{2}\right)e^{-in\overline{\omega}_1 t}\right] \end{aligned}$$

Letting $a_0 = \frac{A_0}{2}$, $a_n = \frac{A_n}{2} - i\frac{B_n}{2}$, and $a_{-n} = \frac{A_n}{2} + i\frac{B_n}{2}$, the above expression becomes

$$P(t) = a_0 + \sum_{n=1}^{\infty}\left[a_n e^{in\overline{\omega}_1 t} + a_{-n}e^{-in\overline{\omega}_1 t}\right] \equiv \sum_{n=-\infty}^{\infty} a_n e^{in\overline{\omega}_1 t} \tag{3.129}$$

In Eq. (3.129), for every positive value of n, say $n = +m$ (m is set to be positive), a_{+m} and a_{-m} are complex conjugate pairs. Rotating a_{+m} and a_{-m} by angle $m\overline{\omega}_1 t$ counterclockwise and clockwise, respectively, leads to other complex conjugate pairs $a_{+m}e^{im\overline{\omega}_1 t}$ and $a_{-m}e^{-im\overline{\omega}_1 t}$, as shown in

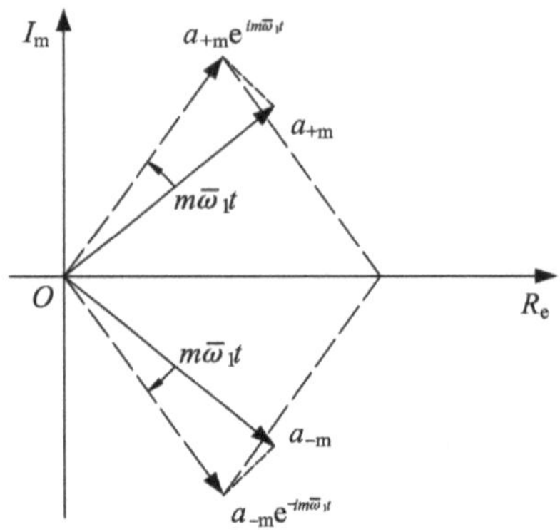

Figure 3.42 Vector representation of exponential load terms.

Fig. 3.42. Since the imaginary parts of $a_{+m}e^{im\overline{\omega}_1 t}$ and $a_{-m}e^{-im\overline{\omega}_1 t}$ cancel each other out, $P(t)$ is a real function.

In Eq. (3.129), a_n, $n = -\infty \to \infty$, are the complex coefficients yet undetermined, and their formula are derived as follows:

Multiplying both sides of Eq. (3.129) by $e^{-in\overline{\omega}_1 t}dt$ and integrating it from 0 to T_P (since n is not a variable here, so n is replaced by m in the summation term), one gets

$$\int_0^{T_P} P(t)e^{-in\overline{\omega}_1 t}dt = \int_0^{T_P}\left(\sum_{m=-\infty}^{\infty} a_m e^{im\overline{\omega}_1 t}\right)e^{-in\overline{\omega}_1 t}dt = \int_0^{T_P}\sum_{m=-\infty}^{\infty} a_m e^{i(m-n)\overline{\omega}_1 t}dt = \sum_{m=-\infty}^{\infty}\int_0^{T_P} a_m e^{i(m-n)\overline{\omega}_1 t}dt$$

when $m \neq n$,

$$\begin{aligned}
\int_0^{T_P} a_m e^{i(m-n)\overline{\omega}_1 t}dt &= \frac{a_m}{i(m-n)\overline{\omega}_1}\int_0^{T_P} e^{i(m-n)\overline{\omega}_1 t}d[i(m-n)\overline{\omega}_1 t] \\
&= \frac{a_m}{i(m-n)\overline{\omega}_1}\left[e^{i(m-n)\overline{\omega}_1 t}\right]_0^{T_P} \\
&= \frac{a_m}{i(m-n)\overline{\omega}_1}\left[e^{i(m-n)\overline{\omega}_1 \frac{2\pi}{\overline{\omega}_1}} - 1\right] \\
&= \frac{a_m}{i(m-n)\overline{\omega}_1}[\cos 2(m-n)\pi + i\sin 2(m-n)\pi - 1] = 0
\end{aligned}$$

thus

$$\int_0^{T_P} P(t)e^{-in\overline{\omega}_1 t}dt = \sum_{m=-\infty}^{\infty}\int_0^{T_P} a_m e^{i(m-n)\overline{\omega}_1 t}dt = \int_0^{T_P} a_n dt = a_n T_P$$

One obtains

$$a_n = \frac{1}{T_P}\int_0^{T_P} P(t)e^{-in\overline{\omega}_1 t}dt \tag{3.130}$$

As shown above, an arbitrary periodic load can be expressed as a Fourier series in the complex form, that is, the sum of many terms of $a_n e^{in\overline{\omega}_1 t}$. a_n is a complex coefficient associated with $P(t)$, and $e^{in\overline{\omega}_1 t}$ represents the force in complex exponential form. If the complex response of the system to the complex load $e^{in\overline{\omega}_1 t}$ is determined, the total response of the linear system can be obtained by means of superposition principle.

Therefore the concept of complex frequency response function $H(\overline{\omega})$ needs to be introduced. Assume that the periodic load lasts long enough that the transient response has disappeared, so only the steady-state response is discussed. Suppose that the system is subjected to a unit harmonic force $e^{i\overline{\omega}t}$. The real and imaginary parts of unit harmonic force $e^{i\overline{\omega}t}$ can be understood as two independent harmonic forces $\cos\overline{\omega}t$ and $\sin\overline{\omega}t$ acting on the system. The complex equation of motion of SDOF systems to unit harmonic force $e^{i\overline{\omega}t}$ is

$$m\ddot{Z} + c\dot{Z} + kZ = e^{i\overline{\omega}t} \tag{3.131}$$

The steady-state response has the form as follows:

$$Z(t) = H(\overline{\omega})e^{i\overline{\omega}t} \tag{3.132}$$

where $H(\overline{\omega})$ is a complex constant. It can be seen later that it is a function of exciting frequency $\overline{\omega}$, hence it is called complex frequency response function. Because load $e^{i\overline{\omega}t}$ is a complex number, the resulting response $Z(t)$ is also a complex number. The real and the imaginary parts of complex response $Z(t)$ correspond to the steady-state response to harmonic force $\cos\overline{\omega}t$ and $\sin\overline{\omega}t$, respectively.

Substituting Eq. (3.132) into Eq. (3.131) leads to

$$H(\overline{\omega}) = \frac{1}{k - \overline{\omega}^2 m + i\overline{\omega}c} = \frac{1}{k\left[1 - \beta^2 + i(2\xi\beta)\right]} \tag{3.133}$$

When $\overline{\omega} = n\overline{\omega}_1$, $\beta = n\beta_1$, and $\beta_1 = \overline{\omega}_1/\omega$,

$$H(n\overline{\omega}_1) = \frac{1}{k\left[1 - n^2\beta_1^2 + i(2\xi n\beta_1)\right]} \tag{3.134}$$

Obviously, when $\overline{\omega} = -n\overline{\omega}_1$,

$$H(-n\overline{\omega}_1) = \frac{1}{k\left[1 - n^2\beta_1^2 - i(2\xi n\beta_1)\right]} \tag{3.135}$$

Thus $H(n\overline{\omega}_1)$ is the complex conjugate of $H(-n\overline{\omega}_1)$.

When $P(t) = \sum\limits_{n=-\infty}^{\infty} a_n e^{in\overline{\omega}_1 t}$, the steady-state response can be expressed as

$$v(t) = \sum_{n=-\infty}^{\infty} a_n H(n\overline{\omega}_1) e^{in\overline{\omega}_1 t} \tag{3.136}$$

Although the load $P(t)$ is expressed as the sum of complex numbers in Eq. (3.129), it is still a real number, and the resulting response $v(t)$ is also a real number.

3.9.2 Fourier integral method

The physical concept of the frequency-domain analysis method is similar to the previous analysis of response to periodic load. Both the procedures involve expressing the applied load in terms of harmonic components, evaluating the response to each component, and superposing the harmonic responses to obtain total structural response. In order to apply the periodic-load technique to arbitrary loads, it is necessary to extend the Fourier series concept so that it will actually represent nonperiodic functions.

The fundamental period T_P and the corresponding frequency $\overline{\omega}_1$ are important parameters in the Fourier series expansion of periodic load $P(t)$. When $P(t)$ is expanded, discrete spectrum with respect to $n\overline{\omega}_1$, $n = 1, 2, \cdots$, is obtained. The characteristic of nonperiodic load is that its waveform never repeats. Thus letting the period T_P approach infinity, that is, $T_P = \infty$, the corresponding frequency $\overline{\omega}_1 = 2\pi/T_P$ will be infinitesimal, that is, $\overline{\omega}_1 \to d\overline{\omega}$, $1/T_P \to d\overline{\omega}/(2\pi)$. The discrete frequency $n\overline{\omega}_1$, $n = -\infty, \cdots, -1, 0, 1, \cdots, \infty$, becomes approximately continuous.

Eq. (3.130) is first rewritten as

$$a_n = \frac{1}{T_P}\int_{-\frac{T_P}{2}}^{\frac{T_P}{2}} P(t) e^{-in\overline{\omega}_1 t} dt, \ n = -\infty, \cdots, -1, 0, 1, \cdots, \infty \tag{3.137}$$

Considering $T_P \to \infty$, and letting $\overline{\omega} = n\overline{\omega}_1$, $d\overline{\omega}/(2\pi) = 1/T_P$, Eq. (3.130) can be further rewritten as

$$a_n = \frac{d\overline{\omega}}{2\pi}\int_{-\infty}^{\infty} P(t) e^{-i\overline{\omega} t} dt \tag{3.138}$$

Substituting Eq. (3.138) into Eq. (3.129), and replacing the summation term in Eq. (3.129) by integral term, the Fourier integral equation can be obtained

$$P(t) = \frac{1}{2\pi}\int_{-\infty}^{\infty}\left[\int_{-\infty}^{\infty} P(t)e^{-i\overline{\omega}t}dt\right]e^{i\overline{\omega}t}d\overline{\omega} = \frac{1}{2\pi}\int_{-\infty}^{\infty}\overline{P}(\overline{\omega})e^{i\overline{\omega}t}d\overline{\omega} \quad (3.139)$$

where

$$\overline{P}(\overline{\omega}) = \int_{-\infty}^{\infty} P(t)e^{-i\overline{\omega}t}dt \quad (3.140)$$

Eqs. (3.139) and (3.140) are called a Fourier transform pair. $\overline{P}(\overline{\omega})$ is known as the Fourier transform of $P(t)$, and $P(t)$ is called the inverse Fourier transform of $\overline{P}(\overline{\omega})$. The necessary condition for the Fourier transforms to exist is that the integral $\int_{-\infty}^{\infty}|P(t)|dt$ is finite. Obviously, this is satisfied as long as the load $P(t)$ acts over a finite period of time.

In Eq. (3.139), $\overline{P}(\overline{\omega})/(2\pi)$ represents the complex amplitude intensity at $\overline{\omega}$. $\overline{P}(\overline{\omega})/(2\pi)e^{i\overline{\omega}t}d\overline{\omega}$ represents a load at $\overline{\omega}$. Eq. (3.139) indicates that $P(t)$ is the sum of infinite loads $\overline{P}(\overline{\omega})/(2\pi)e^{i\overline{\omega}t}d\overline{\omega}$. Since the response of the system to unit harmonic load $e^{i\overline{\omega}t}$ is $H(\overline{\omega})e^{i\overline{\omega}t}$, the steady-state response of a linear system to load $P(t)$ is obtained by applying the superposition principle as follows:

$$v(t) = \frac{1}{2\pi}\int_{-\infty}^{\infty} H(\overline{\omega})\overline{P}(\overline{\omega})e^{i\overline{\omega}t}d\overline{\omega} \quad (3.141)$$

Eq. (3.141) is the basic equation for response analysis with frequency-domain method.

Example 3.11: Analyze the response of a SDOF system to rectangular impulsive load shown in Fig. 3.34. The mass of the system is m, and the stiffness coefficient is k. The undamped and damped frequencies are ω and ω_D, respectively. When $0 < t < t_1$, $P(t) = P_0$; the load is zero otherwise.

Solution:

The Fourier transform of the load function is

$$\overline{P}(\overline{\omega}) = \int_0^{t_1} P_0 e^{-i\overline{\omega}t}dt = -\frac{P_0}{i\overline{\omega}}\left(e^{-i\overline{\omega}t_1} - 1\right)$$

Substituting this load expression together with the complex frequency–response expression of Eq. (3.133) into Eq. (3.141) leads to response in the integral form

$$v(t) = \frac{1}{2\pi}\int_{-\infty}^{\infty}\frac{1}{k\left[1-\beta^2+i(2\xi\beta)\right]}\left[-\frac{P_0}{i\overline{\omega}}\left(e^{-i\overline{\omega}t_1} - 1\right)\right]e^{i\overline{\omega}t}d\overline{\omega}$$

Considering $\overline{\omega} = \beta\omega$ (ω is the natural frequency of the system) and $1-\beta^2+i(2\xi\beta) = -\left(\beta+\sqrt{1-\xi^2}-\xi i\right)\left(\beta-\sqrt{1-\xi^2}-\xi i\right)$, and multiplying the numerator and denominator of the integrand by $i=\sqrt{-1}$, one obtains

$$v(t) = \frac{iP_0}{2\pi k}\left[\int_{-\infty}^{\infty} -\frac{e^{-i\omega\beta(t_1-t)}}{\beta(\beta-\gamma_1)(\beta-\gamma_2)}d\beta + \int_{-\infty}^{\infty}\frac{e^{i\omega\beta t}}{\beta(\beta-\gamma_1)(\beta-\gamma_2)}d\beta\right]$$

where $\gamma_1 = \xi i + \sqrt{1-\xi^2}$, $\gamma_2 = \xi i - \sqrt{1-\xi^2}$.

These two infinite integrals in the above expression can be determined from contour integration in the complex β plane, giving for the case of $0<\xi<1$

$$v(t) = 0, \quad t \le 0$$

$$v(t) = \frac{P_0}{k}\left[1-e^{-\xi\omega t}\left(\cos\omega_D t + \frac{\xi}{\sqrt{1-\xi^2}}\sin\omega_D t\right)\right], \quad 0<t\le t_1$$

$$v(t) = \frac{P_0}{k}e^{-\xi\omega(t-t_1)}\left\{\begin{array}{l}\left[e^{-\xi\omega t_1}\left(\sin\omega_D t_1 - \frac{\xi}{\sqrt{1-\xi^2}}\cos\omega_D t_1\right) + \frac{\xi}{\sqrt{1-\xi^2}}\right]\times\sin\omega_D(t-t_1) + \\ \left[1-e^{-\xi\omega t_1}\left(\cos\omega_D t_1 + \frac{\xi}{\sqrt{1-\xi^2}}\sin\omega_D t_1\right)\right]\times\cos\omega_D(t-t_1)\end{array}\right\}, \quad t>t_1$$

Letting $\xi=0$, the resulting responses will be consistent with those obtained by the time-domain analysis method, as shown in Eqs. (3.114) and (3.115).

The Fourier integral transform of the load function is easy in this example. However the integral to obtain the response is very complex, and contour integration needs to be conducted in the complex plane. If the applied-load function is complex, the integral transform will be cumbersome. The analytical expression for integral transform of the load cannot be obtained, and then it is impossible to obtain the response expression by integral technique. Therefore it is usually impossible to solve these problems analytically in practical engineering. To make the procedure practical, it is necessary to formulate it in terms of a numerical analysis approach, such as DFT and fast Fourier transform (FFT). Detailed information about DFT and FFT can be found in related references.

References

[1] Clough RW, Penzien J. Dynamics of structures. 3rd ed Berkeley, CA: Computers & Structures, Inc; 2003.

[2] Craig Jr RR, Kurdila AJ. Fundamentals of structural dynamics. 2nd ed Hoboken, NJ: John Wiley & Sons. Inc.; 2006.

[3] Zhou Z, Wen Y, Zeng Q. Lectures on dynamics of structures. 2nd ed Beijing, China: Communications Press Co., Ltd; 2017.

Problems

3.1 Why are the natural periods the inherent property of the structures? What quantities of the structures are the natural periods related to?

3.2 What is the critical damping and damping ratio? How is the damping ratio of the system in vibrations measured?

3.3 Analyze the relationship among the work done by external forces, energy dissipation, and system response during resonance.

3.4 Based on the concept of response spectra in this chapter, describe the procedure of developing earthquake response spectra briefly. How can the response spectra be used in engineering design?

3.5 Describe the main idea and conditions for application of the Duhamel integral method in brief. Is the Duhamel integral applicable to the evaluation of the responses of an elastic–plastic system?

3.6 The SDOF frame of Fig. P3.1A is subjected to the blast load history shown in Fig. P3.1B. Evaluate the displacement response for the time $0<t<0.72$ s by the Duhamel integral using an appropriate numerical integral technique.

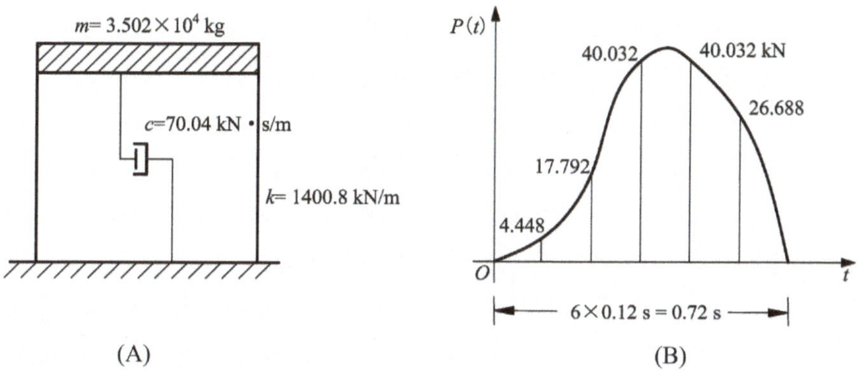

Figure P3.1 Diagram of problem 3.6: (A) SDOF frame; (B) blast load history. *SDOF*, Single-degree-of-freedom.

3.7 As shown in Fig. P3.2, the mass of the object at the end of the beam is M, and the mass of the beam and the spring is ignored. $l = 150$ cm, $M = 897.96$ kg, $EI = 2.93 \times 10^9$ N · cm^2, $k = 3570$ N/cm, the initial displacement $y_0 = 1.3$ cm, and the initial velocity $\dot{y}_0 = 25$ cm/s. Evaluate the natural frequency of this beam, and the displacements and velocities of the object at the instant of $t = 1$ s.

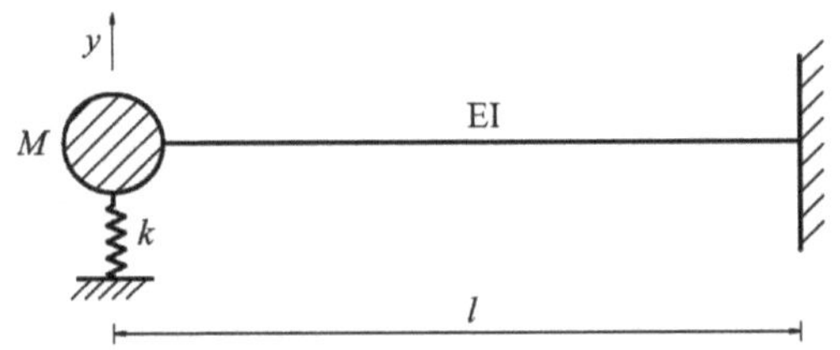

Figure P3.2 Diagram of problem 3.7.

3.8 As shown in Fig. P3.3, the total mass of a worktable equipped with precise instruments is $m = 300$ kg. The worktable is connected to the foundation with springs. The foundation vibrates vertically in simple harmonic form, and the corresponding frequency and amplitude are 10 Hz and 1 cm, respectively. The amplitude of the worktable is required to be less than 0.2 cm, which is measured from static-equilibrium position. Determine the stiffness of the springs required.

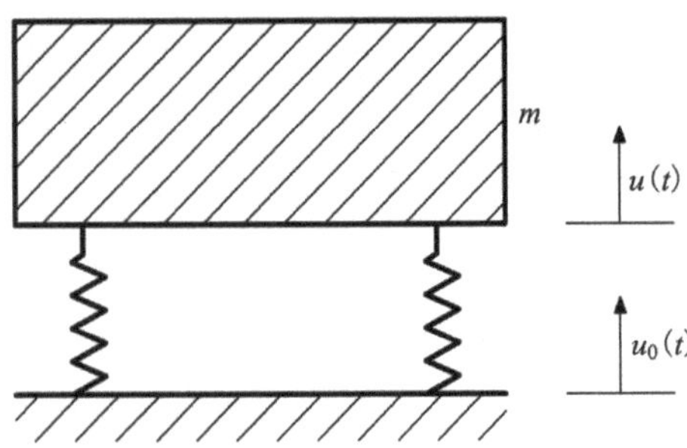

Figure P3.3 Diagram of problem 3.8.

3.9 A SDOF system is subjected to a triangular impulsive load shown in Fig. P3.4. The initial displacement and velocity are zero, the mass of the system is m, the stiffness coefficient is k, and the damping effect is not considered. Evaluate the displacement response of the system using the analytical and Duhamel integral methods, respectively, and formulate the expression of the dynamic magnification factor D.

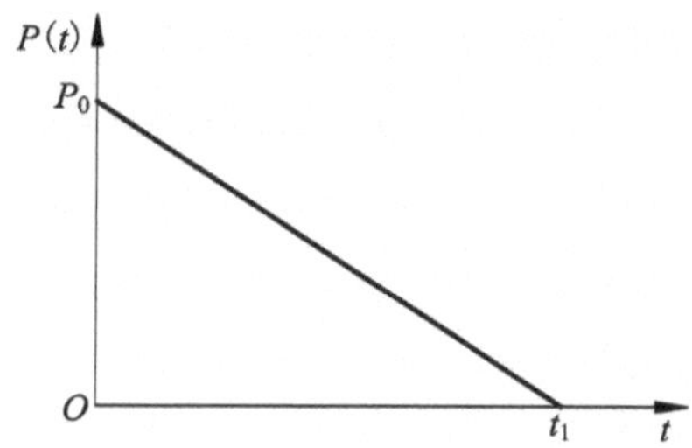

Figure P3.4 Diagram of problem 3.9: triangular impulsive load.

CHAPTER 4

Analysis of dynamic response of MDOF systems: mode superposition method

The mode superposition method is introduced for the analysis of dynamic response of multidegree-of-freedom (MDOF) systems. Natural frequencies and mode shapes of MDOF systems are evaluated by solving the eigenvalue problem. The concept of principal vibration is discussed which will be applied several times in this book. The orthogonality of mode shapes is demonstrated, which is the basis of the mode superposition method. The coupling characteristics of equations of MDOF systems are presented. The equations of motion of MDOF systems are uncoupled into independent differential equations by a linear coordinate transformation. By means of some examples, the application of the mode superposition method is illustrated, and the vibration features of MDOF systems are also discussed.

4.1 Analysis of dynamic properties of multidegree-of-freedom systems

4.1.1 Natural frequencies, mode shapes, and principal vibration

The dynamic response of forced vibration is closely related to the dynamic properties of structures. Thus natural frequencies and mode shapes should be evaluated by free vibration analysis first. Damping has little effect on natural frequencies and mode shapes, and the undamped mode shapes are used by the mode superposition method. Thus the dynamic properties of undamped systems are discussed in this section.

The equations of motion of a freely vibrating undamped system can be obtained by dropping the damping matrix and applied load vector from Eq. (1.1)

$$\boldsymbol{M}\ddot{\boldsymbol{q}} + \boldsymbol{K}\boldsymbol{q} = \boldsymbol{0} \tag{4.1}$$

Fundamentals of Structural Dynamics
DOI: https://doi.org/10.1016/B978-0-12-823704-5.00004-5

The particular solution of Eq. (4.1) is as follows:

$$\boldsymbol{q} = \boldsymbol{A}_i \sin(\omega_i t + \theta_i) \tag{4.2}$$

where $\boldsymbol{A}_i$ is the vector of the displacement amplitudes, and ω_i and θ_i are frequency and phase angle, respectively. Substituting Eq. (4.2) into Eq. (4.1), canceling the common factor $\sin(\omega_i t + \theta_i)$, and letting $\lambda_i = \omega_i^2$ yield

$$(\boldsymbol{K} - \lambda_i \boldsymbol{M})\boldsymbol{A}_i = \boldsymbol{0} \tag{4.3}$$

Eq. (4.3) are a set of homogeneous, linear algebraic equations in $\boldsymbol{A}_i$. Hence, a nontrivial solution exists only if the determinant of the coefficients in Eq. (4.3) vanishes, namely

$$|\boldsymbol{K} - \lambda_i \boldsymbol{M}| = 0 \tag{4.4}$$

Eq. (4.4) is called the frequency or eigenvalue equation of the system. Expanding the determinant will give an n-order algebraic equation in the eigenvalue λ_i (n is the number of DOFs of the system). Here, n roots of λ_i, $i = 1, 2, \cdots, n$, can be solved from the frequency equation, and n natural frequencies of ω_i, $i = 1, 2, \cdots, n$, can be obtained in the order from small to large. The vector, which is made up of the entire set of natural frequencies and arranged in sequence, is called the frequency vector $\boldsymbol{\omega}$

$$\boldsymbol{\omega} = \begin{Bmatrix} \omega_1 \\ \omega_2 \\ \vdots \\ \omega_n \end{Bmatrix}$$

where the lowest frequency ω_1 is called the fundamental frequency.

The obtained λ_i is substituted into Eq. (4.3) to solve for the vector $\boldsymbol{A}_i$, that is, $\boldsymbol{A}_i = \begin{Bmatrix} A_{1i} & A_{2i} & \cdots & A_{ni} \end{Bmatrix}^T$, which is called the ith mode vector (eigenvector) corresponding to ω_i. The deterministic mode vector $\boldsymbol{A}_i$ cannot be solved from Eq. (4.3), and only the proportional relationship among the amplitudes of all the coordinates can be obtained. Therefore $C\boldsymbol{A}_i$ (C is an arbitrary nonzero constant) is also a nontrivial solution of Eq. (4.3), and the ith mode vector of the system. For an n-DOF system, the matrix made up of n mode shapes will be represented by the following mode matrix $\boldsymbol{A}$

$$\boldsymbol{A} = [\boldsymbol{A}_1 \quad \boldsymbol{A}_2 \quad \cdots \quad \boldsymbol{A}_n] = \begin{bmatrix} A_{11} & A_{12} & \cdots & A_{1n} \\ A_{21} & A_{22} & \cdots & A_{2n} \\ \vdots & \vdots & \ddots & \vdots \\ A_{n1} & A_{n2} & \cdots & A_{nn} \end{bmatrix}$$

According to the theory of linear differential equation, the general solution of the free vibration for an undamped system can be written as

$$q = \sum_{i=1}^{n} c_i \boldsymbol{A}_i \sin(\omega_i t + \theta_i) \tag{4.5}$$

Eq. (4.5) contains $2n$ constants: c_i and θ_i, $i = 1, 2, \cdots, n$, and these constants can be determined from the initial displacement and velocity of each coordinate. However, the calculation process is complicated. It will be easier to evaluate the free vibration response by means of the mode superposition method (see Section 4.3).

Under certain initial conditions, all the amplitude constants except a certain c_i in Eq. (4.5) will vanish. In this case, the general solution of free vibration represented by Eq. (4.5) retains only one term, that is,

$$\begin{cases} q_1 = c_i A_{1i} \sin(\omega_i t + \theta_i) \\ q_2 = c_i A_{2i} \sin(\omega_i t + \theta_i) \\ \quad \vdots \\ q_n = c_i A_{ni} \sin(\omega_i t + \theta_i) \end{cases} \tag{4.6}$$

Note that all the coordinates vibrate in a harmonic way with the same frequency ω_i and phase angle θ_i, which pass through the static equilibrium positions and reach the maximum values at the same time. The proportional relationship among all the coordinates remains unchanged, that is,

$$q_1 : q_2 : \cdots : q_n = A_{1i} : A_{2i} : \cdots : A_{ni} \tag{4.7}$$

The free vibration only in the ith natural mode described by Eq. (4.6) is called the ith principal vibration [1]. The coordinate amplitudes and phase angle of the principal vibration are determined from the initial conditions of the system. The shape of principal vibration is completely determined from the proportion relationship among all the components in mode vector $\boldsymbol{A}_i = \{ A_{1i} \quad A_{2i} \quad \cdots \quad A_{ni} \}^T$. The natural frequencies and mode shapes only depend on the stiffness and mass properties of the system and have nothing to do with the initial conditions. The concept of principal vibration provides a theoretical explanation for the phenomenon that a certain principal vibration can be excited in experiments with proper initial conditions. Example 4.2 will illustrate this phenomenon intuitively.

4.1.2 Orthogonality of mode shapes

Letting the ith and jth mode vectors be $\boldsymbol{A}_i$ and $\boldsymbol{A}_j$, respectively, the corresponding eigenvalues are λ_i and λ_j respectively. Substituting the eigenpairs (λ_i, $\boldsymbol{A}_i$) and (λ_j, $\boldsymbol{A}_j$) into Eq. (4.3), respectively, leads to

$$\boldsymbol{K}\boldsymbol{A}_i = \lambda_i \boldsymbol{M}\boldsymbol{A}_i \tag{4.8}$$

$$\boldsymbol{K}\boldsymbol{A}_j = \lambda_j \boldsymbol{M}\boldsymbol{A}_j \tag{4.9}$$

Premultiplying Eq. (4.8) by $\boldsymbol{A}_j^T$ gives

$$\boldsymbol{A}_j^T \boldsymbol{K}\boldsymbol{A}_i = \lambda_i \boldsymbol{A}_j^T \boldsymbol{M}\boldsymbol{A}_i \tag{4.10}$$

Premultiplying Eq. (4.9) by $\boldsymbol{A}_i^T$ gives

$$\boldsymbol{A}_i^T \boldsymbol{K}\boldsymbol{A}_j = \lambda_j \boldsymbol{A}_i^T \boldsymbol{M}\boldsymbol{A}_j \tag{4.11}$$

Considering that $\boldsymbol{K}$ and $\boldsymbol{M}$ are symmetrical matrices, $\boldsymbol{A}_j^T \boldsymbol{K}\boldsymbol{A}_i = \boldsymbol{A}_i^T \boldsymbol{K}\boldsymbol{A}_j$ and $\boldsymbol{A}_j^T \boldsymbol{M}\boldsymbol{A}_i = \boldsymbol{A}_i^T \boldsymbol{M}\boldsymbol{A}_j$ are obtained. Substituting them into Eq. (4.10) yields

$$\boldsymbol{A}_i^T \boldsymbol{K}\boldsymbol{A}_j = \lambda_i \boldsymbol{A}_i^T \boldsymbol{M}\boldsymbol{A}_j \tag{4.12}$$

Subtracting Eq. (4.11) from Eq. (4.12) leads to

$$(\lambda_i - \lambda_j)\boldsymbol{A}_i^T \boldsymbol{M}\boldsymbol{A}_j = 0 \tag{4.13}$$

When $\lambda_i \neq \lambda_j$, one obtains

$$\boldsymbol{A}_i^T \boldsymbol{M}\boldsymbol{A}_j = 0 \tag{4.14}$$

$$\boldsymbol{A}_i^T \boldsymbol{K}\boldsymbol{A}_j = 0 \tag{4.15}$$

Eqs. (4.14) and (4.15) indicate that the modes of the system are orthogonal with respect to $\boldsymbol{M}$ and $\boldsymbol{K}$. When $\lambda_i = \lambda_j$, the orthogonality conditions of the modes may not be satisfied. When $i = j$, Eq. (4.12) can be rearranged as follows:

$$\lambda_i = \frac{\boldsymbol{A}_i^T \boldsymbol{K}\boldsymbol{A}_i}{\boldsymbol{A}_i^T \boldsymbol{M}\boldsymbol{A}_i} = \frac{K_i}{M_i} \tag{4.16}$$

where $M_i = \boldsymbol{A}_i^T \boldsymbol{M}\boldsymbol{A}_i$ and $K_i = \boldsymbol{A}_i^T \boldsymbol{K}\boldsymbol{A}_i$, which are called the ith generalized mass and stiffness corresponding to $\boldsymbol{A}_i$, respectively.

Dividing each $\boldsymbol{A}_i$, $i = 1, 2, \cdots, n$, by $\sqrt{M_i}$ yields a new set of mode vectors $\overline{\boldsymbol{A}}_i = {}^{\boldsymbol{A}_i}/_{\sqrt{M_i}}$, $i = 1, 2, \cdots, n$, which are called normal mode vectors, and the corresponding normal mode matrix is denoted by $\overline{\boldsymbol{A}}$. Thus

$$\overline{\boldsymbol{A}}_i^T \boldsymbol{M} \overline{\boldsymbol{A}}_i = \overline{M}_i \tag{4.17}$$

$$\overline{\boldsymbol{A}}_i^T \boldsymbol{K} \overline{\boldsymbol{A}}_i = \overline{K}_i \tag{4.18}$$

where $\overline{M}_i$ and $\overline{K}_i$ are the ith generalized mass and stiffness corresponding to $\overline{\boldsymbol{A}}_i$, respectively. It is easily demonstrated that $\overline{M}_i = 1$ and $\overline{K}_i = \lambda_i$.

The above process to determine normal mode vectors is one of the normalizing procedures. The normalizing procedures also include (1) setting a specified coordinate to be unity and determining other elements according to the proportional relationship; and (2) setting the maximum absolute value of coordinates to unity.

Example 4.1: As shown in Fig. 4.1, the system vibrates freely along one horizontal direction on a frictionless plane. Here, $m_1 = m_2 = m_3 = m$, and $k_1 = k_2 = k_3 = k$ are known. Evaluate the natural frequencies and mode shapes of the system.

Solution:

The equations of motion in free vibration for the system are given as follows:

$$\boldsymbol{M}\ddot{\boldsymbol{q}} + \boldsymbol{K}\boldsymbol{q} = \boldsymbol{0}$$

where $\ddot{\boldsymbol{q}} = \{\ddot{v}_1 \quad \ddot{v}_2 \quad \ddot{v}_3\}^T$, $\boldsymbol{q} = \{v_1 \quad v_2 \quad v_3\}^T$

$$\boldsymbol{M} = \begin{bmatrix} m & 0 & 0 \\ 0 & m & 0 \\ 0 & 0 & m \end{bmatrix}, \ \boldsymbol{K} = \begin{bmatrix} 2k & -k & 0 \\ -k & 2k & -k \\ 0 & -k & k \end{bmatrix}$$

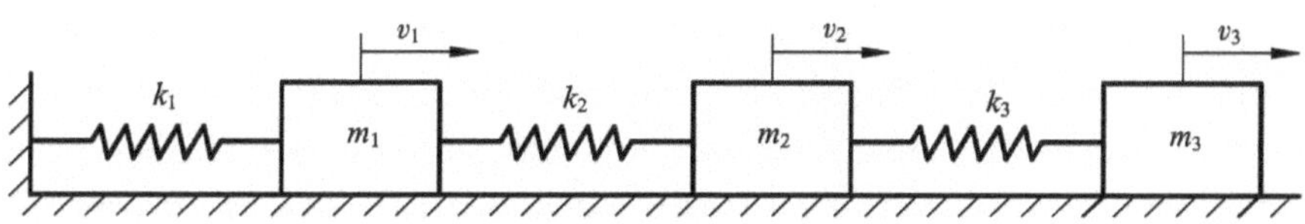

Figure 4.1 Schematic diagram of a multimass–spring system.

from which

$$\boldsymbol{K}-\lambda_i\boldsymbol{M}=\begin{bmatrix}2k-\lambda_i m & -k & 0\\ -k & 2k-\lambda_i m & -k\\ 0 & -k & k-\lambda_i m\end{bmatrix}$$

The solutions of the frequency equation $|\boldsymbol{K}-\lambda_i\boldsymbol{M}|=0$ are as follows:

$$\lambda_1=\frac{0.198k}{m},\quad \lambda_2=\frac{1.555k}{m},\quad \lambda_3=\frac{3.247k}{m}$$

The corresponding natural frequencies are $\omega_1=0.445\sqrt{k/m}$, $\omega_2=1.247\sqrt{k/m}$ and $\omega_3=1.802\sqrt{k/m}$, respectively.

Substituting each eigenvalue λ_i into the equation $(\boldsymbol{K}-\lambda_i\boldsymbol{M})\boldsymbol{A}_i=\boldsymbol{0}$, respectively, the mode vectors can be solved

$$\boldsymbol{A}_1=\{\,1.000\quad 1.802\quad 2.247\,\}^T$$

$$\boldsymbol{A}_2=\{\,1.000\quad 0.445\quad -0.802\,\}^T$$

$$\boldsymbol{A}_3=\{\,1.000\quad -1.247\quad 0.555\,\}^T$$

From $M_i=\boldsymbol{A}_i^T\boldsymbol{M}\boldsymbol{A}_i$, the generalized masses can be evaluated

$$M_1=9.296m,\quad M_2=1.841m,\quad M_3=2.863m$$

From $\overline{\boldsymbol{A}}_i=\boldsymbol{A}_i/\sqrt{M_i}$, the normal mode vectors are given as follows:

$$\overline{\boldsymbol{A}}_1=\frac{1}{\sqrt{m}}\{\,0.328\quad 0.591\quad 0.737\,\}^T$$

$$\overline{\boldsymbol{A}}_2=\frac{1}{\sqrt{m}}\{\,0.737\quad 0.328\quad -0.591\,\}^T$$

$$\overline{\boldsymbol{A}}_3=\frac{1}{\sqrt{m}}\{\,0.591\quad -0.737\quad 0.328\,\}^T$$

When the number of DOFs of the system is very large, the computational effort required to solve Eqs. (4.4) and (4.3) to obtain all the natural frequencies and modes of the system will be very great. However, only the first few frequencies and modes are required in the dynamic analysis of practical structures. Therefore some approximate methods were developed to evaluate lower frequencies and modes (see Chapter 6: Approximate evaluation of natural frequencies and mode shapes, for details).

4.1.3 Repeated frequency case

When the natural frequencies and mode shapes of structures are evaluated, it occasionally happens that a system has a repeated frequency [1]. The mode vectors associated with the repeated frequency may not necessarily satisfy the orthogonality conditions. A 2-DOF system is shown in Fig. 4.2, where the mass m is supported by two springs in the horizontal and vertical directions, respectively. The spring stiffness coefficients are both k. The small-amplitude free vibration around the static equilibrium position is performed, ignoring the influence of gravity. Taking the horizontal and vertical displacements u and v of mass m as generalized coordinates, the equation of motion is as follows:

$$\begin{bmatrix} m & 0 \\ 0 & m \end{bmatrix} \begin{Bmatrix} \ddot{u} \\ \ddot{v} \end{Bmatrix} + \begin{bmatrix} k & 0 \\ 0 & k \end{bmatrix} \begin{Bmatrix} u \\ v \end{Bmatrix} = \begin{Bmatrix} 0 \\ 0 \end{Bmatrix}$$

Substituting the mass and stiffness matrices into Eq. (4.4) yields

$$\begin{vmatrix} k - \lambda_i m & 0 \\ 0 & k - \lambda_i m \end{vmatrix} = 0$$

where $\lambda_1 = \lambda_2 = k/m$ can be solved from the above equation, and the corresponding natural frequencies of the system are $\omega_1 = \omega_2 = \sqrt{k/m}$.

Substituting $\lambda_1 = \lambda_2 = k/m$ into Eq. (4.3) yields

$$\begin{bmatrix} 0 & 0 \\ 0 & 0 \end{bmatrix} \begin{Bmatrix} A_{1i} \\ A_{2i} \end{Bmatrix} = \begin{Bmatrix} 0 \\ 0 \end{Bmatrix}$$

Obviously, two mode vectors can be chosen arbitrarily, which can be taken as $\boldsymbol{A}_1 = \begin{Bmatrix} 1 \\ 1 \end{Bmatrix}$ and $\boldsymbol{A}_2 = \begin{Bmatrix} 2 \\ 1 \end{Bmatrix}$ without the loss of generality. The

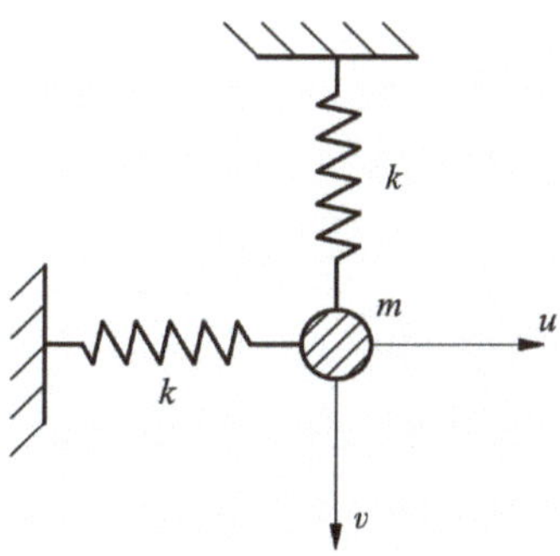

Figure 4.2 Mass–spring system with repeated natural frequency.

arbitrary linear combination of the two mode vectors can also be used as the mode vector of the system.

In this example, the two natural frequencies coincide with each other, and the two mode vectors are not orthogonal with respect to $\boldsymbol{M}$ and $\boldsymbol{K}$. In this case, the equations cannot be uncoupled by the following mode superposition method.

In accordance with the theorem of linear algebra, an n-DOF system has n orthogonal mode vectors regardless of whether the natural frequencies overlap. These mode vectors are orthogonal with respect to $\boldsymbol{M}$ and $\boldsymbol{K}$. Next, the procedure is introduced to find a set of orthogonal mode vectors for the case of repeated frequency, which is the preparation for the following mode superposition method.

When $\lambda_i = \lambda_j$, the mode vectors $\boldsymbol{A}_i$ and $\boldsymbol{A}_j$ may not satisfy the orthogonality conditions. When $\boldsymbol{A}_i$ and $\boldsymbol{A}_j$ are not orthogonal, it is necessary to find a set of orthogonalized mode vectors by the orthogonalizing process, which is referred to as a space complete basis in mathematics. The orthogonalizing procedure is given as follows:

1. Select the mode vectors $\boldsymbol{A}_i$ and $\boldsymbol{A}_j$ arbitrarily, when $\lambda_i = \lambda_j$.
2. Let $\boldsymbol{A}_j^* = \boldsymbol{A}_i + c\boldsymbol{A}_j$, and determine an appropriate coefficient c to make $\boldsymbol{A}_i$ and $\boldsymbol{A}_j^*$ satisfy the orthogonality conditions.
3. $\boldsymbol{A}_i$ and $\boldsymbol{A}_j^*$ satisfy the orthogonality conditions, if the following relation holds, that is,

$$\boldsymbol{A}_i^T\boldsymbol{M}\boldsymbol{A}_j^* = \boldsymbol{A}_i^T\boldsymbol{M}(\boldsymbol{A}_i + c\boldsymbol{A}_j) = \boldsymbol{A}_i^T\boldsymbol{M}\boldsymbol{A}_i + c\boldsymbol{A}_i^T\boldsymbol{M}\boldsymbol{A}_j = 0$$

   from which

$$c = -\frac{\boldsymbol{A}_i^T\boldsymbol{M}\boldsymbol{A}_i}{\boldsymbol{A}_i^T\boldsymbol{M}\boldsymbol{A}_j}.$$

4. Keep $\boldsymbol{A}_i$ unchanged and replace the original mode vector $\boldsymbol{A}_j$ with $\boldsymbol{A}_j^*$ ($\boldsymbol{A}_j^*$ is still denoted as $\boldsymbol{A}_j$ for the convenience in the expression). Thus all the mode vectors of the system are orthogonal to each other.

The obtained n orthogonal mode vectors $\boldsymbol{A}_1, \boldsymbol{A}_2, \cdots, \boldsymbol{A}_n$ are linearly independent, which is proven below:

Let n coefficients C_i, $i = 1, 2, \cdots, n$, satisfy

$$\sum_{i=1}^{n} C_i\boldsymbol{A}_i = \mathbf{0} \tag{4.19}$$

Premultiplying Eq. (4.19) by $\boldsymbol{A}_j^T\boldsymbol{M}$ leads to

$$\sum_{i=1}^{n} C_i\boldsymbol{A}_j^T\boldsymbol{M}\boldsymbol{A}_i = 0 \tag{4.20}$$

According to the orthogonality conditions of mode vectors, one gets $C_j = 0$. Let $j = 1, 2, \cdots, n$, thus $C_1 = C_2 = \cdots = C_n = 0$ is obtained, with which $\boldsymbol{A}_1, \boldsymbol{A}_2, \cdots, \boldsymbol{A}_n$ are proven to be linearly independent. Therefore the orthogonal mode vector set $\boldsymbol{A}_1, \boldsymbol{A}_2, \cdots, \boldsymbol{A}_n$constitutes the complete basis of n-dimensional space. With the substituting process above, the new mode vector set is linearly independent and also satisfies the orthogonality conditions, which satisfies the requirement of the mode superposition method.

4.2 Coupling characteristics and uncoupling procedure of equations of MDOF systems

4.2.1 Coupling characteristics of equations of MDOF systems

In general, the equations of motion for MDOF systems have coupling terms [1]. When the equations of motion are expressed in matrix form, coupling terms can be shown in the off-diagonal elements. If the mass matrix is not a diagonal matrix, the equations of motion are coupled with respect to the mass, which is called inertial coupling or dynamic coupling. If the stiffness matrix is not diagonal, the equations of motion are coupled with respect to the stiffness, which is called elastic coupling or static coupling. In these situations, n equations must be solved simultaneously to obtain the system responses, and the solving process becomes complicated due to the coupling characteristics. Whether the equations of motion of systems are coupled or not is related to the selection of the generalized coordinates. For example, when the translation or rotation of the mass center of an object is selected as generalized coordinates, the diagonal mass matrix can be obtained. When the generalized coordinates are selected properly, the equations of motion will be uncoupled. The coupling characteristics with respect to the damping and processing procedure will be introduced in Section 4.5.

A 2-DOF system is taken as an example here, and three different sets of generalized coordinates are selected to discuss the coupling characteristics of equations of motion. As shown in Fig. 4.3, the system consists of a rigid rod with mass m. Points A and D of the rigid rod are supported by

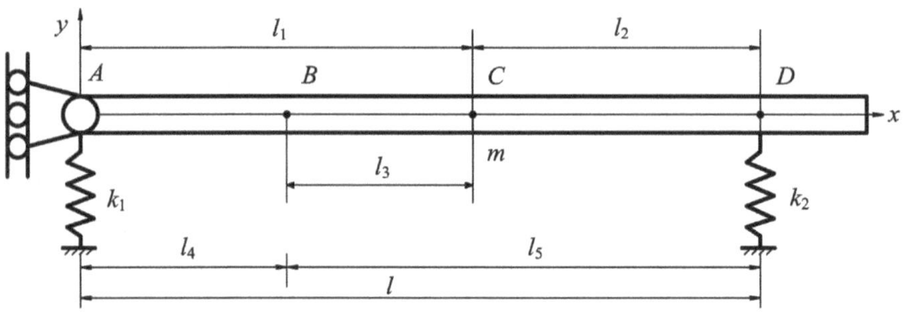

Figure 4.3 2-DOF undamped system. *DOF*, Degree-of-freedom.

springs with stiffness coefficients k_1 and k_2, respectively. The constraint of the support at point A only allows the rigid rod to move in the $x-y$ plane, and the translation of point A along the x axis is restricted. Point C is the mass center of the rigid rod and J_C represents the moment of inertia about the z axis passing through point C (perpendicular to the $x-y$ plane, not shown). Point B is a special point, with which the relationship $k_1 l_4 = k_2 l_5$ holds. If a force acts at point B along the y axis, only translation of the system will occur, and the rotation will vanish. If a moment acts about point B, the system will only rotate and not translate. Since the acting positions and types of external loads affect only the load vector, rather than the coupling characteristics of the equations of the system, the free vibration equations in terms of different generalized coordinates are formulated below.

Once the translation y_A at point A and the rotation θ_A of the rigid rod about point A are selected as the displacement coordinates of the system, the equations of motion can be formulated as follows:

$$\begin{bmatrix} m & ml_1 \\ ml_1 & ml_1^2 + J_C \end{bmatrix} \begin{Bmatrix} \ddot{y}_A \\ \ddot{\theta}_A \end{Bmatrix} + \begin{bmatrix} k_1 + k_2 & k_2 l \\ k_2 l & k_2 l^2 \end{bmatrix} \begin{Bmatrix} y_A \\ \theta_A \end{Bmatrix} = \begin{Bmatrix} 0 \\ 0 \end{Bmatrix} \tag{4.21}$$

In Eq. (4.21) the off-diagonal elements in mass and stiffness matrices are not zero, and both inertial and elastic coupling appear. The phenomenon of inertial coupling indicates that the two accelerations are not independent of each other, that is, the system is coupled with respect to the mass. The phenomenon of elastic coupling indicates that a displacement not only causes a reactive force corresponding to itself, but also causes a reactive force corresponding to another displacement coordinate, and the equations of the system are coupled with respect to the stiffness.

Once the translation y_B at point B and the rotation θ_B of the rigid rod about point B are taken as the displacement coordinates of the system, the equations of motion may be formulated

$$\begin{bmatrix} m & ml_3 \\ ml_3 & ml_3^2 + J_C \end{bmatrix} \begin{Bmatrix} \ddot{y}_B \\ \ddot{\theta}_B \end{Bmatrix} + \begin{bmatrix} k_1 + k_2 & 0 \\ 0 & k_1 l_4^2 + k_2 l_5^2 \end{bmatrix} \begin{Bmatrix} y_B \\ \theta_B \end{Bmatrix} = \begin{Bmatrix} 0 \\ 0 \end{Bmatrix} \tag{4.22}$$

where $\boldsymbol{K}$ is a diagonal matrix and $\boldsymbol{M}$ is not a diagonal matrix. It is shown that only inertial coupling occurs in Eq. (4.22), rather than elastic coupling.

Once the translation y_C at point C and the rotation θ_C of the rigid rod about point C are selected as the displacement coordinates of the system, the equations of motion are as follows:

$$\begin{bmatrix} m & 0 \\ 0 & J_C \end{bmatrix} \begin{Bmatrix} \ddot{y}_C \\ \ddot{\theta}_C \end{Bmatrix} + \begin{bmatrix} k_1 + k_2 & k_2 l_2 - k_1 l_1 \\ k_2 l_2 - k_1 l_1 & k_1 l_1^2 + k_2 l_2^2 \end{bmatrix} \begin{Bmatrix} y_C \\ \theta_C \end{Bmatrix} = \begin{Bmatrix} 0 \\ 0 \end{Bmatrix} \tag{4.23}$$

where $\boldsymbol{M}$ is a diagonal matrix, $\boldsymbol{K}$ is not a diagonal matrix, and only elastic coupling and no inertial coupling occurs.

It is obviously shown from the above three cases that the coupling characteristics of the equations of motion depend on the selection of coordinates rather than the characteristics of systems. Theoretically, as long as the displacement coordinates are selected properly, neither inertial nor elastic coupling will occur in the equations of motion, and the equations of motion are independent of each other.

4.2.2 Uncoupling procedure of equations of MDOF systems

As discussed above, elastic or inertial uncoupling (see Section 4.5 for damping uncoupling) can be achieved by selecting displacement coordinates for a system properly. The generalized coordinates are called the principal coordinates (modal coordinates), denoted as $T_1, T_2, \cdots, T_n$, with which elastic and inertial coupling in the equations of motion will not appear. The transformation between the original geometric coordinates, $q_1, q_2, \cdots, q_n$, and the principal coordinates, $T_1, T_2, \cdots, T_n$, can be carried out by using the orthogonalized mode matrix $\boldsymbol{A}$, which is called principal coordinate transformation, that is,

$$\boldsymbol{q} = \boldsymbol{AT} \tag{4.24}$$

where

$$\boldsymbol{q} = \left\{ q_1 \quad q_2 \quad \cdots \quad q_n \right\}^T$$

$$\boldsymbol{A} = \left[\boldsymbol{A}_1 \quad \boldsymbol{A}_2 \quad \cdots \quad \boldsymbol{A}_n \right]$$

$$\boldsymbol{A}_i = \left\{ A_{1i} \quad A_{2i} \quad \ldots \quad A_{ni} \right\}^T, \; i = 1, 2, \cdots, n$$

$$\boldsymbol{T} = \left\{ T_1 \quad T_2 \quad \cdots \quad T_n \right\}^T$$

Eq. (4.24) indicates the mode superposition principle for linear vibration analysis, that is, any displacements of n-DOF systems can be expressed as a linear combination of n orthogonal mode shapes.

In Chapter 2, Formulation of equations of motion of systems, the linear vibration equations of motion for n-DOF undamped systems were given as follows:

$$\boldsymbol{M}\ddot{\boldsymbol{q}} + \boldsymbol{K}\boldsymbol{q} = \mathbf{Q} \tag{4.25}$$

where $\mathbf{Q}$ represents the generalized load vector corresponding to generalized coordinate vector $\boldsymbol{q}$. When $\mathbf{Q} = \mathbf{0}$, this is the case of free vibration.

Substituting Eq. (4.24) into Eq. (4.25) leads to

$$\boldsymbol{M}\boldsymbol{A}\ddot{\boldsymbol{T}} + \boldsymbol{K}\boldsymbol{A}\boldsymbol{T} = \mathbf{Q} \tag{4.26}$$

Premultiplying Eq. (4.26) by $\boldsymbol{A}^T$ gives

$$\boldsymbol{A}^T\boldsymbol{M}\boldsymbol{A}\ddot{\boldsymbol{T}} + \boldsymbol{A}^T\boldsymbol{K}\boldsymbol{A}\boldsymbol{T} = \boldsymbol{A}^T\mathbf{Q} \tag{4.27}$$

According to the orthogonality conditions, Eq. (4.27) can be written as follows:

$$\boldsymbol{M}^*\ddot{\boldsymbol{T}} + \boldsymbol{K}^*\boldsymbol{T} = \boldsymbol{P}^* \tag{4.28}$$

where $\boldsymbol{M}^* = \text{diag}(M_1, M_2, \cdots, M_n)$, which is called generalized mass matrix; $\boldsymbol{K}^* = \text{diag}(K_1, K_2, \cdots, K_n)$, which is called generalized stiffness matrix; $\boldsymbol{P}^* = \boldsymbol{A}^T\mathbf{Q} = \left\{ P_1 \quad P_2 \quad \cdots \quad P_n \right\}^T$, which is called generalized load vector; and

$$M_i = \boldsymbol{A}_i^T\boldsymbol{M}\boldsymbol{A}_i, \quad K_i = \boldsymbol{A}_i^T\boldsymbol{K}\boldsymbol{A}, \quad P_i = \boldsymbol{A}_i^T\mathbf{Q}, \quad i = 1, 2, \cdots, n$$

Eq. (4.28) can be written in the form of components

$$M_i\ddot{T}_i + K_iT_i = P_i, \; i = 1, 2, \cdots, n \tag{4.29}$$

Eq. (4.29) are the uncoupled equations of motion in principal coordinates.

If the mode matrix $\boldsymbol{A}$ is replaced by the normal mode matrix $\overline{\boldsymbol{A}}$ in principal coordinate transform, one obtains

$$q = \overline{\boldsymbol{A}}\overline{\boldsymbol{T}} \tag{4.30}$$

where

$$q = \{ q_1 \quad q_2 \quad \cdots \quad q_n \}^T$$

$$\overline{\boldsymbol{A}} = [\overline{\boldsymbol{A}}_1 \quad \overline{\boldsymbol{A}}_2 \quad \cdots \quad \overline{\boldsymbol{A}}_n]$$

$$\overline{\boldsymbol{A}}_i = \{ \overline{A}_{1i} \quad \overline{A}_{2i} \quad \cdots \quad \overline{A}_{ni} \}^T, \quad i = 1, 2, \cdots, n$$

$$\overline{\boldsymbol{T}} = \{ \overline{T}_1 \quad \overline{T}_2 \quad \cdots \quad \overline{T}_n \}^T$$

Similarly, substituting Eq. (4.30) into Eq. (4.25) leads to

$$\boldsymbol{M}\overline{\boldsymbol{A}}\ddot{\overline{\boldsymbol{T}}} + \boldsymbol{K}\overline{\boldsymbol{A}}\overline{\boldsymbol{T}} = \mathbf{Q} \tag{4.31}$$

Premultiplying Eq. (4.31) by $\overline{\boldsymbol{A}}^T$ gives

$$\overline{\boldsymbol{A}}^T\boldsymbol{M}\overline{\boldsymbol{A}}\ddot{\overline{\boldsymbol{T}}} + \overline{\boldsymbol{A}}^T\boldsymbol{K}\overline{\boldsymbol{A}}\overline{\boldsymbol{T}} = \overline{\boldsymbol{A}}^T\mathbf{Q} \tag{4.32}$$

According to the orthogonality conditions, Eq. (4.32) can be written as follows:

$$\ddot{\overline{\boldsymbol{T}}} + \boldsymbol{\lambda}\overline{\boldsymbol{T}} = \overline{\boldsymbol{P}} \tag{4.33}$$

where $\overline{\boldsymbol{P}} = \overline{\boldsymbol{A}}^T\mathbf{Q} = \{ \overline{P}_1 \quad \overline{P}_2 \quad \cdots \quad \overline{P}_n \}^T$, $\overline{P}_i = \overline{\boldsymbol{A}}_i^T\mathbf{Q}$, $i = 1, 2, \cdots, n$, $\boldsymbol{\lambda} = \text{diag}(\lambda_1, \lambda_2, \cdots, \lambda_n)$, and λ_i can be found in Eq. (4.16).

Eq. (4.33) can be written in the form of components

$$\ddot{\overline{T}}_i + \lambda_i\overline{T}_i = \overline{P}_i, \; i = 1, 2, \cdots, n \tag{4.34}$$

Eq. (4.34) are essentially equivalent to Eq. (4.29), and only the mode matrices of different forms are used for the coordinate transformation. Eq. (4.30) is known as the normal coordinate transformation, and $\overline{T}_1, \overline{T}_2, \cdots, \overline{T}_n$ are called the normal coordinates. Hence, Eq. (4.34) is the equation of motion in normal coordinates. According to the above analysis, the normal mode is a special mode, and the corresponding normal coordinates are also a special form of principal coordinates.

The dynamic displacements of the system are expressed as the linear combination of all n mode vectors shown above. It should be noted that for most types of loads, the displacement contributions generally are greatest for the lower modes and tend to decrease for the higher modes (as illustrated by Example 4.4). Consequently, it usually is not necessary to include the contribution of all the higher modes in the superposition process. The series can be truncated when the response has been obtained with desired accuracy. Moreover, the mathematical idealization of any complex structural system also tends to be less reliable in predicting the higher modes. For these reasons, the number of modes considered in dynamic response analysis should be limited [2]. The dynamic response of n-DOF systems can be approximately expressed as the linear combination of the first N mode vectors ($N < n$). Therefore Eqs. (4.24) and (4.30) can be approximately written respectively as follows:

$$\boldsymbol{q} \approx \boldsymbol{A}_{n\times N}\boldsymbol{T}_N \tag{4.35}$$

$$\boldsymbol{q} \approx \overline{\boldsymbol{A}}_{n\times N}\overline{\boldsymbol{T}}_N \tag{4.36}$$

where

$$\boldsymbol{A}_{n\times N} = \begin{bmatrix} \boldsymbol{A}_1 & \boldsymbol{A}_2 & \cdots & \boldsymbol{A}_N \end{bmatrix}$$

$$\boldsymbol{T}_N = \begin{Bmatrix} T_1 & T_2 & \cdots & T_N \end{Bmatrix}^T$$

$$\overline{\boldsymbol{A}}_{n\times N} = \begin{bmatrix} \overline{\boldsymbol{A}}_1 & \overline{\boldsymbol{A}}_2 & \cdots & \overline{\boldsymbol{A}}_N \end{bmatrix}$$

$$\overline{\boldsymbol{T}}_N = \begin{Bmatrix} \overline{T}_1 & \overline{T}_2 & \cdots & \overline{T}_N \end{Bmatrix}^T$$

Substituting Eqs. (4.35) and (4.36) into Eq. (4.25), respectively, leads to

$$M_i\ddot{T}_i + K_iT_i = P_i, \quad i = 1, 2, \cdots, N \tag{4.37}$$

$$\ddot{\overline{T}}_i + \lambda_i\overline{T}_i = \overline{P}_i, \quad i = 1, 2, \cdots, N \tag{4.38}$$

To sum up, coupled equations of motion for n-DOF systems can be uncoupled into n or N independent differential equations by linear coordinate transformation. Thus the problem of coupled MDOF systems is transformed into n or Nindependent problems of single-degree-of-freedom (SDOF) systems. The next task is to solve n (or N) independent equations respectively using the methods introduced in Chapter 3, Analysis of dynamic response of SDOF systems. Then, the system responses expressed by the original geometric coordinates can be obtained by the coordinate transformation of these independent solutions (see the following three sections for details). The method mentioned above is called the mode superposition method or the mode synthesis method for dynamic response analysis of systems.

4.3 Analysis of free vibration response of undamped systems

According to the previous discussion, the equations of motion for n-DOF systems can be uncoupled into n independent equations by means of normal coordinate transformation. The corresponding free vibration equations of motion in normal coordinates are as follows:

$$\ddot{\overline{T}}_i(t) + \omega_i^2 \overline{T}_i(t) = 0, \ i = 1, 2, \cdots, n \tag{4.39}$$

If the initial conditions in terms of the normal coordinate $\overline{T}_i(t)$ are known as $\overline{T}_i(0)$ and $\dot{\overline{T}}_i(0)$, the free vibration response of the ith normal coordinate can be evaluated as follows:

$$\overline{T}_i(t) = \overline{T}_i(0) \cos \omega_i t + \frac{\dot{\overline{T}}_i(0)}{\omega_i} \sin\omega_i t, \quad i = 1, 2, \cdots, n \tag{4.40}$$

Premultiplying Eq. (4.30) by $\overline{\boldsymbol{A}}^T \boldsymbol{M}$, considering the orthogonality conditions, and letting $t = 0$ leads to

$$\overline{\boldsymbol{T}}(0) = \overline{\boldsymbol{A}}^T \boldsymbol{M} \boldsymbol{q}_0 \tag{4.41}$$

Similarly, the following equation can be obtained

$$\dot{\overline{\boldsymbol{T}}}(0) = \overline{\boldsymbol{A}}^T \boldsymbol{M} \dot{\boldsymbol{q}}_0 \tag{4.42}$$

where $\boldsymbol{q}_0 = \{ q_{01} \quad q_{02} \quad \cdots \quad q_{0n} \}^T$ and $\dot{\boldsymbol{q}}_0 = \{ \dot{q}_{01} \quad \dot{q}_{02} \quad \cdots \quad \dot{q}_{0n} \}^T$ are the initial conditions expressed in terms of original geometric coordinates, and $\overline{\boldsymbol{T}}(0) = \{ \overline{T}_1(0) \quad \overline{T}_2(0) \quad \cdots \quad \overline{T}_n(0) \}^T$ and $\dot{\overline{\boldsymbol{T}}}(0) = \{ \dot{\overline{T}}_1(0) \dot{\overline{T}}_2(0) \cdots \dot{\overline{T}}_n(0) \}^T$ are the initial conditions expressed in terms of normal coordinates.

When a rigid body mode is present for a system, the corresponding natural frequency $\omega_i = 0$, and Eq. (4.39) becomes

$$\ddot{\overline{T}}_i(t) = 0 \tag{4.43}$$

Integrating Eq. (4.43) twice with respect to time t yields

$$\overline{T}_i(t) = \overline{T}_i(0) + \dot{\overline{T}}_i(0)t \tag{4.44}$$

The response of the rigid body mode represented by normal coordinates can be evaluated from Eq. (4.44).

Based on the obtained responses of normal coordinates $\overline{T}_i$, $i = 1, 2, \ldots, n$, the free vibration responses of the system expressed in terms of the original geometric coordinates can be evaluated by Eq. (4.30).

Example 4.2: Consider the system given in Example 4.1, assume $k_1 = 0$, and evaluate the free vibration responses of the system.

Solution: According to the analysis in Example 4.1, the free vibration equations of motion of the system are as follows:

$$\boldsymbol{M}\ddot{\boldsymbol{q}} + \boldsymbol{K}\boldsymbol{q} = \boldsymbol{0}$$

where $\ddot{\boldsymbol{q}} = \left\{ \ddot{v}_1 \quad \ddot{v}_2 \quad \ddot{v}_3 \right\}^T$, $\boldsymbol{q} = \left\{ v_1 \quad v_2 \quad v_3 \right\}^T$

$$\boldsymbol{M} = \begin{bmatrix} m & 0 & 0 \\ 0 & m & 0 \\ 0 & 0 & m \end{bmatrix}, \quad \boldsymbol{K} = \begin{bmatrix} k & -k & 0 \\ -k & 2k & -k \\ 0 & -k & k \end{bmatrix}$$

Obviously, $|\boldsymbol{K}| = 0$ because the stiffness matrix is singular.

$$\boldsymbol{K} - \lambda_i \boldsymbol{M} = \begin{bmatrix} k - \lambda_i m & -k & 0 \\ -k & 2k - \lambda_i m & -k \\ 0 & -k & k - \lambda_i m \end{bmatrix}$$

From $|\boldsymbol{K} - \lambda_i \boldsymbol{M}| = 0$, the eigenvalues can be solved as follows:

$$\lambda_1 = 0, \quad \lambda_2 = \frac{k}{m}, \quad \lambda_3 = \frac{3k}{m}$$

The corresponding natural frequencies are $\omega_1 = 0$, $\omega_2 = \sqrt{k/m}$, and $\omega_3 = \sqrt{3k/m}$.

The eigenvalues λ_i, $i = 1, 2, 3$, are substituted into $(\boldsymbol{K} - \lambda_i \boldsymbol{M})\boldsymbol{A}_i = \boldsymbol{0}$, respectively, and three mode vectors of the system are solved as follows:

$$\boldsymbol{A}_1 = \left\{ 1 \quad 1 \quad 1 \right\}^T$$

$$\boldsymbol{A}_2 = \left\{ 1 \quad 0 \quad -1 \right\}^T$$

$$\boldsymbol{A}_3 = \left\{ 1 \quad -2 \quad 1 \right\}^T$$

Note that $\boldsymbol{A}_1$ is the rigid body mode corresponding to $\omega_1 = 0$.

Assuming that the system is in static state at the beginning, the mass m_1 is suddenly struck to get the initial velocity $\dot{v}_{01}$. The responses of the system caused by this impact will be evaluated below.

The generalized mass of the system is calculated as follows:

$$M_1 = \boldsymbol{A}_1^T \boldsymbol{M} \boldsymbol{A}_1 = m(1^2 + 1^2 + 1^2) = 3m$$

$$M_2 = \boldsymbol{A}_2^T \boldsymbol{M} \boldsymbol{A}_2 = m(1^2 + 1^2) = 2m$$

$$M_3 = \boldsymbol{A}_3^T \boldsymbol{M} \boldsymbol{A}_3 = m(1^2 + 2^2 + 1^2) = 6m$$

Thus the normal mode matrix can be written as

$$\overline{\boldsymbol{A}} = \begin{bmatrix} \dfrac{\boldsymbol{A}_1}{\sqrt{M_1}} & \dfrac{\boldsymbol{A}_2}{\sqrt{M_2}} & \dfrac{\boldsymbol{A}_3}{\sqrt{M_3}} \end{bmatrix} = \frac{1}{\sqrt{6m}} \begin{bmatrix} \sqrt{2} & \sqrt{3} & 1 \\ \sqrt{2} & 0 & -2 \\ \sqrt{2} & -\sqrt{3} & 1 \end{bmatrix}$$

from which

$$\overline{\boldsymbol{A}}^T \boldsymbol{M} = \frac{1}{\sqrt{6m}} \begin{bmatrix} \sqrt{2} & \sqrt{2} & \sqrt{2} \\ \sqrt{3} & 0 & -\sqrt{3} \\ 1 & -2 & 1 \end{bmatrix} \begin{bmatrix} m & 0 & 0 \\ 0 & m & 0 \\ 0 & 0 & m \end{bmatrix} = \sqrt{\frac{m}{6}} \begin{bmatrix} \sqrt{2} & \sqrt{2} & \sqrt{2} \\ \sqrt{3} & 0 & -\sqrt{3} \\ 1 & -2 & 1 \end{bmatrix}$$

The initial condition vectors of the system expressed in terms of original geometric coordinates are written as

$$\boldsymbol{q}_0 = \begin{Bmatrix} 0 \\ 0 \\ 0 \end{Bmatrix}, \quad \dot{\boldsymbol{q}}_0 = \begin{Bmatrix} \dot{v}_{01} \\ 0 \\ 0 \end{Bmatrix} \tag{4.45}$$

Then, Eq. (4.45) is transformed into the initial condition vectors in terms of normal coordinates.

$$\overline{\boldsymbol{T}}(0) = \overline{\boldsymbol{A}}^T \boldsymbol{M} \boldsymbol{q}_0 = \begin{Bmatrix} 0 \\ 0 \\ 0 \end{Bmatrix}$$

$$\dot{\overline{\boldsymbol{T}}}(0) = \overline{\boldsymbol{A}}^T \boldsymbol{M} \dot{\boldsymbol{q}}_0 = \sqrt{\frac{m}{6}} \begin{bmatrix} \sqrt{2} & \sqrt{2} & \sqrt{2} \\ \sqrt{3} & 0 & -\sqrt{3} \\ 1 & -2 & 1 \end{bmatrix} \begin{Bmatrix} \dot{v}_{01} \\ 0 \\ 0 \end{Bmatrix} = \dot{v}_{01} \sqrt{\frac{m}{6}} \begin{Bmatrix} \sqrt{2} \\ \sqrt{3} \\ 1 \end{Bmatrix} \tag{4.46}$$

Substituting the first row of Eq. (4.46) into Eq. (4.44) and substituting the second and third rows into Eq. (4.40), respectively, leads to

$$\overline{\boldsymbol{T}}(t) = \dot{v}_{01}\sqrt{\frac{m}{6}}\begin{Bmatrix} \sqrt{2}t \\ (\sqrt{3}\sin\omega_2 t)/\omega_2 \\ (\sin\omega_3 t)/\omega_3 \end{Bmatrix} \tag{4.47}$$

Transforming $\overline{\boldsymbol{T}}(t)$ into original geometric coordinates gives

$$\begin{aligned} \boldsymbol{q} = \overline{\boldsymbol{A}}\overline{\boldsymbol{T}}(t) &= \frac{1}{\sqrt{6m}}\begin{bmatrix} \sqrt{2} & \sqrt{3} & 1 \\ \sqrt{2} & 0 & -2 \\ \sqrt{2} & -\sqrt{3} & 1 \end{bmatrix} \dot{v}_{01}\sqrt{\frac{m}{6}}\begin{Bmatrix} \sqrt{2}t \\ (\sqrt{3}\sin\omega_2 t)/\omega_2 \\ (\sin\omega_3 t)/\omega_3 \end{Bmatrix} \\ &= \frac{\dot{v}_{01}}{6}\begin{Bmatrix} 2t + (3\sin\omega_2 t)/\omega_2 + (\sin\omega_3 t)/\omega_3 \\ 2t - (2\sin\omega_3 t)/\omega_3 \\ 2t - (3\sin\omega_2 t)/\omega_2 + (\sin\omega_3 t)/\omega_3 \end{Bmatrix} \end{aligned} \tag{4.48}$$

In Eq. (4.48), the terms of rigid displacements in each response component are all equal to $\dot{v}_{01}t/3$.

Once all the masses of the system have the same initial velocity $\dot{v}_0$, the initial velocity vector is $\dot{\boldsymbol{q}}_0 = \{\dot{v}_0 \quad \dot{v}_0 \quad \dot{v}_0\}^T$. Then, Eqs. (4.46)–(4.48) can be rewritten as

$$\overline{\boldsymbol{T}}(0) = \begin{Bmatrix} 0 \\ 0 \\ 0 \end{Bmatrix}, \quad \dot{\overline{\boldsymbol{T}}}(0) = \dot{v}_0\sqrt{3m}\begin{Bmatrix} 1 \\ 0 \\ 0 \end{Bmatrix}, \quad \overline{\boldsymbol{T}}(t) = \dot{v}_0 t\sqrt{3m}\begin{Bmatrix} 1 \\ 0 \\ 0 \end{Bmatrix}, \quad \boldsymbol{q} = \dot{v}_0 t\begin{Bmatrix} 1 \\ 1 \\ 1 \end{Bmatrix} \tag{4.49}$$

Note that only rigid body translations are present and no oscillation occurs.

If the initial velocities of masses m_1, m_2, and m_3 are $\dot{v}_0$, 0, and $-\dot{v}_0$, respectively, and the initial displacements are all zero, the initial velocity vector is $\dot{\boldsymbol{q}}_0 = \{\dot{v}_0 \quad 0 \quad -\dot{v}_0\}^T$. Then, Eqs. (4.46)–(4.48) can be written as

$$\overline{\boldsymbol{T}}(0) = \begin{Bmatrix} 0 \\ 0 \\ 0 \end{Bmatrix}, \quad \dot{\overline{\boldsymbol{T}}}(0) = \dot{v}_0\sqrt{2m}\begin{Bmatrix} 0 \\ 1 \\ 0 \end{Bmatrix}, \quad \overline{\boldsymbol{T}}(t) = \frac{\sqrt{2m}\dot{v}_0\sin\omega_2 t}{\omega_2}\begin{Bmatrix} 0 \\ 1 \\ 0 \end{Bmatrix}, \quad \boldsymbol{q} = \frac{\dot{v}_0\sin\omega_2 t}{\omega_2}\begin{Bmatrix} 1 \\ 0 \\ -1 \end{Bmatrix} \tag{4.50}$$

In this case, the initial conditions only excite the second principal vibration of the system. Thus a certain principal vibration of the system may occur independently under proper initial conditions, which is also known as a simple harmonic vibration. This characteristic is consistent with the theoretical analysis in Section 4.1, which will be applied to verify the orthogonality of modes and evaluate the free vibration of continuous systems, etc.

4.4 Response of undamped systems to arbitrary dynamic loads

Eq. (4.34) have given the equations of motion in normal coordinates for undamped systems, which are directly referenced as follows:

$$\ddot{\overline{T}}_i(t) + \lambda_i \overline{T}_i(t) = \overline{P}_i, \quad i = 1, 2, \cdots, n, \quad \lambda_i = \omega_i^2 \tag{4.51}$$

Considering the initial conditions, the solution of Eq. (4.51) is obtained by the Duhamel integral

$$\overline{T}_i(t) = \overline{T}_i(0) \cos \omega_i t + \frac{\dot{\overline{T}}_i(0)}{\omega_i} \sin \omega_i t + \frac{1}{\omega_i} \int_0^t \overline{P}_i(\tau) \sin \omega_i (t - \tau) d\tau \tag{4.52}$$

where the initial conditions $\overline{T}_i(0)$ and $\dot{\overline{T}}_i(0)$ are determined from Eqs. (4.41) and (4.42), respectively.

When the ith mode of the system is rigid body mode, that is, $\lambda_i = 0$, the corresponding equation of motion is given as follows:

$$\ddot{\overline{T}}_i(t) = \overline{P}_i \tag{4.53}$$

Integrating Eq. (4.53) twice with respect to t leads to

$$\overline{T}_i(t) = \int_0^t \left[\int_0^t \overline{P}_i dt \right] dt + C_1 t + C_2 \tag{4.54}$$

where the integral constants C_1 and C_2 are determined from the initial conditions of the rigid body motion. Therefore instead of Eq. (4.52), Eq. (4.54) is used to evaluate the normal coordinate response when the rigid body mode is present.

When the applied load is a simple harmonic load, say $\overline{P}_i = P_{i0} \sin \overline{\omega} t$, the corresponding response can be obtained from Eq. (4.52)

$$\overline{T}_i(t) = \overline{T}_i(0) \cos \omega_i t + \frac{\dot{\overline{T}}_i(0)}{\omega_i} \sin \omega_i t - \frac{P_{i0} \overline{\omega}}{\omega_i (\omega_i^2 - \overline{\omega}^2)} \sin \omega_i t + \frac{P_{i0}}{\omega_i^2 - \overline{\omega}^2} \sin \overline{\omega} t \tag{4.55}$$

In Eq. (4.55), the first three terms vibrate at the natural frequency ω_i. The first two terms are determined from the initial conditions. The third term is called associated free vibration, and its amplitude is related to the applied load. In practical engineering, the first three terms will decay rapidly due to damping. The last term vibrates at the frequency of the simple harmonic load $\overline{\omega}$ and continues with the action of load, which is known as the steady-state response.

Example 4.3: A 3-DOF system is shown in Fig. 4.4, and $m_1 = 2m$, $m_2 = 1.5m$, $m_3 = m$, $k_1 = 3k$, $k_2 = 2k$, and $k_3 = k$ are known. Assume that the simple harmonic load $P_2(t) = P\sin\overline{\omega}t$ is applied to mass m_2. Evaluate the steady-state response of the system.

Solution: The mass matrix $\boldsymbol{M}$ and stiffness matrix $\boldsymbol{K}$ can be obtained as follows:

$$\boldsymbol{M} = m\begin{bmatrix} 2 & 0 & 0 \\ 0 & 1.5 & 0 \\ 0 & 0 & 1 \end{bmatrix}, \quad \boldsymbol{K} = k\begin{bmatrix} 5 & -2 & 0 \\ -2 & 3 & -1 \\ 0 & -1 & 1 \end{bmatrix}$$

The natural frequencies and mode matrix are thus obtained, respectively,

$$\omega_1^2 = 0.3515\frac{k}{m}, \quad \omega_2^2 = 1.6066\frac{k}{m}, \quad \omega_3^2 = 3.5419\frac{k}{m},$$

$$\boldsymbol{A} = \begin{bmatrix} 0.3018 & -0.6790 & -0.9598 \\ 0.6485 & -0.6066 & 1.0000 \\ 1.0000 & 1.0000 & -0.3934 \end{bmatrix}$$

Then, the normal mode matrix is

$$\overline{\boldsymbol{A}} = \frac{1}{\sqrt{m}}\begin{bmatrix} 0.2242 & -0.4317 & -0.5132 \\ 0.4816 & -0.3857 & 0.5348 \\ 0.7427 & 0.6358 & -0.2104 \end{bmatrix}$$

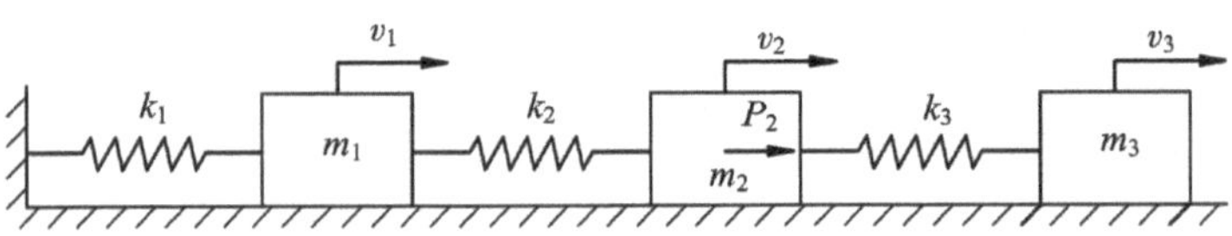

Figure 4.4 3-DOF mass–spring system. *DOF*, Degree-of-freedom.

from which

$$\overline{\boldsymbol{P}} = \overline{\boldsymbol{A}}^T \boldsymbol{P} = \frac{1}{\sqrt{m}} \begin{bmatrix} 0.2242 & 0.4816 & 0.7427 \\ -0.4317 & -0.3857 & 0.6358 \\ -0.5132 & 0.5348 & -0.2104 \end{bmatrix} \begin{Bmatrix} 0 \\ P\sin\overline{\omega}t \\ 0 \end{Bmatrix} = \frac{P\sin\overline{\omega}t}{\sqrt{m}} \begin{Bmatrix} 0.4816 \\ -0.3857 \\ 0.5348 \end{Bmatrix}$$

The steady-state responses in terms of normal coordinates may be evaluated by the Duhamel integral

$$\overline{\boldsymbol{T}}(t) = \begin{Bmatrix} \overline{T}_1(t) \\ \overline{T}_2(t) \\ \overline{T}_3(t) \end{Bmatrix} = \frac{P\sin\overline{\omega}t}{\sqrt{m}} \begin{Bmatrix} \dfrac{0.4816D_1}{\omega_1^2} \\ \dfrac{-0.3857D_2}{\omega_2^2} \\ \dfrac{0.5348D_3}{\omega_3^2} \end{Bmatrix}$$

where $D_i = 1/\left(1 - \overline{\omega}^2/\omega_i^2\right)$, $i = 1, 2, 3$.

Introducing $\omega_1^2 = 0.3515k/m$, $\omega_2^2 = 1.6066k/m$, and $\omega_3^2 = 3.5419k/m$ into the above expression yields

$$\overline{\boldsymbol{T}}(t) = \frac{P\sin\overline{\omega}t\sqrt{m}}{k} \begin{Bmatrix} 1.3701D_1 \\ -0.2401D_2 \\ 0.1510D_3 \end{Bmatrix}$$

By means of normal coordinate transformation, one gets

$$q = \begin{Bmatrix} v_1 \\ v_2 \\ v_3 \end{Bmatrix} = \overline{\boldsymbol{A}}\overline{\boldsymbol{T}}(t) = \frac{P\sin\overline{\omega}t}{k} \begin{Bmatrix} 0.3072D_1 + 0.1037D_2 - 0.0775D_3 \\ 0.6598D_1 + 0.0926D_2 + 0.0808D_3 \\ 1.0176D_1 + 0.1527D_2 - 0.0318D_3 \end{Bmatrix}$$

It should be noted that resonances will occur in the system when the frequency of simple harmonic load $\overline{\omega}$ is equal to any natural frequency of the system.

4.5 Response of damped systems to arbitrary dynamic loads

Damping is present in any practical structure, but it is not necessary to take it into consideration in all cases. The vibration analysis of SDOF systems shows that damping can be neglected due to the short duration when the excitation is an impulsive load. When a periodic load is exerted upon a linearly elastic system, it can be expanded into the sum of many harmonic components by the Fourier series, and the response excited by each harmonic component can be evaluated separately. The damping must be considered when the exciting frequency is close to a natural frequency (especially lower natural frequencies), which will be illustrated by Example 4.4 in this section. No matter whether the analytical or numerical methods are adopted, the damping should be taken into account for general dynamic loads with a long duration.

The uncoupled equations of motion and the response analysis for damped systems will be introduced. The forced vibration equation of motion of n-DOF damped systems is given as follows:

$$\boldsymbol{M}\ddot{\boldsymbol{q}} + \boldsymbol{C}\dot{\boldsymbol{q}} + \boldsymbol{K}\boldsymbol{q} = \boldsymbol{Q} \tag{4.56}$$

From the introduction in Section 4.1, the modes are orthogonal to each other with respect to the mass matrix $\boldsymbol{M}$ and stiffness matrix $\boldsymbol{K}$, rather than the damping matrix $\boldsymbol{C}$. Unless proper assumptions are adopted, the uncoupled equations of motion of damped systems cannot be obtained. Therefore Rayleigh damping is given as

$$\boldsymbol{C} = a_0\boldsymbol{M} + a_1\boldsymbol{K} \tag{4.57}$$

where a_0 and a_1 are the constants yet undetermined. Here, $\boldsymbol{C}$ is expressed as the linear combination of $\boldsymbol{M}$ and $\boldsymbol{K}$, which is also referred to as proportional damping.

Substituting Eqs. (4.57) and (4.30) into Eq. (4.56), premultiplying Eq. (4.56) by $\overline{\boldsymbol{A}}^T$, and considering the orthogonality conditions of the modes, leads to

$$\ddot{\overline{\boldsymbol{T}}}(t) + (a_0\boldsymbol{I} + a_1\boldsymbol{\lambda})\dot{\overline{\boldsymbol{T}}}(t) + \boldsymbol{\lambda}\overline{\boldsymbol{T}}(t) = \overline{\boldsymbol{P}} \tag{4.58}$$

where

$$\overline{\boldsymbol{T}} = \left\{ \overline{T}_1 \quad \overline{T}_2 \quad \cdots \quad \overline{T}_n \right\}^T$$

$$\overline{\boldsymbol{P}} = \overline{\boldsymbol{A}}^T\boldsymbol{Q} = \left\{ \overline{P}_1 \quad \overline{P}_2 \quad \cdots \quad \overline{P}_n \right\}^T, \quad \overline{P}_i = \overline{\boldsymbol{A}}_i^T\boldsymbol{Q}, \quad i = 1, 2, \cdots, n$$

$$\boldsymbol{\lambda} = \text{diag}(\lambda_1, \lambda_2, \cdots, \lambda_n), \quad \lambda_i = \omega_i^2, \quad i = 1, 2, \cdots, n$$

$\boldsymbol{I}$ is an $n \times n$ identity matrix.

Eq. (4.58) is written in the form of components

$$\ddot{\overline{T}}_i(t) + (a_0 + a_1\omega_i^2)\dot{\overline{T}}_i(t) + \omega_i^2\overline{T}_i(t) = \overline{P}_i, \quad i = 1, 2, \cdots, n \tag{4.59}$$

Let

$$a_0 + a_1\omega_i^2 = 2\xi_i\omega_i \tag{4.60}$$

where ξ_i is the damping ratio of the system corresponding to the ith mode. Thus Eq. (4.59) is the equations of motion of SDOF systems with viscous damping in normal coordinates, which can be rewritten as

$$\ddot{\overline{T}}_i(t) + 2\xi_i\omega_i\dot{\overline{T}}_i(t) + \omega_i^2\overline{T}_i(t) = \overline{P}_i, \quad i = 1, 2, \cdots, n \tag{4.61}$$

Solving Eq. (4.61) yields

$$\begin{aligned}\overline{T}_i(t) = e^{-\xi_i\omega_i t}&\left[\frac{\dot{\overline{T}}_i(0) + \overline{T}_i(0)\xi_i\omega_i}{\omega_{Di}}\sin\omega_{Di}t + \overline{T}_i(0)\cos\omega_{Di}t\right] \\ &+ \frac{1}{\omega_{Di}}\int_0^t \overline{P}_i(\tau)e^{-\xi_i\omega_i(t-\tau)}\sin\omega_{Di}(t-\tau)d\tau\end{aligned} \tag{4.62}$$

where the initial conditions $\overline{T}_i(0)$ and $\dot{\overline{T}}_i(0)$ are determined from Eqs. (4.41) and (4.42), respectively.

When $\overline{P}_i(t) = P_{i0}\sin\overline{\omega}t$, the general solution of Eq. (4.61) can be obtained directly from Eq. (3.43), that is,

$$\begin{aligned}\overline{T}_i(t) = \mathrm{e}^{-\xi_i\omega_i t}&\left[\frac{\dot{\overline{T}}_i(0) + \overline{T}_i(0)\xi_i\omega_i}{\omega_{Di}}\sin\omega_{Di}t + \overline{T}_i(0)\cos\omega_{Di}t\right] \\ &+ \rho_i\mathrm{e}^{-\xi_i\omega_i t}\left[\sin\theta_i\cos\omega_{Di}t + \frac{\xi_i\omega_i\sin\theta_i - \overline{\omega}\cos\theta_i}{\omega_{Di}}\sin\omega_{Di}t\right] \\ &+ \rho_i\sin(\overline{\omega}t - \theta_i)\end{aligned} \tag{4.63}$$

where

$$\theta_i = \tan^{-1}\left(\frac{2\xi_i\overline{\omega}/\omega_i}{1 - \overline{\omega}^2/\omega_i^2}\right) \tag{4.64}$$

$$\rho_i = \frac{P_{i0}}{\overline{K}_i}D_i \tag{4.65}$$

$$D_i = \frac{1}{\sqrt{\left(1 - \frac{\overline{\omega}^2}{\omega_i^2}\right)^2 + \left(\frac{2\xi_i\overline{\omega}}{\omega_i}\right)^2}} \tag{4.66}$$

$\overline{K}_i = \omega_i^2$ (since $\overline{M}_i = 1$, $\overline{K}_i = \omega_i^2\overline{M}_i = \omega_i^2$)

When the responses in terms of normal coordinates $\overline{\boldsymbol{T}}(t)$ have been evaluated, the responses in terms of original geometric coordinates $\boldsymbol{q}(t)$ can be obtained by Eq. (4.30). It has been proven that the first mode generally contributes the most to the total response. Some examples show that the distribution of the first mode accounts for more than 90% of the total response, the contributions of the second, third, and so on become less and less, and the contributions of the higher modes almost vanish. Therefore when the mode superposition method is applied in practice, only the contributions of the lower modes are considered (see Eqs. 4.35 and 4.36). A tentative calculation can be carried out to determine the proper number of required modes. The contribution of the first N modes is calculated first, and then the contribution of the first $(N+1)$ modes is also calculated. When the calculated results are close, this means that the first N modes are enough to satisfy the desired accuracy.

Based on Rayleigh's assumption, the uncoupled equations of motion of a damped system have been obtained, and the proportional constants a_0 and a_1 need to be determined. If any two natural frequencies of the system are known (say ω_1, ω_2, and $\omega_1 < \omega_2$) and the corresponding damping ratios ξ_1 and ξ_2 are given, the constants a_0 and a_1 can be obtained from $a_0 + a_1\omega_i^2 = 2\xi_i\omega_i$ as follows:

$$a_0 = \frac{2\xi_1\omega_1\omega_2^2 - 2\xi_2\omega_1^2\omega_2}{\omega_2^2 - \omega_1^2}$$

$$a_1 = \frac{2\xi_2\omega_2 - 2\xi_1\omega_1}{\omega_2^2 - \omega_1^2}$$

In practical engineering, the damping ratios ξ_1 and ξ_2 are usually determined according to empirical or measured data, and ξ_2 is often set to equal ξ_1. The recommended damping ratios ξ for bridge dynamic analysis by Eurocode 1 (EN1991-2) are listed in Table 4.1.

Table 4.1 Recommended values of damping ratio ξ (%) [3].

Bridge type	**Bridge span (m)** $L < 20$	$L \geq 20$
Steel or composite beam	$0.5 + 0.125(20 - L)$	0.5
Prestressed concrete beam	$1.0 + 0.07(20 - L)$	1.0
Reinforced concrete beam	$1.5 + 0.07(20 - L)$	1.5

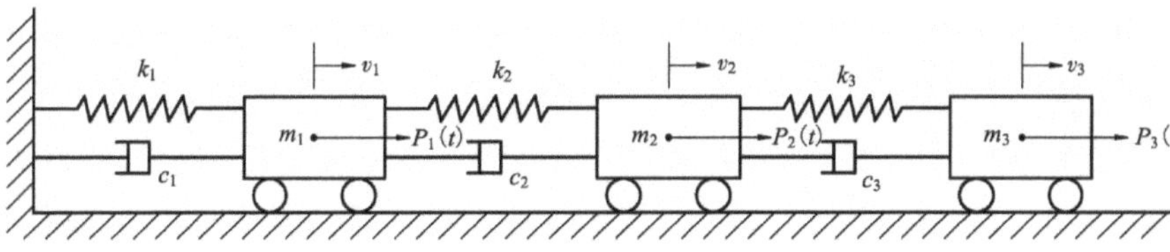

Figure 4.5 Schematic diagram of a multimass–spring-damper system.

Example 4.4: As shown in Fig. 4.5, the three masses are subjected to simple harmonic loads $P_1(t) = P_2(t) = P_3(t) = P\sin\overline{\omega}t$, respectively, and $\overline{\omega} = 1.25\sqrt{k/m}$. Rayleigh's assumption is applied to the damping of the system. Here, $m_1 = m_2 = m_3 = m$ and $k_1 = k_2 = k_3 = k$ are known. The damping ratios ξ_i, $i = 1, 2, 3$, are all set to be 0.01. The steady-state response of the system will be evaluated below [4].

Solution: Based on the system shown in Example 4.1, three dampers and applied loads are added to the system, as shown in Fig. 4.5. The corresponding external load vector is as follows:

$$\mathbf{Q} = P\sin\overline{\omega}t\begin{Bmatrix} 1 \\ 1 \\ 1 \end{Bmatrix}$$

The normal mode matrix has been obtained in Example 4.1 as follows:

$$\overline{\boldsymbol{A}} = \frac{1}{\sqrt{m}}\begin{bmatrix} 0.328 & 0.737 & 0.591 \\ 0.591 & 0.328 & -0.737 \\ 0.737 & -0.591 & 0.328 \end{bmatrix}$$

from which

$$\overline{\mathbf{P}} = \overline{\boldsymbol{A}}^T\mathbf{Q} = \frac{1}{\sqrt{m}}\begin{bmatrix} 0.328 & 0.591 & 0.737 \\ 0.737 & 0.328 & -0.591 \\ 0.591 & -0.737 & 0.328 \end{bmatrix} P\sin\overline{\omega}t\begin{Bmatrix} 1 \\ 1 \\ 1 \end{Bmatrix} = \frac{P\sin\overline{\omega}t}{\sqrt{m}}\begin{Bmatrix} 1.656 \\ 0.474 \\ 0.182 \end{Bmatrix}$$

The natural frequencies have been evaluated in Example 4.1 as follows:

$$\omega_1 = 0.445\sqrt{\frac{k}{m}}, \quad \omega_2 = 1.247\sqrt{\frac{k}{m}}, \quad \omega_3 = 1.802\sqrt{\frac{k}{m}}$$

One can obtain

$$\overline{K}_1 = \omega_1^2 = 0.198\frac{k}{m}, \quad \overline{K}_2 = \omega_2^2 = 1.555\frac{k}{m}, \quad \overline{K}_3 = \omega_3^2 = 3.247\frac{k}{m}$$

$$D_1 = \frac{1}{\sqrt{\left(1-\frac{\overline{\omega}^2}{\omega_1^2}\right)^2 + \left(\frac{2\xi_1\overline{\omega}}{\omega_1}\right)^2}} = \frac{1}{\sqrt{\left(1-\frac{1.5625}{0.198}\right)^2 + \left(2\times 0.01\times \frac{1.25}{0.445}\right)^2}} = 0.1451$$

$$D_2 = \frac{1}{\sqrt{\left(1-\frac{\overline{\omega}^2}{\omega_2^2}\right)^2 + \left(\frac{2\xi_2\overline{\omega}}{\omega_2}\right)^2}} = \frac{1}{\sqrt{\left(1-\frac{1.5625}{1.555}\right)^2 + \left(2\times 0.01\times \frac{1.25}{1.247}\right)^2}} = 48.496$$

$$D_3 = \frac{1}{\sqrt{\left(1-\frac{\overline{\omega}^2}{\omega_3^2}\right)^2 + \left(\frac{2\xi_3\overline{\omega}}{\omega_3}\right)^2}} = \frac{1}{\sqrt{\left(1-\frac{1.5625}{3.247}\right)^2 + \left(2\times 0.01\times \frac{1.25}{1.802}\right)^2}} = 1.9269$$

$$\theta_1 = \tan^{-1}\left(\frac{2\xi_1\overline{\omega}/\omega_1}{1-\overline{\omega}^2/\omega_1^2}\right) = \tan^{-1}\left(\frac{2\times 0.01\times\dfrac{1.25}{0.445}}{1-\dfrac{1.5625}{0.198}}\right)$$

$$= \tan^{-1}(-0.008152) = -0.00814 \text{ rad } \theta_2 = \tan^{-1}\left(\frac{2\xi_2\overline{\omega}/\omega_2}{1-\overline{\omega}^2/\omega_2^2}\right)$$

$$= \tan^{-1}\left(\frac{2\times 0.01\times\dfrac{1.25}{1.247}}{1-\dfrac{1.5625}{1.555}}\right) = \tan^{-1}(-4.15677)$$

$$= -1.3336 \text{ rad } \theta_3 = \tan^{-1}\left(\frac{2\xi_3\overline{\omega}/\omega_3}{1-\overline{\omega}^2/\omega_3^2}\right) = \tan^{-1}\left(\frac{2\times 0.01\times\dfrac{1.25}{1.802}}{1-\dfrac{1.5625}{3.247}}\right)$$

$$= \tan^{-1}(0.02674) = 0.0267 \text{ rad}$$

The steady-state responses of the system in terms of the normal coordinates (the third term of Eq. 4.63) are thus obtained

$$\overline{T}_1(t) = \frac{P_{10}D_1}{\overline{K}_1}\sin(\overline{\omega}t-\theta_1) = \frac{1.656\times 0.1451}{0.198\dfrac{k}{m}}\frac{P}{\sqrt{m}}\sin(\overline{\omega}t-\theta_1) = 1.214\frac{P\sqrt{m}}{k}\sin(\overline{\omega}t-\theta_1)$$

$$\overline{T}_2(t) = \frac{P_{20}D_2}{\overline{K}_2}\sin(\overline{\omega}t-\theta_2) = \frac{0.474\times 48.496}{1.555\dfrac{k}{m}}\frac{P}{\sqrt{m}}\sin(\overline{\omega}t-\theta_2) = 14.783\frac{P\sqrt{m}}{k}\sin(\overline{\omega}t-\theta_2)$$

$$\overline{T}_3(t) = \frac{P_{30}D_3}{\overline{K}_3}\sin(\overline{\omega}t-\theta_3) = \frac{0.182\times 1.9269}{3.247\dfrac{k}{m}}\frac{P}{\sqrt{m}}\sin(\overline{\omega}t-\theta_3) = 0.108\frac{P\sqrt{m}}{k}\sin(\overline{\omega}t-\theta_3)$$

The steady-state responses in terms of the original geometric coordinates can be evaluated

$$q = \begin{Bmatrix} v_1 \\ v_2 \\ v_3 \end{Bmatrix} = \overline{\boldsymbol{A}\boldsymbol{T}} = \frac{1}{\sqrt{m}} \begin{bmatrix} 0.328 & 0.737 & 0.591 \\ 0.591 & 0.328 & -0.737 \\ 0.737 & -0.591 & 0.328 \end{bmatrix}$$

$$\begin{Bmatrix} 1.214\dfrac{P\sqrt{m}}{k}\sin(\overline{\omega}t - \theta_1) \\[2ex] 14.783\dfrac{P\sqrt{m}}{k}\sin(\overline{\omega}t - \theta_2) \\[2ex] 0.108\dfrac{P\sqrt{m}}{k}\sin(\overline{\omega}t - \theta_3) \end{Bmatrix} = \frac{P}{k}$$

$$\begin{Bmatrix} 0.398\sin(\overline{\omega}t - \theta_1) + 10.895\sin(\overline{\omega}t - \theta_2) + 0.064\sin(\overline{\omega}t - \theta_3) \\ 0.717\sin(\overline{\omega}t - \theta_1) + 4.849\sin(\overline{\omega}t - \theta_2) - 0.080\sin(\overline{\omega}t - \theta_3) \\ 0.895\sin(\overline{\omega}t - \theta_1) - 8.737\sin(\overline{\omega}t - \theta_2) + 0.035\sin(\overline{\omega}t - \theta_3) \end{Bmatrix}$$

In this example, the system is subjected to simple harmonic applied loads, and $\overline{T}_i(t)$ is also a simple harmonic quantity. The frequency of $\overline{T}_i(t)$ is the same as the exciting frequency, and its amplitude is $P_{i0}D_i/\overline{K}_i$, which accounts for the contribution of the ith mode to the total response. Here, D_i is the dynamic magnification factor. When the natural frequency ω_i of the system is close to the exciting frequency $\overline{\omega}$, the system resonates and D_i becomes very large. When ω_i is much larger than $\overline{\omega}$, D_i approaches unity. Here, P_{i0} is the amplitude of the generalized applied load, which is determined from the amplitudes of original external loads and their spatial distribution (i.e., it is determined from the original applied load vector and the mode matrix). The value of P_{i0} does not continuously increase as the mode number i increases. However, the generalized stiffness $\overline{K}_i$ continuously increases with the increase of the mode number i. Thus the generalized static displacement $P_{i0}/\overline{K}_i$ generally tends to be very small with the increase of the mode number i. Therefore, when ω_i is far larger than $\overline{\omega}$, the amplitude $P_{i0}D_i/\overline{K}_i$ of $\overline{T}_i(t)$ generally becomes less and less, and the contribution of this mode to the total response also reduces. In this example, the contribution of the second mode to the total response is largest, and the contribution of the third mode is least. That is because the exciting frequency $\overline{\omega} = 1.25\sqrt{k/m}$ is close to the second natural frequency $\omega_2 = 1.247\sqrt{k/m}$.

If the damping is neglected, one obtains

$$D_1 = \frac{1}{\sqrt{\left(1 - \frac{\overline{\omega}^2}{\omega_1^2}\right)^2}} = \frac{1}{\sqrt{\left(1 - \frac{1.5625}{0.198}\right)^2}} = 0.145$$

$$D_2 = \frac{1}{\sqrt{\left(1 - \frac{\overline{\omega}^2}{\omega_2^2}\right)^2}} = \frac{1}{\sqrt{\left(1 - \frac{1.5625}{1.555}\right)^2}} = 207.33$$

$$D_3 = \frac{1}{\sqrt{\left(1 - \frac{\overline{\omega}^2}{\omega_3^2}\right)^2}} = \frac{1}{\sqrt{\left(1 - \frac{1.5625}{3.247}\right)^2}} = 1.928$$

Note that the damping has a significant influence on the response in the resonance zone, while there is a minor influence on the response off the resonance zone.

This example only discusses the vibration characteristics of a MDOF system subjected to simple harmonic loads with a single frequency. According to the concept of frequency domain analysis in Section 3.9, an arbitrary dynamic load can be approximately regarded as a periodic load with a very large period, so it can be expressed as the sum of many simple harmonic loads. The effect of each simple harmonic load is consistent with the characteristics discussed in this example. It is shown from this example that the displacement contributions of the lower modes are dominant, and the contributions of the higher modes tend to decrease for most types of loads. In practical analysis, the required number of modes can be determined according to the desired accuracy by trial and error.

References

[1] Wen B, Liu S, Chen Z, Li H. Theory of the mechanical vibration and its applications. Beijing: Higher Education Press; 2009.
[2] Clough RW, Penzien J. Dynamics of structures. 3rd ed. Berkeley, CA: Computers & Structures, Inc; 2003.
[3] Liu J, Du X. Dynamics of structures. Beijing: China Machine Press; 2007.
[4] Zhou Z, Wen Y, Zeng Q. Lectures on dynamics of structures. 2nd ed. Beijing: China Communications Press Co., Ltd; 2017.

Problems

4.1 What is the difference between the concept of principal vibration and mode shape?

4.2 When the mode superposition method is applied to analyze the response of MDOF systems or continuous systems, the lower modes of the system are employed to express the displacement of systems. Explain the reasons for doing so.

4.3 The superposition principle is used in the mode superposition method. Explain the conditions with which this method can be applied to structural dynamic analysis.

4.4 Does the superposition method apply to static analysis? If it does, describe the process briefly.

4.5 For an n-DOF system, explain the conditions of resonance and the corresponding reasons.

4.6 A two-story rigid frame is shown in Fig. P4.1. The masses of the floors are $m_1 = 120$ t and $m_2 = 100$ t, respectively. The mass of all columns is concentrated on the floors, and the flexural stiffness of all the columns is $EI = 80\,\text{MN}\cdot\text{m}^2$. The height of all columns is 4m, and the stiffness of the girder is infinite. Evaluate the natural frequencies and mode shapes of the structure.

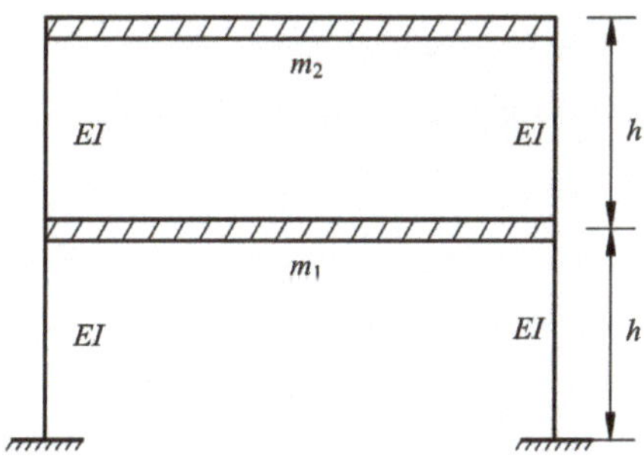

Figure P4.1 Figure of problem 4.6.

4.7 The system, as shown in Problem 4.6, is subjected to two types of impulsive loads respectively at the top floor along the horizontal direction, namely, (1) a triangular impulse load with the duration of $2t_1$ and a maximum value P_0 and (2) a rectangular impulse load with the duration of t_1 and a maximum value of P_0. The total impulse of the two kinds of loads is equal. Use the mode superposition method to evaluate the dynamic response, compare the maximum displacements of the two responses, and discuss the effect of t_1/T (T is the natural period of the system) on the resulting responses.

CHAPTER 5

Analysis of dynamic response of continuous systems: straight beam

Structures are discretized into multidegree-of-freedom (MDOF) systems, and then dynamic responses can be analyzed by using the method in Chapter 4, Analysis of dynamic response of MDOF systems: mode superposition method. However, actual structures have continuously distributed properties, say distributed elasticity and mass, which are called continuous or distributed parameter systems. For example, bridge structures consisting of plates, beams, and rods are all continuous systems. Strictly speaking, infinite generalized coordinates are required to describe the configuration of such systems at any instant of time, which are also called infinite-degree-of-freedom (IDOF) systems. If the motions of these systems are described by a finite number of coordinates, only approximate results of actual dynamic behavior can be achieved. Increasing the number of DOFs considered in the analysis can lead to higher accuracy of the results as required. However, for a real structure having distributed properties, in principle, an infinite number of coordinates are required to converge to the exact results. Therefore it is obviously impossible to obtain an exact solution for continuous systems by means of approaches with finite DOFs.

To describe the vibration of an IDOF system completely, it is necessary to establish a continuous displacement function of the time and spatial position. Therefore the equations of motion describing the IDOF system have the form of partial differential equations. However, the equations of motion of complex systems can only be solved by numerical methods. In most cases, a discrete-coordinate formulation is preferable to a continuous-coordinate formulation. For this reason, the present treatment will be limited to simple systems. In the present chapter, the fundamental procedure of deriving and solving the partial

Fundamentals of Structural Dynamics
DOI: https://doi.org/10.1016/B978-0-12-823704-5.00005-7

differential equations of continuous systems is illustrated by using a bending straight beam (Bernoulli–Euler beam) as an example [1].

5.1 Differential equations of motion of undamped straight beam

A straight nonuniform beam is shown in Fig. 5.1A and only the case of beam flexure is discussed in this chapter. Here, it is assumed that the main physical properties of this beam, the flexural stiffness $EI(x)$ and the mass per unit length $m(x)$ vary arbitrarily with position x along the span. The influence of damping is not considered temporarily (damping effect will be discussed in Section 5.5). Assuming that a transverse load $p(x, t)$ varies arbitrarily with position and time, the transverse displacement $v(x, t)$ is also a function of these variables. The end support conditions for the beam are arbitrary, although they are pictured as simple supports for illustrative purposes here. In Chapter 2, Formulation of equations of motion of systems, Hamilton's principle and the principle of total potential energy with a stationary value in elastic system dynamics are applied to formulate the differential equation of motion of an undamped straight beam, and the direct equilibrium method is used to derive the above equation again in this section.

Considering the dynamic equilibrium of forces acting on the differential segment of the beam shown in Fig. 5.1B, the dynamic equilibrium equations of this simple system can be easily derived. Summing all the forces acting vertically leads to the first dynamic equilibrium relationship

$$V(x,t) + p(x,t)dx - \left[V(x,t) + \frac{\partial V(x,t)}{\partial x}dx\right] - f_I(x,t)dx = 0 \quad (5.1)$$

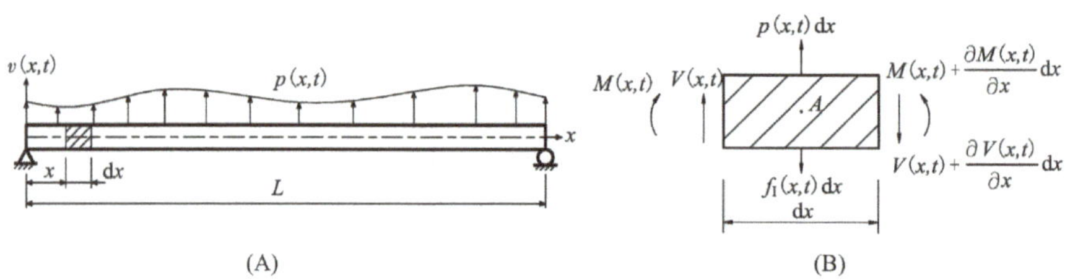

Figure 5.1 Simply supported beam subjected to dynamic load: (A) beam properties and coordinates; (B) resultant forces acting on differential segment.

where $V(x, t)$ is the vertical force acting on the cut section, and $f_I(x, t)dx$ is the resultant transverse inertial force equal to the mass of the segment multiplied by its transverse acceleration, that is,

$$f_I(x, t)dx = m(x)\frac{\partial^2 v(x, t)}{\partial t^2}dx \tag{5.2}$$

Substituting Eq. (5.2) into Eq. (5.1) and dividing the resulting equation by dx yields

$$\frac{\partial V(x, t)}{\partial x} = p(x, t) - m(x)\frac{\partial^2 v(x, t)}{\partial t^2} \tag{5.3}$$

This equation is similar to the standard static relationship between the shear and transverse load, but with the load now being the resultant of the applied load and inertial force.

A second equilibrium relationship is obtained by summing the moments about point A on the elastic axis. Ignoring the second-order term of the moment involving the inertial force and applied load, one obtains

$$M(x, t) + V(x, t)dx - \left[M(x, t) + \frac{\partial M(x, t)}{\partial x}dx\right] = 0 \tag{5.4}$$

Because the rotational inertia is neglected, the equation above may be directly simplified to a standard static relationship between shear and moment

$$\frac{\partial M(x, t)}{\partial x} = V(x, t) \tag{5.5}$$

Differentiating Eq. (5.5) with respect to x, and substituting the result into Eq. (5.3) gives

$$\frac{\partial^2 M(x, t)}{\partial x^2} + m(x)\frac{\partial^2 v(x, t)}{\partial t^2} = p(x, t) \tag{5.6}$$

Introducing the basic relationship between the moment and curvature of the beam, $M = EI\frac{\partial^2 v}{\partial x^2}$, Eq. (5.6) becomes

$$\frac{\partial^2}{\partial x^2}\left[EI(x)\frac{\partial^2 v(x, t)}{\partial x^2}\right] + m(x)\frac{\partial^2 v(x, t)}{\partial t^2} = p(x, t) \tag{5.7}$$

Eq. (5.7) is the partial differential equation of motion of a beam to an arbitrary distributed load. To avoid the mathematical complexity of dealing with systems of variable nature, the following discussion will be limited to beams of constant nature along the length, that is, $EI(x) = EI$ and $m(x) = \overline{m}$. However, this is not a necessary limitation because it is more efficient to adopt the approach with discrete coordinates (such as the finite element method) to model systems with variable nature.

5.2 Modal expansion of displacement and orthogonality of mode shapes of straight beam

Similar to the mode superposition analysis of MDOF systems, the amplitudes of the mode shape response can be chosen as generalized coordinates to determine the structural response of continuous (distributed parameter) systems.

Because a continuous system is essentially a vibrating system with infinite DOFs, there exist infinite modes and generalized coordinates. By analogy with the mode superposition analysis of MDOF systems, the displacement of a straight beam can be expressed as follows [2]:

$$v(x, t) = \sum_{i=1}^{\infty} \varphi_i(x) T_i(t) \tag{5.8}$$

Corresponding to the physical quantities used in Eq. (4.24), $\varphi_i(x)$ represents the ith mode shape of the straight beam, which is a continuous function of the position x along the elastic axis. Here, $T_i(t)$ is the ith principal coordinate (modal coordinate) of the system. Eq. (5.8) is simply a statement that any physically permissible displacement pattern can be made up by superposing appropriate amplitudes of the mode shapes for the structure. This principle can be illustrated by the simple example shown in Fig. 5.2, in which the arbitrary displacement of a beam with an overhanging end is expressed as the sum of a set of modal components.

The orthogonality relationships also apply to the mode shapes of the distributed parameter beam, which are equivalent to those defined previously for MDOF systems. The orthogonality can be proven by means of Betti's law (reciprocal theorem of work). The simply supported beam has stiffness and mass varying arbitrarily along its length. Fig. 5.3 shows the mth and nth principal vibrations of the beam. For each principal vibration, the displacement and the inertial force producing this displacement are shown in Fig. 5.3A and B, respectively.

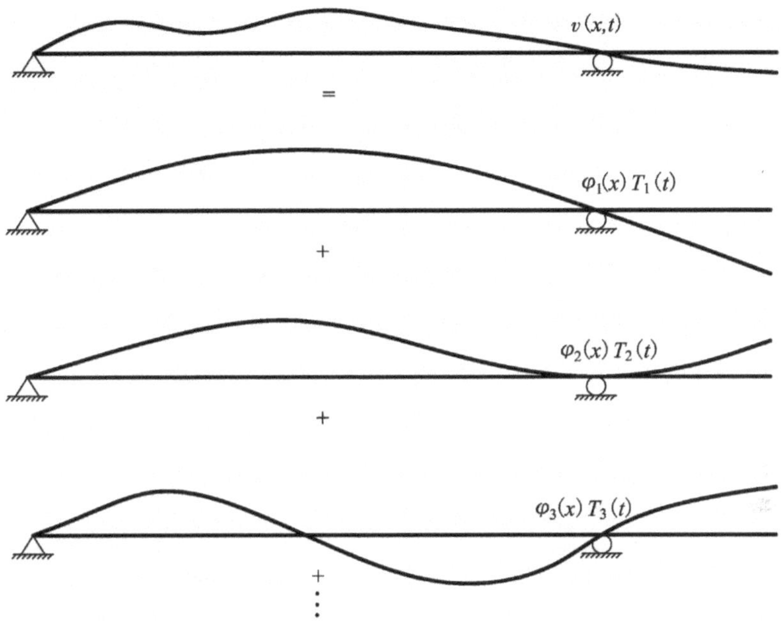

Figure 5.2 Arbitrary displacement of a beam with an overhanging end represented by modal coordinates.

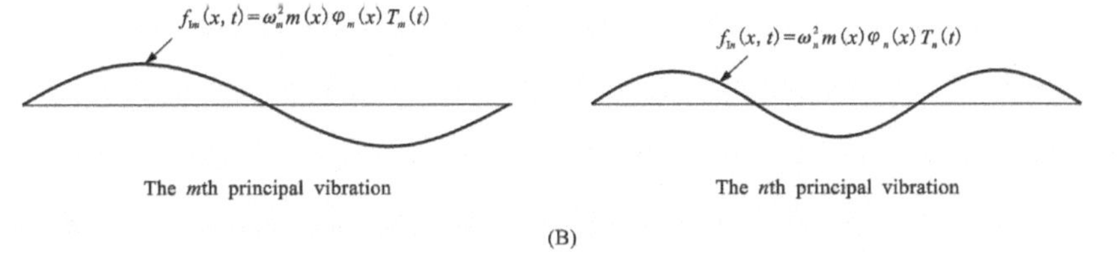

Figure 5.3 Two principal vibrations of a simply supported beam: (A) displacements corresponding to different principal vibrations; (B) inertial forces corresponding to different principal vibrations.

Betti's law is applied to the two principal vibrations, which means that the work done by the inertial forces of the nth principal vibration acting on the displacement of the mth principal vibration are equal to the work by the forces of the mth principal vibration acting on the displacement of the nth principal vibration, that is,

$$\int_0^L v_m(x,t) f_{In}(x,t)dx = \int_0^L v_n(x,t) f_{Im}(x,t)dx \tag{5.9}$$

Furthermore, the following relationships apply to the two principal vibrations:

$$\begin{cases} v_m(x,t) = \varphi_m(x)T_m(t) \\ v_n(x,t) = \varphi_n(x)T_n(t) \\ f_{Im}(x,t) = -m(x)\ddot{v}_m(x,t) = m(x)\omega_m^2 T_m(t)\varphi_m(x) \\ f_{In}(x,t) = -m(x)\ddot{v}_n(x,t) = m(x)\omega_n^2 T_n(t)\varphi_n(x) \end{cases} \tag{5.10}$$

where $v_m(x, t)$ and $v_n(x, t)$ are the displacements of the mth and the nth principal vibrations, respectively. By analogy with Eq. (4.6), the principal vibrations are simple harmonic, and the vibration frequencies of the two principal vibrations are ω_m and ω_n, respectively, which will also be illustrated in Section 5.3. Therefore the latter two in Eq. (5.10) hold. Substituting Eq. (5.10) into Eq. (5.9) yields

$$T_m(t)T_n(t)\omega_n^2 \int_0^L \varphi_m(x)m(x)\varphi_n(x)dx = T_m(t)T_n(t)\omega_m^2 \int_0^L \varphi_n(x)m(x)\varphi_m(x)dx \tag{5.11}$$

which can be rewritten as

$$(\omega_n^2 - \omega_m^2)T_m(t)T_n(t) \int_0^L \varphi_m(x)m(x)\varphi_n(x)dx = 0 \tag{5.12}$$

When the frequencies of the two principal vibrations are different, their mode shapes must satisfy the following orthogonality condition:

$$\int_0^L \varphi_m(x)m(x)\varphi_n(x)dx = 0, \quad \omega_m \neq \omega_n \tag{5.13}$$

Eq. (5.13) is the orthogonality condition of the distributed parameter beam involving the mass as the weighting parameter. Obviously, the orthogonality condition of continuous systems is equivalent to that of the MDOF systems indicated in Eq. (4.14). When the two principal vibrations have the same frequency, the orthogonality condition may not apply. However, this does not often occur in ordinary structural problems.

In addition, a second orthogonality condition, involving the stiffness property rather than the mass as the weighting parameter, can be derived for continuous systems, as it was earlier for discrete MDOF systems indicated in Eq. (4.15). From Eq. (5.7), the free-vibration equation of motion of the straight beam is obtained as follows:

$$\frac{\partial^2}{\partial x^2}\left[EI(x)\frac{\partial^2 v(x,t)}{\partial x^2}\right]+m(x)\frac{\partial^2 v(x,t)}{\partial t^2}=0 \tag{5.14}$$

When only the mth principal vibration occurs in the straight beam, $v(x,t)=v_m(x,t)$. Considering Eq. (5.14), the inertial force $f_{Im}(x,t)$ in Eq. (5.10) can be written as

$$f_{Im}(x,t)=-m(x)\frac{\partial^2 v_m(x,t)}{\partial t^2}=\frac{\partial^2}{\partial x^2}\left[EI(x)\frac{\partial^2 v_m(x,t)}{\partial x^2}\right] \tag{5.15}$$

Combining Eqs. (5.10) and (5.15), one obtains

$$\frac{\partial^2}{\partial x^2}\left[EI(x)\frac{\partial^2 v_m(x,t)}{\partial x^2}\right]=m(x)\omega_m^2 v_m(x,t) \tag{5.16}$$

Substituting the first expression of Eq. (5.10) into Eq. (5.16) yields

$$m(x)\varphi_m(x)=\frac{1}{\omega_m^2}\frac{d^2}{dx^2}\left[EI(x)\frac{d^2\varphi_m(x)}{dx^2}\right] \tag{5.17}$$

Substituting Eq. (5.17) into Eq. (5.13) yields

$$\int_0^L \varphi_n(x)\frac{d^2}{dx^2}\left[EI(x)\frac{d^2\varphi_m(x)}{dx^2}\right]dx=0,\quad \omega_m\neq\omega_n \tag{5.18}$$

Eq. (5.18) is the orthogonality condition of the distributed parameter beam involving the stiffness as the weighting parameter. By integrating Eq. (5.18) twice by parts, a more convenient form of this orthogonality condition can be obtained

$$\varphi_n(x)\overline{V}_m(x)|_0^L-\varphi_n^{'}(x)\overline{M}_m(x)|_0^L+\int_0^L \varphi_m^{''}(x)\varphi_n^{''}(x)EI(x)dx=0,\quad \omega_m\neq\omega_n \tag{5.19}$$

where

$$\overline{V}_m(x)=\frac{d}{dx}\left[EI(x)\frac{d^2\varphi_m(x)}{dx^2}\right],\quad \overline{M}_m(x)=EI(x)\frac{d^2\varphi_m(x)}{dx^2}$$

Eq. (5.19) is the orthogonality condition involving stiffness as the weighting parameter under general boundary conditions. Multiplying both sides of Eq. (5.19) by $T_m(t)T_n(t)$ yields

$$\varphi_n(x)T_n(t)\cdot\overline{V}_m(x)|_0^L T_m(t) - \varphi_n'(x)T_n(t)\cdot\overline{M}_m(x)|_0^L T_m(t) + \int_0^L \varphi_m''(x)T_m(t)\cdot\varphi_n''(x)T_n(t)EI(x)\,dx = 0, \quad \omega_m \neq \omega_n \tag{5.20}$$

Considering $v_n(x,t) = \varphi_n(x)T_n(t)$, $v_m(x,t) = \varphi_m(x)T_m(t)$, $v_n''(x,t) = \varphi_n''(x)T_n(t)$ and $v_m''(x,t) = \varphi_m''(x)T_m(t)$, one gets

$$M_m(x,t) = EI(x)\frac{d^2 v_m(x,t)}{dx^2} = EI(x)\frac{d^2\varphi_m(x)}{dx^2}T_m(t) = \overline{M}_m(x)T_m(t),$$

$$V_m(x,t) = \frac{dM_m(x,t)}{dx} = \frac{d}{dx}\left[EI(x)\frac{d^2\varphi_m(x)}{dx^2}T_m(t)\right] = \overline{V}_m(x)T_m(t)$$

Then, Eq. (5.20) can be rewritten as

$$v_n(x,t)\cdot V_m(x,t)|_0^L - v_n'(x,t)\cdot M_m(x,t)|_0^L + \int_0^L v_m''(x,t)\cdot v_n''(x,t)EI(x)dx = 0, \quad \omega_m \neq \omega_n \tag{5.21}$$

The first two terms in Eq. (5.21) represent the work done by the boundary vertical section force of the mth principal vibration acting on the end displacements of the nth principal vibration and the work done by the end moment of the mth principal vibration acting on the corresponding rotations of the nth principal vibration, respectively. For the standard clamped-end, hinged-end, or free-end conditions, these terms will vanish, and the corresponding orthogonality condition involving the stiffness as the weighting parameter can be simplified as follows:

$$\int_0^L \varphi_m''(x)\varphi_n''(x)EI(x)dx = 0, \quad \omega_m \neq \omega_n \tag{5.22}$$

However, they contribute to the orthogonality relationship if the beam has elastic supports or if it has a lumped mass at its end. Therefore they must be retained in the expression when considering such cases.

5.3 Free vibration analysis of undamped straight beam

It is assumed that the nature of the straight beam along its length is constant, that is, $EI(x) = EI$ and $m(x) = \overline{m}$. From Eq. (5.7), the undamped free-vibration equation of motion for the straight beam can be obtained as follows:

$$EI\frac{\partial^4 v(x,t)}{\partial x^4} + \overline{m}\frac{\partial^2 v(x,t)}{\partial t^2} = 0 \tag{5.23}$$

Considering the ith principal vibration of the beam, $v(x,t) = \varphi_i(x)T_i(t)$ and substituting it into Eq. (5.23) gives

$$EI\frac{d^4\varphi_i(x)}{dx^4}T_i(t) + \overline{m}\frac{d^2T_i(t)}{dt^2}\varphi_i(x) = 0 \tag{5.24}$$

Dividing Eq. (5.24) by $EI\varphi_i(x)T_i(t)$, the variables can be separated as follows:

$$\frac{\varphi_i^{\mathrm{IV}}(x)}{\varphi_i(x)} + \frac{\overline{m}}{EI}\frac{\ddot{T}_i(t)}{T_i(t)} = 0 \tag{5.25}$$

where the dot notation ".." indicates a derivative with respect to variable t twice, and the superscript "IV" indicates a derivative with respect to variable x four times. Since the first term in this equation is a function of x only and the second term is a functions of t only, the entire equation can be satisfied for arbitrary values of x and t, only if each term is a constant, that is,

$$\frac{\varphi_i^{\mathrm{IV}}(x)}{\varphi_i(x)} = -\frac{\overline{m}}{EI}\frac{\ddot{T}_i(t)}{T_i(t)} = a^4 \tag{5.26}$$

where the single constant involved is designated in the form of a^4 for later mathematical convenience. Two ordinary differential equations can be obtained from Eq. (5.26)

$$\ddot{T}_i(t) + \omega_i^2 T_i(t) = 0 \tag{5.27}$$

$$\varphi_i^{\mathrm{IV}}(x) - a^4\varphi_i(x) = 0 \tag{5.28}$$

where

$$\omega_i^2 = \frac{a^4 EI}{\overline{m}}, \quad a^4 = \frac{\omega_i^2\overline{m}}{EI} \tag{5.29}$$

First, Eq. (5.28) can be solved in the usual way by introducing a solution of the form

$$\varphi_i(x) = Ge^{Sx} \tag{5.30}$$

where G and S are complex constants yet undetermined. Substituting Eq. (5.30) into Eq. (5.28) gives

$$(S^4 - a^4)Ge^{Sx} = 0 \tag{5.31}$$

from which

$$S_{1,2} = \pm ia, \quad S_{3,4} = \pm a \tag{5.32}$$

By incorporating each of these roots into Eq. (5.30) separately and adding the resulting four terms, the complete solution can be obtained as follows:

$$\varphi_i(x) = G_1 e^{iax} + G_2 e^{-iax} + G_3 e^{ax} + G_4 e^{-ax} \tag{5.33}$$

where G_1, G_2, G_3, and G_4 are complex constants. Expressing the exponential functions in terms of their trigonometric and hyperbolic equivalents, considering that $\varphi_i(x)$ must be real and setting the entire imaginary part of the right-hand side of this equation to be zero leads to

$$\varphi_i(x) = A_1 \cos(ax) + A_2 \sin(ax) + A_3 \cosh(ax) + A_4 \sinh(ax) \tag{5.34}$$

where A_1, A_2, A_3, and A_4 are real constants which can be expressed by G_1, G_2, G_3, and G_4. These four real constants are determined from the known boundary conditions (displacement, rotation, moment or shear) at the beam ends. From the boundary conditions and Eq. (5.34), homogeneous algebraic equations containing four unknown real constants can be obtained. According to the condition that the obtained equations have nontrivial solutions, the coefficient determinant must be zero. An equation about a (the frequency equation) can be obtained, from which the frequency parameter a may be determined. Then, the homogeneous equations can be used to determine the relative relationship among the four real constants and the mode function $\varphi_i(x)$ can be determined.

Second, Eq. (5.27) is solved. Eq. (5.27) is the free vibration equation of motion for an undamped single-degree-of-freedom (SDOF) system, which has the following solution (see Eq. 3.2):

$$T_i = C_{1i} \cos \omega_i t + C_{2i} \sin \omega_i t \tag{5.35}$$

where the constants C_{1i} and C_{2i} depend on the initial displacement $T_i(0)$ and velocity $\dot{T}_i(0)$, that is,

$$T_i(t) = T_i(0)\cos\omega_i t + \frac{\dot{T}_i(0)}{\omega_i}\sin\omega_i t \tag{5.36}$$

When the value of frequency parameter a is solved, and the initial conditions (the value of $T_i(0)$ and $\dot{T}_i(0)$) of the ith principal vibration are known, the time history of the principal vibration can be obtained from Eq. (5.36). When the free vibration occurs under specified initial conditions (specified $v(x,0)$ and $\dot{v}(x,0)$), the initial condition of the ith principal vibration can be obtained from the expansion theorem.

$$v(x,0) = \sum_{i=1}^{\infty}\varphi_i(x)T_i(0), \quad \dot{v}(x,0) = \sum_{i=1}^{\infty}\varphi_i(x)\dot{T}_i(0) \tag{5.37}$$

Multiplying both sides of Eq. (5.37) by $\varphi_i(x)$, integrating the resulting equations, and considering the orthogonality conditions leads to

$$T_i(0) = \frac{\int_0^L \varphi_i(x)v(x,0)\,dx}{\int_0^L \varphi_i^2(x)\,dx}, \quad \dot{T}_i(0) = \frac{\int_0^L \varphi_i(x)\dot{v}(x,0)\,dx}{\int_0^L \varphi_i^2(x)\,dx} \tag{5.38}$$

It has been previously stated that mode functions $\varphi_i(x)$ are not uniquely defined. Therefore the values of the initial conditions of the ith principal vibration are associated with a specified $\varphi_i(x)$.

Substituting Eqs. (5.31) and (5.36) into Eq. (5.8) gives

$$v(x,t) = \sum_{i=1}^{\infty}\left\{\begin{bmatrix} A_1\cos(ax) + A_2\sin(ax) + A_3\cosh(ax) + \\ A_4\sinh(ax) \end{bmatrix}\cdot\left[T_i(0)\cos\omega_i t + \frac{\dot{T}_i(0)}{\omega_i}\sin\omega_i t\right]\right\} \tag{5.39}$$

Apparently, after the mode functions $\varphi_i(x)$ are determined, and the response of generalized coordinates $T_i(t)$ is solved from Eq. (5.27), the free vibration response can be obtained from Eq. (5.8). It is shown from Eq. (5.39) that the principal vibrations contained in the free vibrations are simple harmonic vibrations, which can occur under appropriate initial conditions independently. This concept is also the basis for proving the orthogonality of mode shapes in Section 5.2.

Example 5.1: Considering the uniform simply supported beam shown in Fig. 5.4A, the four known boundary conditions of this beam are given as follows:

$$v(0,t)=0,\quad M(0,t)=EIv''(0,t)=0,\quad v(L,t)=0,\quad M(L,t)=EIv''(L,t)=0$$

Considering that a certain principal vibration can occur independently under appropriate initial conditions, letting $v(x,t)=\varphi_i(x)T_i(t)$ in this case, and substituting it into the boundary conditions above, one obtains

$$\varphi_i(0)=0,\quad \varphi_i''(0)=0 \tag{5.40}$$

$$\varphi_i(L)=0,\quad \varphi_i''(L)=0 \tag{5.41}$$

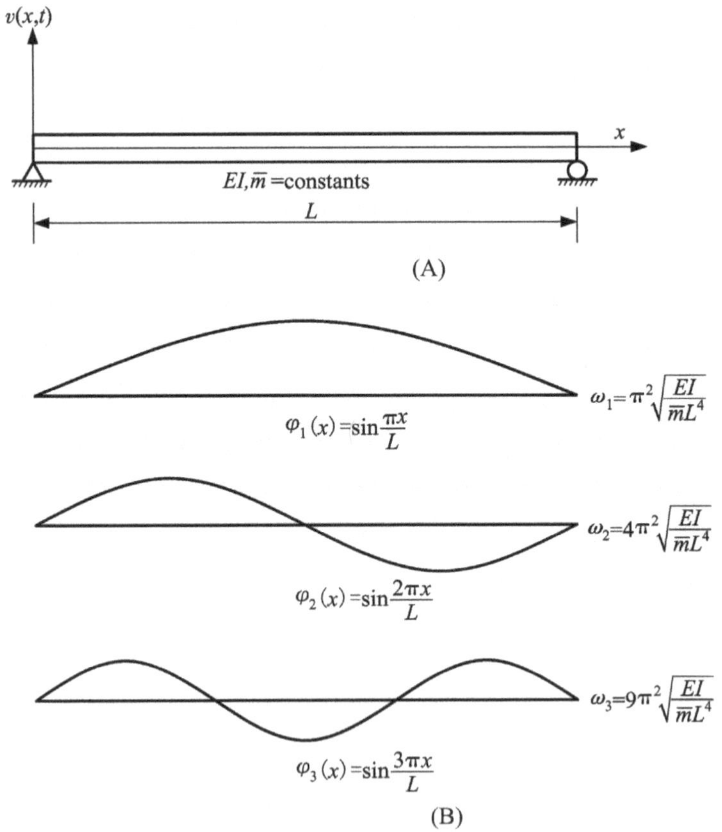

Figure 5.4 Analysis of the natural modes and frequencies of a simply supported beam: (A) basic properties of a simply supported beam; (B) first three natural modes and frequencies of a simply supported beam.

Substituting Eqs. (5.40) and (5.41) into Eq. (5.34) and its second derivative and simplifying leads to a homogeneous equation in a matrix form as follows:

$$\begin{bmatrix} 1 & 0 & 1 & 0 \\ -1 & 0 & 1 & 0 \\ \cos aL & \sin aL & \cosh aL & \sinh aL \\ -\cos aL & -\sin aL & \cosh aL & \sinh aL \end{bmatrix} \begin{Bmatrix} A_1 \\ A_2 \\ A_3 \\ A_4 \end{Bmatrix} = \mathbf{0} \tag{5.42}$$

Considering that the above equations have nontrivial solutions, the determinant of the coefficient matrix must equal zero, thus giving the frequency equation

$$\begin{vmatrix} 1 & 0 & 1 & 0 \\ -1 & 0 & 1 & 0 \\ \cos aL & \sin aL & \cosh aL & \sinh aL \\ -\cos aL & -\sin aL & \cosh aL & \sinh aL \end{vmatrix} = 0$$

from which

$$\sin aL \cdot \sinh aL = 0$$

Since $\sinh aL \neq 0$, thus

$$\sin aL = 0 \tag{5.43}$$

Solving Eq. (5.43) leads to

$$a = i\pi/L, \quad i = 1, 2, \cdots \tag{5.44}$$

Substituting Eq. (5.44) into the first expression of Eq. (5.29) and taking the square root of both sides yields the frequency expression

$$\omega_i = i^2\pi^2\sqrt{\frac{EI}{\overline{m}L^4}}, \quad i = 1, 2, \cdots$$

Substituting $a = i\pi/L$ into Eq. (5.42), one obtains $A_1 = A_3 = A_4 = 0$ easily, and the value of A_2 is indeterminate. Substituting them into Eq. (5.34) yields the mode functions of the simply supported beam

$$\varphi_i(x) = A_2 \sin\frac{i\pi x}{L}, \quad i = 1, 2, \cdots$$

The first three modes are shown in Fig. 5.4B along with the corresponding circular frequencies.

Example 5.2: The uniform bar of length L shown in Fig. 5.5, is lifted from its right-hand support as indicated and then dropped producing a rotation about its left-hand pinned support. Assuming it rotates as a rigid body, the initial velocity distribution upon initial impact is given as follows:

$$\dot{v}(x,0) = \frac{x}{L}\dot{v}_t \tag{5.45}$$

where $\dot{v}_t$ represents the tip velocity. Corresponding to the rigid body rotation concept, the initial displacement is $v(x,0) = 0$.

Solution: The ith mode shape of this simply supported beam is given by

$$\varphi_i(x) = \sin\frac{i\pi x}{L} \tag{5.46}$$

Substituting $v(x,0)$ and $\dot{v}(x,0)$ into Eq. (5.38) yields

$$T_i(0) = 0, \quad \dot{T}_i(0) = \begin{cases} \dfrac{2\dot{v}_t}{i\pi}, & i = \text{odd} \\ -\dfrac{2\dot{v}_t}{i\pi}, & i = \text{even} \end{cases} \tag{5.47}$$

The generalized coordinates of the ith principal vibration can be obtained from Eq. (5.36)

$$T_i(t) = \begin{cases} \dfrac{2\dot{v}_t}{i\pi\omega_i}\sin\omega_i t, & i = \text{odd} \\ -\dfrac{2\dot{v}_t}{i\pi\omega_i}\sin\omega_i t, & i = \text{even} \end{cases} \tag{5.48}$$

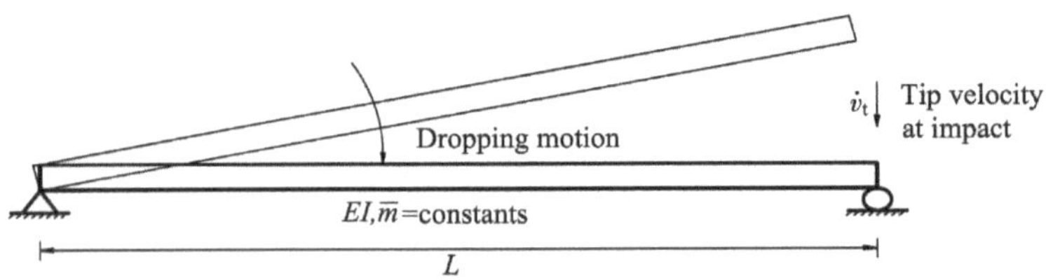

Figure 5.5 Analysis of free vibration of simply supported beam.

Substituting Eqs. (5.46) and (5.48) into Eq. (5.8) gives

$$v(x,t) = \frac{2\dot{v}_t}{\pi}\left(\frac{1}{\omega_1}\sin\frac{\pi x}{L}\sin\omega_1 t - \frac{1}{2\omega_2}\sin\frac{2\pi x}{L}\sin\omega_2 t + \cdots\right) \tag{5.49}$$

5.4 Forced vibration analysis of undamped straight beam

Two orthogonality conditions (Eqs. 5.13 and 5.18) also provide the means for uncoupling the equations of motion of distributed parameter systems, as it was earlier for discrete parameter systems. Eq. (5.7) is the forced vibration equation of motion for an undamped straight beam. Substituting Eq. (5.8) into Eq. (5.7) yields

$$\sum_{j=1}^{\infty} m(x)\varphi_j(x)\ddot{T}_j(t) + \sum_{j=1}^{\infty}\frac{d^2}{dx^2}\left[EI(x)\frac{d\varphi_j(x)}{dx^2}\right]T_j(t) = p(x,t) \tag{5.50}$$

Note that the subscript i in summations has been replaced by j for convenience. Multiplying each term by $\varphi_i(x)$ and integrating gives

$$\sum_{j=1}^{\infty}\ddot{T}_j(t)\int_0^L m(x)\varphi_j(x)\varphi_i(x)dx + \sum_{j=1}^{\infty}T_j(t)\int_0^L \varphi_i(x)\frac{d^2}{dx^2}\left[EI(x)\frac{d^2\varphi_j(x)}{dx^2}\right]dx = \int_0^L \varphi_i(x)p(x,t)dx \tag{5.51}$$

When the two orthogonality relationships are applied to the first two terms, it is obvious that all the terms in the series expansions, except the ith one, vanish, thus

$$\ddot{T}_i(t)\int_0^L m(x)\varphi_i^2(x)dx + T_i(t)\int_0^L \varphi_i(x)\frac{d^2}{dx^2}\left[EI(x)\frac{d^2\varphi_i(x)}{dx^2}\right]dx = \int_0^L \varphi_i(x)p(x,t)\,dx \tag{5.52}$$

Replacing the subscript m by i in Eq. (5.17), multiplying it by $\varphi_i(x)$, and integrating yields

$$\int_0^L \varphi_i(x)\frac{d^2}{dx^2}\left[EI(x)\frac{d^2\varphi_i(x)}{dx^2}\right]dx = \omega_i^2\int_0^L \varphi_i^2(x)m(x)dx \tag{5.53}$$

The integral on the right-hand side of this equation is the generalized mass of the *i*th mode

$$M_i = \int_0^L \varphi_i^2(x) m(x) dx \tag{5.54}$$

Considering Eqs. (5.53) and (5.54), Eq. (5.52) can be written in the following form:

$$M_i \ddot{T}_i(t) + \omega_i^2 M_i T_i(t) = P_i(t) \tag{5.55}$$

where

$$P_i(t) = \int_0^L \varphi_i(x) p(x,t) dx \tag{5.56}$$

which is the generalized load associated with $\varphi_i(x)$.

An equation of the type of Eq. (5.55) can be established for each mode of the structure, when its generalized mass and load are evaluated by using Eqs. (5.54) and (5.56), respectively. These are the equations of motion in modal coordinates for an undamped straight beam. Solving Eq. (5.55) leads to the time history of the steady-state response $T_i(t)$ of each modal coordinate, and the transient response of each modal coordinate can be obtained from the initial conditions. Then, the response of dynamic displacement, $v(x, t)$, can be evaluated from Eq. (5.8).

Example 5.3: In order to illustrate the above mode-superposition analysis procedure, the steady-state response of a uniform simply supported beam subjected to a central step-function load as shown in Fig. 5.6A and B, will be evaluated.

Solution: The natural frequencies and mode shapes of the simply supported beam evaluated in Example 5.1 are as follows:

$$\omega_i = i^2\pi^2\sqrt{\frac{EI}{\overline{m}L^4}}, \quad i = 1, 2, \cdots \tag{5.57}$$

$$\varphi_i(x) = \sin\frac{i\pi x}{L}, \quad i = 1, 2, \cdots \tag{5.58}$$

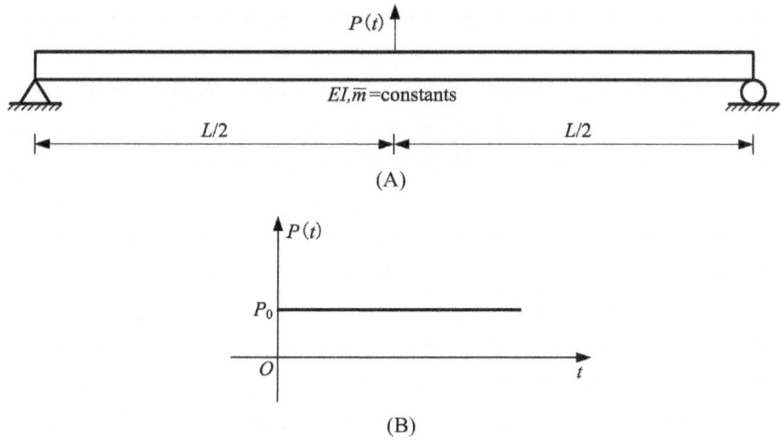

Figure 5.6 Example of dynamic response analysis for an undamped straight beam: (A) arrangement of beam and load; (B) applied step-function load.

From Eqs. (5.54) and (5.56), the generalized mass and load are evaluated to be, respectively,

$$M_i = \int_0^L \varphi_i^2(x)m(x)dx = \overline{m}\int_0^L \sin^2\left(\frac{i\pi x}{L}\right)dx = \frac{\overline{m}L}{2} \tag{5.59}$$

$$P_i(t) = \int_0^L \varphi_i(x)p(x,t)dx = P_0\varphi_i\left(\frac{L}{2}\right) = \alpha_i P_0 \tag{5.60}$$

where

$$\alpha_i = \begin{cases} 1, & i = 1, 5, 9 \\ -1, & i = 3, 7, 11, \cdots \\ 0, & i = \text{even} \end{cases}$$

Solving Eq. (5.55) with the Duhamel integral method yields

$$\begin{aligned} T_i(t) &= \frac{1}{M_i\omega_i}\int_0^t P_i(\tau)\sin\omega_i(t-\tau)d\tau \\ &= \frac{2\alpha_i P_0}{\overline{m}L\omega_i}\int_0^t \sin\omega_i(t-\tau)d\tau = \frac{2\alpha_i P_0}{\overline{m}L\omega_i^2}(1-\cos\omega_i t) \end{aligned} \tag{5.61}$$

Substituting Eqs. (5.46) and (5.61) into Eq. (5.8), and considering $\omega_i^2 = i^4\pi^4 EI/(\overline{m}L^4)$, one obtains the steady-state response

$$v(x,t) = \sum_{i=1}^{\infty} \varphi_i(x)T_i(t) = \frac{2P_0L^3}{\pi^4 EI}\sum_{i=1}^{\infty}\frac{\alpha_i}{i^4}(1-\cos\omega_i t)\sin\frac{i\pi x}{L} \tag{5.62}$$

The moment and shear response of the straight beam can be further evaluated

$$M(x,t) = EI\frac{\partial^2 v(x,t)}{\partial x^2} = -\frac{2P_0L}{\pi^2}\sum_{i=1}^{\infty}\frac{\alpha_i}{i^2}(1-\cos\omega_i t)\sin\frac{i\pi x}{L} \tag{5.63}$$

$$V(x,t) = EI\frac{\partial^3 v(x,t)}{\partial x^3} = -\frac{2P_0}{\pi}\sum_{i=1}^{\infty}\frac{\alpha_i}{i}(1-\cos\omega_i t)\cos\frac{i\pi x}{L} \tag{5.64}$$

Note that the higher modes contribute an insignificant amount to displacement due to the position of i^4 in Eq. (5.62). However, their contributions become more significant for the moment response and even more significant for shear. In other words, the series in Eq. (5.64) converges much more slowly with the mode number i than the series in Eq. (5.63) does, which in turn converges much more slowly than the series in Eq. (5.62). Therefore proper selection of that number depends upon the response quantities being evaluated.

5.5 Forced vibration analysis of damped straight beam

In the preceding formulation of the partial differential equations of motion for the straight beam (Fig. 5.7A), no damping was included. Now distributed viscous damping of two types will be included, as shown in Fig. 5.7C. One is the resistance of external media such as water, air, soil, etc., which is called external damping, while the other is the distributed damping stress along the height caused by the repeated straining of fibers in the structural section, which is called internal damping. Both of these types of damping are assumed to be viscous damping. Therefore it is convenient to separately consider the effects of the two forms of viscous damping above in equations of motion [3].

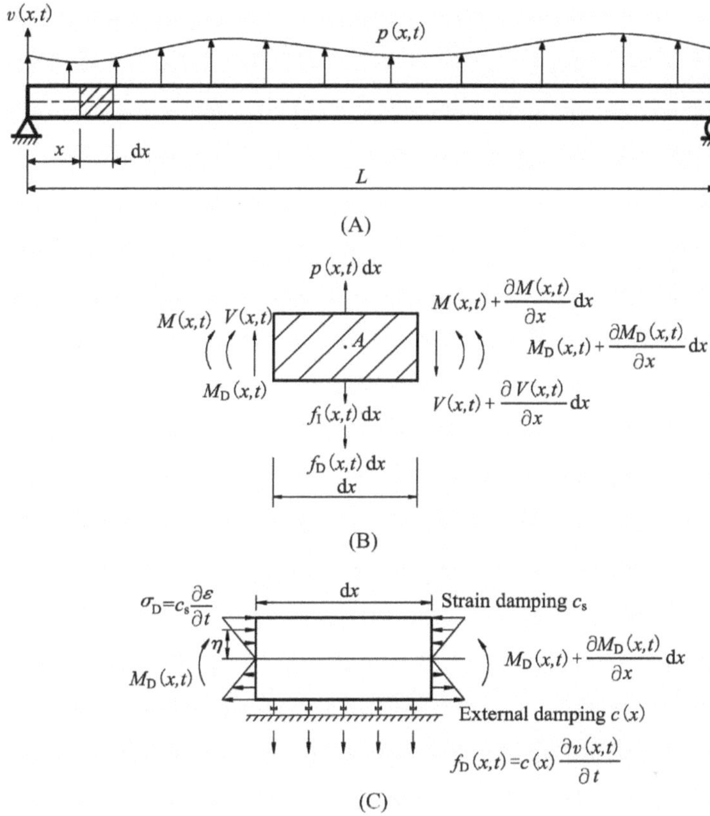

Figure 5.7 Damped simply supported beam: (A) distributed parameter beam with arbitrary loads; (B) dynamic equilibrium of differential segment; (C) damping force model of differential segment.

The damping force generated by the external damping is proportional to the vertical velocity of the beam. For the differential segment shown in Fig. 5.7B, the external damping force is given as follows:

$$f_D(x) = c(x)\frac{\partial v(x,t)}{\partial t} \tag{5.65}$$

The damping stress generated by internal damping is related to the strain velocity of material, that is,

$$\sigma_D(x,\eta,t) = c_s\frac{\partial \varepsilon(x,\eta,t)}{\partial t} = c_s\eta\frac{\partial}{\partial t}\left[\frac{M(x,t)}{EI(x)}\right] \tag{5.66}$$

where $\sigma_D(x,\eta,t)$ is the strain damping stress, c_s is the strain damping coefficient, and $\varepsilon(x,\eta,t)$ is the strain at the point with the distance η from the neutral axis on the cross section. Assuming that the stress is linearly distributed along the height on the cross section, these damping stresses produce a damping moment, that is,

$$M_D(x,t) = \int_A \sigma_D(x,\eta,t)\eta dA = \int_A c_s\eta^2 \frac{\partial}{\partial t}\left[\frac{M(x,t)}{EI(x)}\right] dA \tag{5.67}$$

where η is the distance between any point and the neutral axis on the cross section, and A is the area of the cross section.

Introducing the basic relationship between the moment and curvature of the beam, $M = EI\partial^2 v/\partial x^2$, Eq. (5.67) becomes

$$M_D(x,t) = c_s I(x)\frac{\partial^3 v(x,t)}{\partial x^2 \partial t} \tag{5.68}$$

where $I(x) = \int_A \eta^2 dA$.

Considering the dynamic equilibrium of the differential segment shown in Fig. 5.7B, the dynamic equilibrium equation of the damped straight beam can also be derived. Summing all the forces acting vertically yields the first dynamic equilibrium relationship

$$\begin{aligned} V(x,t) + p(x,t)dx - \left[V(x,t) + \frac{\partial V(x,t)}{\partial x}dx\right] - m(x)\frac{\partial^2 v(x,t)}{\partial t^2}dx \\ - c(x)\frac{\partial v(x,t)}{\partial t}dx = 0 \end{aligned} \tag{5.69}$$

Simplifying Eq. (5.69) gives

$$\frac{\partial V(x,t)}{\partial x} = p(x,t) - m(x)\frac{\partial^2 v(x,t)}{\partial t^2} - c(x)\frac{\partial v(x,t)}{\partial t} \tag{5.70}$$

The second equilibrium relationship is obtained by summing all the moments about point A on the elastic axis. Dropping the second-order moment terms involving the inertial force, external damping force, and external load yields

$$\begin{aligned} M(x,t) + M_D(x,t) + V(x,t)dx \\ - \left[M(x,t) + \frac{\partial M(x,t)}{\partial x}dx + M_D(x,t) + \frac{\partial M_D(x,t)}{\partial x}dx\right] = 0 \end{aligned} \tag{5.71}$$

Simplifying Eq. (5.71) gives

$$\frac{\partial M(x,t)}{\partial x} + \frac{\partial M_D(x,t)}{\partial x} = V(x,t) \tag{5.72}$$

By taking the derivative of Eq. (5.72) with respect to x and substituting the resulting equation into Eq. (5.70), one obtains

$$\frac{\partial^2}{\partial x^2}[M(x,t)+M_D(x,t)]+m(x)\frac{\partial^2 v(x,t)}{\partial t^2}+c(x)\frac{\partial v(x,t)}{\partial t}=p(x,t) \quad (5.73)$$

Introducing $M = EI\partial^2 v/\partial x^2$ and Eq. (5.68), the equation of motion for the distributed parameter beam, considering both internal and external damping, is obtained from Eq. (5.73)

$$\frac{\partial^2}{\partial x^2}\left[EI(x)\frac{\partial^2 v(x,t)}{\partial x^2}+c_sI(x)\frac{\partial^3 v(x,t)}{\partial x^2\partial t}\right]+m(x)\frac{\partial^2 v(x,t)}{\partial t^2}+c(x)\frac{\partial v(x,t)}{\partial t}=p(x,t) \quad (5.74)$$

Substituting Eq. (5.8) into Eq. (5.74) leads to

$$\sum_{j=1}^{\infty}m(x)\varphi_j(x)\ddot{T}_j(t)+\sum_{j=1}^{\infty}c(x)\varphi_j(x)\dot{T}_j(t)+\sum_{j=1}^{\infty}\frac{d^2}{dx^2}\left[c_sI(x)\frac{d^2\varphi_j(x)}{dx^2}\right]\dot{T}_j(t)$$
$$+\sum_{j=1}^{\infty}\frac{d^2}{dx^2}\left[EI(x)\frac{d^2\varphi_j(x)}{dx^2}\right]T_j(t)=p(x,t) \quad (5.75)$$

Note that the subscript i in summations has been replaced by j for convenience. Multiplying each term of Eq. (5.75) by $\varphi_i(x)$, integrating and using the orthogonality conditions, one obtains

$$M_i\ddot{T}_i(t)+\sum_{j=1}^{\infty}\dot{T}_j(t)\int_0^L\varphi_i(x)\left\{c(x)\varphi_j(x)+\frac{d^2}{dx^2}\left[c_sI(x)\frac{d^2\varphi_j(x)}{dx^2}\right]\right\}dx$$
$$+\omega_i^2M_iT_i(t)=P_i(t) \quad (5.76)$$

where the meaning and expression of M_i and $P_i(t)$ can be found in the previous section.

Obviously, the terms related to damping in Eq. (5.76) are generally coupled to each other, so these equations in modal coordinates need to be solved simultaneously. Assuming $c(x) = a_0m(x)$ and $c_sI(x) = a_1EI(x)$

(i.e., $c_s = a_1 E$), Eq. (5.76) can be uncoupled by means of the orthogonality conditions (see Eqs. 5.13 and 5.18) as follows:

$$M_i \ddot{T}_i(t) + (a_0 M_i + a_1 \omega_i^2 M_i) \dot{T}_i(t) + \omega_i^2 M_i T_i(t) = P_i(t) \tag{5.77}$$

By letting $C_i = a_0 M_i + a_1 \omega_i^2 M_i$, and introducing the damping ratio ξ_i of ith principal vibration, that is,

$$\xi_i = \frac{C_i}{2M_i \omega_i} = \frac{a_0}{2\omega_i} + \frac{a_1 \omega_i}{2}$$

Eq. (5.77) can be simplified to

$$\ddot{T}_i(t) + 2\xi_i \omega_i \dot{T}_i(t) + \omega_i^2 T_i(t) = \frac{P_i(t)}{M_i}, \quad i = 1, 2, \cdots \tag{5.78}$$

Eq. (5.78) is the equations of motion in modal coordinates for the forced vibration of a damped straight beam. It is shown from the analysis above that for a distributed parameter system, the partial differential equation of motion can be transformed into an infinite number of uncoupled equations of motion in modal coordinates. Each equation of motion in modal coordinates contains only one modal coordinate.

As mentioned above, damping is assumed to be proportional to the stiffness or mass, which is the classical Rayleigh damping assumption. Similar to the discrete parameter systems, the Rayleigh damping assumption can also be used to achieve the uncoupling of equations of the distributed parameter system. The parameters a_0 and a_1 can be determined from the damping ratios and the corresponding natural frequencies.

In principle, the total response of the system is the superposition of all modal component responses. Similar to the discrete MDOF systems, the lower modes generally contribute to the total response more than higher modes for most types of load. Therefore it is usually not necessary to include all the higher modes in the superposition process. When the response has been obtained with desired accuracy, the series can be truncated, and computational effort can be reduced greatly. In addition, the mathematical idealization of any complex structure also tends to be less reliable in predicting the higher modes; for this reason, too, it is well to limit the number of modes considered in dynamic response analysis.

Therefore the equations of motion in modal coordinates can be solved according to the procedure used for SDOF systems to obtain the required time history of the steady-state response $T_i(t)$. Transient response of the corresponding modal coordinates can be evaluated from the initial conditions. Finally, the total response in terms of the original geometric coordinates can be obtained from Eq. (5.8).

References

[1] Clough RW, Penzien J. Dynamics of structures. 3rd ed. Berkeley, CA: Computers & Structures, Inc; 2003.
[2] Zhou Z, Wen Y, Zeng Q. Lectures on dynamics of structures. 2nd ed. Beijing: China Communications Press Co., Ltd; 2017.
[3] Liu J, Du X. Dynamics of structures. Beijing: China Machine Press; 2007.

Problems

5.1 Compare the equations of motion, orthogonality conditions, and expansion expressions between finite-DOF and continuous systems.

5.2 When the beam shown in Fig. 5.1 is subjected to an axial load, analyze the relationship between natural frequencies and axial load.

5.3 Evaluate the first three frequencies of the cantilever beam with a mass at the end shown in Fig. P5.1, when the end lumped mass $M = 2\overline{m}L$ and the moment of inertia of the lumped mass is ignored. In Fig. P5.1, EI is the flexural stiffness of the beam, and $\overline{m}$ is the mass of the beam per unit length. Plot the shape of these three modes.

5.4 The uniform simply supported beam is subjected to a transverse, concentrated load P at midspan. Analyze the free vibration of the beam when the load P is removed suddenly.

5.5 The uniform simply supported beam shown in Fig. P5.2 is subjected to a transverse load $P(x, t) = \delta(x - a)\delta(t)$, where $\delta(x - a)$ and $\delta(t)$ are

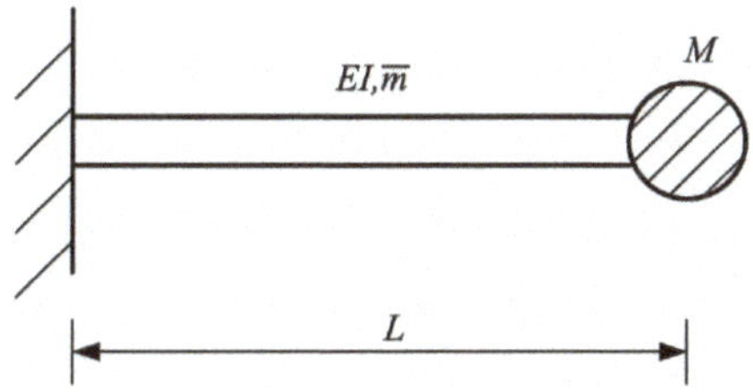

Figure P5.1 Figure of problem 5.3.

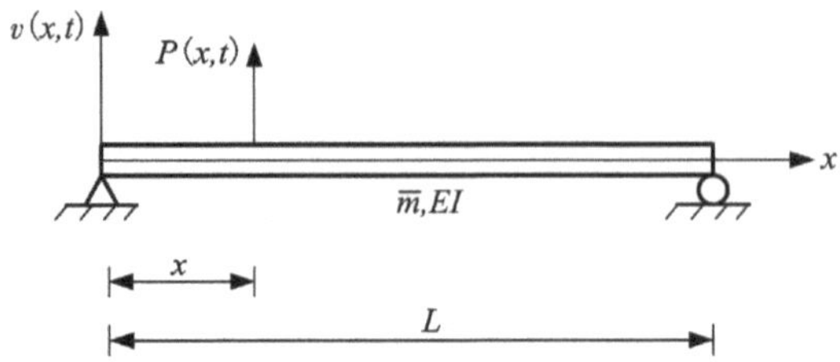

Figure P5.2 Figure of problem 5.5.

Dirac delta functions. Using elementary beam theory and the mode superposition method, determine the series expressions for transverse deflection $v(x, t)$, internal moment $M(x, t)$, and internal shear $V(x, t)$ caused by the load $P(x, t)$ defined above. In Fig. P5.2, EI is the flexural stiffness of the beam, and $\overline{m}$ is the mass of the beam per unit length. Discuss the rates of convergence of these series expressions.

CHAPTER 6

Approximate evaluation of natural frequencies and mode shapes

The determination of natural frequencies and mode shapes is an important step in the dynamic response analysis of linear systems by means of the mode superposition method. When the number of degrees of freedom (DOFs) of the system is larger than three ($n > 3$), it is difficult to manually evaluate all frequencies and mode shapes. As discussed in Chapter 4, Analysis of dynamic response of MDOF systems: mode superposition method, the structural response is mainly contributed by the first few modes, and the contribution of higher modes is negligibly small. Some practical methods, such as Rayleigh energy method, Rayleigh–Ritz method, matrix iteration method, subspace iteration method, and so on, were developed to evaluate the first few natural frequencies and modes approximately. These methods can be used to easily evaluate the first few frequencies and modes of the system by using computer programs or software.

According to the principle of mode superposition, the displacement response of the system can be expressed approximately by the product of the first few modes and the corresponding generalized coordinates (the modal coordinates). The above process is one of the few ways to reduce the DOF of the system. Therefore the common methods of reducing DOF in dynamic analysis are further discussed at the end of this chapter, as well as the relationship between dynamic and static DOF.

6.1 Rayleigh energy method

The Rayleigh energy method is one of the most effective and simplest methods used to evaluate the fundamental frequency of a system. The frequency equation can be derived according to the law of conservation of energy or equation of motion.

Fundamentals of Structural Dynamics
DOI: https://doi.org/10.1016/B978-0-12-823704-5.00006-9

As discussed in Chapter 4, Analysis of dynamic response of MDOF systems: mode superposition method, a certain principal vibration of the system can be excited by appropriate initial conditions, which is a free vibration at a specified frequency. When the conservative system vibrates freely at a certain frequency, the energy input and dissipation are not present in accordance with the law of the conservation of energy. Thus the total mechanical energy E remains constant, that is,

$$E = T + V = E_0 \tag{6.1}$$

where T is the kinetic energy of the freely vibrating system at any instant of time, V is the corresponding potential energy, and E_0 is a constant.

When the displacements of the vibrating system reach their maximum values, the kinetic energy is zero and the potential energy reaches its maximum value $V_{\max}$. When the system passes through the static-equilibrium position, the kinetic energy reaches its maximum value $T_{\max}$ and the potential energy is zero. According to the law of the conservation of mechanical energy, one obtains

$$T_{\max} = V_{\max} \tag{6.2}$$

The frequency equation can be derived from Eq. (6.2) as follows:

The free-vibration equation of motion for the undamped n-DOF system, which is a conservative system, is given by

$$\boldsymbol{M}\ddot{\boldsymbol{q}} + \boldsymbol{K}\boldsymbol{q} = \boldsymbol{0} \tag{6.3}$$

From Eq. (4.6), the ith principal vibration of the system can be expressed as

$$\boldsymbol{q}_i = c_i \boldsymbol{A}_i \sin(\omega_i t + \theta_i) \tag{6.4}$$

The kinetic energy of the small-amplitude vibration of the undamped system is $T = \frac{1}{2}\dot{\boldsymbol{q}}_i^T \boldsymbol{M}\dot{\boldsymbol{q}}_i = \frac{1}{2}c_i^2 \boldsymbol{A}_i^T \boldsymbol{M}\boldsymbol{A}_i \omega_i^2 \cos^2(\omega_i t + \theta_i)$. When $\cos^2(\omega_i t + \theta_i) = 1$, $T_{\max} = \frac{1}{2}c_i^2 \omega_i^2 \boldsymbol{A}_i^T \boldsymbol{M}\boldsymbol{A}_i$. The potential energy of the same system is $V = \frac{1}{2}\boldsymbol{q}_i^T \boldsymbol{K}\boldsymbol{q}_i = \frac{1}{2}c_i^2 \boldsymbol{A}_i^T \boldsymbol{K}\boldsymbol{A}_i \sin^2(\omega_i t + \theta_i)$, so the maximum potential energy is $V_{\max} = \frac{1}{2}c_i^2 \boldsymbol{A}_i^T \boldsymbol{K}\boldsymbol{A}_i$.

Substituting $T_{\max}$ and $V_{\max}$ into Eq. (6.2) leads to

$$\omega_i^2 = \frac{\boldsymbol{A}_i^T \boldsymbol{K}\boldsymbol{A}_i}{\boldsymbol{A}_i^T \boldsymbol{M}\boldsymbol{A}_i} \equiv R_{\mathrm{I}}(\boldsymbol{A}_i) \tag{6.5}$$

Eq. (6.5) is the frequency equation of Rayleigh energy method, in which $R_{\mathrm{I}}(\boldsymbol{A}_i)$ is called the first Rayleigh quotient.

In addition, Eq. (6.5) can also be derived directly from the equation of motion as follows:

Substituting the ith principal vibration $\boldsymbol{q}_i = c_i\boldsymbol{A}_i\sin(\omega_i t + \theta_i)$ into Eq. (6.3) leads to

$$\boldsymbol{K}\boldsymbol{A}_i = \omega_i^2\boldsymbol{M}\boldsymbol{A}_i$$

Premultiplying both sides of the above equation by $\boldsymbol{A}_i^T$, one obtains

$$\boldsymbol{A}_i^T\boldsymbol{K}\boldsymbol{A}_i = \omega_i^2\boldsymbol{A}_i^T\boldsymbol{M}\boldsymbol{A}_i \tag{6.6}$$

Then, Eq. (6.5) can also be obtained from Eq. (6.6).

When the flexibility matrix $\boldsymbol{R}$ of the system is known instead of the stiffness matrix $\boldsymbol{K}$, the frequency equation can be alternatively obtained as follows:

When the system vibrates freely at the ith natural frequency, the inertial force acting on it can be expressed as

$$\boldsymbol{f}_I = -\boldsymbol{M}\ddot{\boldsymbol{q}}_i$$

Considering $\boldsymbol{q}_i = c_i\boldsymbol{A}_i\sin(\omega_i t + \theta_i)$, $\dot{\boldsymbol{q}}_i = \boldsymbol{A}_i c_i\omega_i\cos(\omega_i t + \theta_i)$ and $\ddot{\boldsymbol{q}}_i = -\boldsymbol{A}_i c_i\omega_i^2\sin(\omega_i t + \theta_i) = -\omega_i^2\boldsymbol{q}_i$, the inertial force $\boldsymbol{f}_I$ can be rewritten as

$$\boldsymbol{f}_I = \omega_i^2\boldsymbol{M}\boldsymbol{q}_i$$

The displacement of the system produced by the inertial force $\boldsymbol{f}_I$ is

$$\overline{\boldsymbol{q}}_i = \boldsymbol{R}\boldsymbol{f}_I = \omega_i^2\boldsymbol{R}\boldsymbol{M}\boldsymbol{q}_i$$

The work done by inertial force $\boldsymbol{f}_I$ is transformed into the potential energy of the system, that is,

$$V = \frac{1}{2}\boldsymbol{f}_I^T\overline{\boldsymbol{q}}_i = \frac{1}{2}\omega_i^4\boldsymbol{q}_i^T\boldsymbol{M}\boldsymbol{R}\boldsymbol{M}\boldsymbol{q}_i$$

Note that the symmetry of $\boldsymbol{M}$ has been used in the above equation.

Therefore the maximum potential energy of the system is $V_{\max} = \frac{1}{2}c_i^2\omega_i^4\boldsymbol{A}_i^T\boldsymbol{M}\boldsymbol{R}\boldsymbol{M}\boldsymbol{A}_i$. The expression of the maximum kinetic energy of the system is still $T_{\max} = \frac{1}{2}c_i^2\omega_i^2\boldsymbol{A}_i^T\boldsymbol{M}\boldsymbol{A}_i$. Similarly, substituting $T_{\max}$ and $V_{\max}$ into Eq. (6.2) leads to

$$\omega_i^2 = \frac{\boldsymbol{A}_i^T\boldsymbol{M}\boldsymbol{A}_i}{\boldsymbol{A}_i^T\boldsymbol{M}\boldsymbol{R}\boldsymbol{M}\boldsymbol{A}_i} \equiv R_{\text{II}}(\boldsymbol{A}_i) \tag{6.7}$$

Eq. (6.7) is another frequency equation of Rayleigh energy method, in which $R_{\text{II}}(\boldsymbol{A}_i)$ is called the second Rayleigh quotient.

Similarly, Eq. (6.7) can also be derived directly from the equation of motion as follows:

From Eq. (6.3), one obtains

$$\boldsymbol{q} = -\boldsymbol{RM}\ddot{\boldsymbol{q}} \tag{6.8}$$

Substituting $\boldsymbol{q}_i = c_i\boldsymbol{A}_i\sin(\omega_i t + \theta_i)$ into Eq. (6.8) yields

$$\boldsymbol{A}_i = \omega_i^2\boldsymbol{RMA}_i$$

Premultiplying both sides of the above equation by $\boldsymbol{A}_i^T\boldsymbol{M}$ leads to

$$\boldsymbol{A}_i^T\boldsymbol{MA}_i = \omega_i^2\boldsymbol{A}_i^T\boldsymbol{MRMA}_i \tag{6.9}$$

Obviously, Eq. (6.7) can also be obtained from Eq. (6.9).

The first and second Rayleigh quotients have the following properties [1]: (1) when $\boldsymbol{A}_i$ is a precise mode, the Rayleigh quotient is equal to the true value of ω_i^2; (2) if $\boldsymbol{A}_i$ is an approximation to a certain mode with an error that is a first-order infinitesimal, the Rayleigh quotient is an approximation to the true value of ω_i^2 with an error which is a second-order infinitesimal, that is, Rayleigh quotient is stationary value in the neighborhood of the true value of ω_i^2; and (3) the Rayleigh quotient is bounded between ω_1^2 and ω_n^2, the squares of the lowest and highest natural frequencies, that is, it provides an upper bound for ω_1^2 and a lower bound for ω_n^2.

To solve for ω_i^2 from Eq. (6.5) or Eq. (6.7), the approximate mode vector $\boldsymbol{A}_i$ must be selected. The fundamental mode $\boldsymbol{A}_1$ can be generally assumed conveniently, whereas it is difficult to estimate the higher modes. Therefore the Rayleigh energy method can only be applied to evaluate the fundamental frequency ω_1. The accuracy of the Rayleigh energy method completely depends on the assumed approximate mode $\boldsymbol{A}_1$. When the approximate mode $\boldsymbol{A}_i$ satisfies the displacement (geometric) and force boundary conditions simultaneously, a relatively accurate frequency can be obtained. If the two types of boundary conditions cannot be satisfied simultaneously, the displacement boundary condition should be satisfied at least, otherwise the result will have a large error. According to the third property mentioned above, the fundamental frequency corresponding to the true mode is the lower bound of the evaluated frequencies by Rayleigh energy method. Therefore when the approximations obtained by this method are judged, the lowest evaluated frequency is most accurate.

As explained above, the free-vibration displacement of the system is produced by the inertial force, and the inertial force is proportional to the mass of the system. It can be deduced that the fundamental mode of the system is close to the deflection curve produced by its self-weight. If the horizontal vibration is investigated, the gravity should act in the horizontal direction. Therefore the self-weight deflection curve of the system is generally adopted as the assumed fundamental mode. If other similar curves are adopted, they must satisfy the displacement boundary condition of the system (the deflection curve produced by self-weight satisfies this condition naturally).

The deflection curve of the system due to self-weight can be used as the assumed mode of $\boldsymbol{A}_1$ to obtain a good estimate of ω_1. When a new displacement given by $\omega_1^2\boldsymbol{RMA}_1$, which is produced by inertial force corresponding to the assumed mode $\boldsymbol{A}_1$, is used to express the maximum potential energy $V_{\max}$, rather than the assumed mode $\boldsymbol{A}_1$, a more accurate value of ω_1 can be obtained. If the maximum potential energy $V_{\max}$ and the maximum kinetic energy $T_{\max}$ are both expressed by the new displacement $\omega_1^2\boldsymbol{RMA}_1$, the result will be more accurate than those given by previous methods (see the improved Rayleigh method in Ref. [2] for details).

$V_{\max}$ and $T_{\max}$ are evaluated using the assumed mode $\boldsymbol{A}_i$ in the expression of $R_{\mathrm{I}}(\boldsymbol{A}_i)$. In the expression of $R_{\mathrm{II}}(\boldsymbol{A}_i)$, $V_{\max}$ and $T_{\max}$ are evaluated by the new displacement $\omega_i^2\boldsymbol{RMA}_i$ and the assumed mode $\boldsymbol{A}_i$, respectively. The above analysis shows that for any assumed mode $\boldsymbol{A}_i$, $R_{\mathrm{II}}(\boldsymbol{A}_i)$ is closer to the square of true frequency of the structure than $R_{\mathrm{I}}(\boldsymbol{A}_i)$, giving

$$R_{\mathrm{I}}(\boldsymbol{A}_i) \geq R_{\mathrm{II}}(\boldsymbol{A}_i) \tag{6.10}$$

Eqs. (6.5) and (6.7) represent two forms of the Rayleigh energy method, and each has its own advantages. The former applies to the case with known stiffness matrix, and the latter is applicable to the case with known flexibility matrix. Generally, the former is simpler, whereas the latter is more accurate. In addition, only the formulae leading to approximate natural frequencies of discrete systems are given in this section. The application of Rayleigh energy method for continuous systems is essentially consistent with discrete systems. The expressions of $T_{\max}$ and $V_{\max}$ can be written in integral form according to the assumed mode functions, and the corresponding frequency equation can be obtained by setting $T_{\max}$ to equal $V_{\max}$. The detailed process can be found in Ref. [2].

Example 6.1: As shown in Fig. 6.1A, three disks are connected to a rotating shaft. The moment of inertia of each disk is J, the torsional

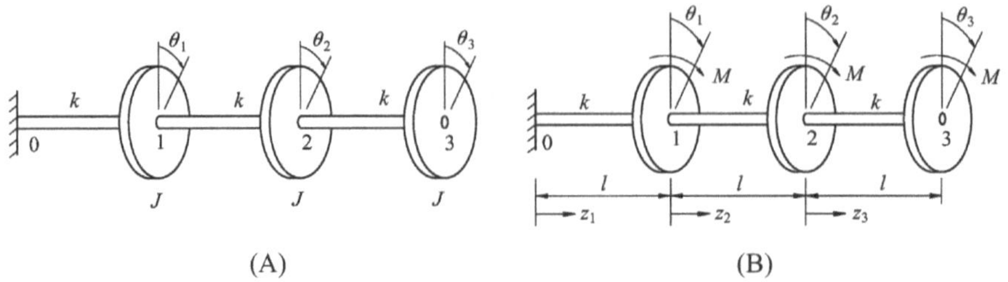

Figure 6.1 Diagram of vibrating system of rotating shaft: (A) diagram of dynamic property; (B) diagram of analysis of static–displacement curve.

stiffness of each shaft segment is k, and the mass of the shaft is ignored. Evaluate the fundamental frequency of this system.

Solution:

The mass, stiffness, and flexibility matrices of the system, respectively, are

$$\boldsymbol{M} = J\begin{bmatrix} 1 & 0 & 0 \\ 0 & 1 & 0 \\ 0 & 0 & 1 \end{bmatrix}, \boldsymbol{K} = k\begin{bmatrix} 2 & -1 & 0 \\ -1 & 2 & -1 \\ 0 & -1 & 1 \end{bmatrix} \text{ and } \boldsymbol{R} = \boldsymbol{K}^{-1} = \frac{1}{k}\begin{bmatrix} 1 & 1 & 1 \\ 1 & 2 & 2 \\ 1 & 2 & 3 \end{bmatrix}$$

The fundamental mode is assumed to be $\boldsymbol{A}_1 = \{1 \quad 1 \quad 1\}^T$ and one obtains $\boldsymbol{A}_1^T\boldsymbol{M}\boldsymbol{A}_1 = 3J$, $\boldsymbol{A}_1^T\boldsymbol{K}\boldsymbol{A}_1 = k$, and $\boldsymbol{A}_1^T\boldsymbol{M}\boldsymbol{R}\boldsymbol{M}\boldsymbol{A}_1 = 14J^2/k$.

From Eq. (6.5), $\omega_1^2 = k/(3J) = 0.333k/J$ is solved, and $\omega_1^2 = 3J/\left(14J^2/k\right) = 0.214k/J$ is obtained from Eq. (6.7).

If the assumed mode is determined according to the deflection curve due to self-weight (the calculation process is attached below), $\boldsymbol{A}_1 = \{3 \quad 5 \quad 6\}^T$ is selected.

Therefore $\boldsymbol{A}_1^T\boldsymbol{M}\boldsymbol{A}_1 = 70J$, $\boldsymbol{A}_1^T\boldsymbol{K}\boldsymbol{A}_1 = 14k$, and $\boldsymbol{A}_1^T\boldsymbol{M}\boldsymbol{R}\boldsymbol{M}\boldsymbol{A}_1 = 353\,J^2/k$, and then one gets $\omega_1^2 = 0.200k/J$ from Eq. (6.5) and $\omega_1^2 = 0.1983k/J$ from Eq. (6.7).

The exact value of the fundamental frequency of the system is $\omega_1 = \sqrt{0.1981k/J}$, and the evaluated results are consistent with the concepts in this section.

Appendix 1: Derivation of stiffness matrix $\boldsymbol{K}$ and mass matrix $\boldsymbol{M}$

The total potential energy of the system is

$$\Pi_d = \frac{1}{2}k\left[\theta_1^2 + (\theta_2-\theta_1)^2 + (\theta_3-\theta_2)^2\right] + J\ddot{\theta}_1\theta_1 + J\ddot{\theta}_2\theta_2 + J\ddot{\theta}_3\theta_3$$

Its variation with respect to displacements is

$$\begin{aligned}\delta_\varepsilon \Pi_d &= k[\theta_1\delta\theta_1 + (\theta_2-\theta_1)(\delta\theta_2-\delta\theta_1) + (\theta_3-\theta_2)(\delta\theta_3-\delta\theta_2)] \\ &\quad + J\ddot{\theta}_1\delta\theta_1 + J\ddot{\theta}_2\delta\theta_2 + J\ddot{\theta}_3\delta\theta_3 \\ &= \delta\theta_1 k(2\theta_1-\theta_2) + \delta\theta_2 k(-\theta_1+2\theta_2-\theta_3) + \delta\theta_3 k(-\theta_2+\theta_3) \\ &\quad + J\ddot{\theta}_1\delta\theta_1 + J\ddot{\theta}_2\delta\theta_2 + J\ddot{\theta}_3\delta\theta_3\end{aligned}$$

According to the principle of total potential energy with a stationary value in elastic system dynamics and the "set-in-right-position" rule for assembling system matrices, one obtains

$$\boldsymbol{K} = \begin{array}{c} \\ \delta\theta_1 \\ \delta\theta_2 \\ \delta\theta_3 \end{array}\begin{array}{c} \begin{array}{ccc}\theta_1 & \theta_2 & \theta_3\end{array} \\ \begin{bmatrix} 2k & -k & 0 \\ -k & 2k & -k \\ 0 & -k & k \end{bmatrix}\end{array}, \quad \boldsymbol{M} = \begin{array}{c} \\ \delta\theta_1 \\ \delta\theta_2 \\ \delta\theta_3 \end{array}\begin{array}{c} \begin{array}{ccc}\ddot{\theta}_1 & \ddot{\theta}_2 & \ddot{\theta}_3\end{array} \\ \begin{bmatrix} J & 0 & 0 \\ 0 & J & 0 \\ 0 & 0 & J \end{bmatrix}\end{array}$$

Appendix 2: Analysis of the deflection curve due to self-weight

Assuming that a torque M is exerted on each disk, the rotation angle of the rotating shaft is denoted by θ, as shown in Fig. 6.1B, and the torsional differential equation of the system is

$$\frac{d\theta}{dz} = \frac{M_T}{k}$$

For segment 0–1: $\theta' = 3M/k$ and $\theta = 3Mz_1/k + C_1$ are solved.

From the boundary conditions: $z_1 = 0$, $\theta_0 = 0$, $C_1 = 0$, and $\theta = 3Mz_1/k$ are obtained.

Therefore when $z_1 = l$, $\theta = \theta_1 = 3Ml/k$.

For segment 1–2: $\theta' = 2M/k$ and $\theta = 2Mz_2/k + C_2$ are solved.

From the boundary conditions: $z_2 = 0$, $\theta = \theta_1 = 3Ml/k$, $C_2 = 3Ml/k$, and $\theta = 2Mz_2/k + 3Ml/k$ are obtained.

Therefore when $z_2 = l$, $\theta = \theta_2 = 5Ml/k$.

For segment 2–3: $\theta' = M/k$ and $\theta = Mz_3/k + C_3$ are solved.

From the boundary conditions: $z_3 = 0$, $\theta = \theta_2 = 5Ml/k$, $C_3 = 5Ml/k$, and $\theta = Mz_3/k + 5Ml/k$ are obtained.

Therefore when $z_3 = l$, $\theta = \theta_3 = 6Ml/k$ is obtained.

Finally, the assumed fundamental mode is selected as $\boldsymbol{A}_1 = \{\theta_1 \quad \theta_2 \quad \theta_3\}^T = \{3 \quad 5 \quad 6\}^T$ according to the deflection curve of the whole shaft due to self-weight (only the relative values are taken).

6.2 Rayleigh–Ritz method

Although the fundamental frequency of the system can be effectively estimated, the higher frequencies cannot be obtained using the Rayleigh energy method. Higher frequencies as well as the fundamental frequency are often required in practical analysis. Ritz solved this problem by using the variational principle as follows:

Based on the first Rayleigh quotient, the exact natural frequencies ω_i, $i = 1, 2, \cdots, s$, would be evaluated using Eq. (6.5), if the exact natural modes $\boldsymbol{A}_i$, $i = 1, 2, \cdots, s$, could be selected. Since $\boldsymbol{A}_i$ is unknown, it is impossible to directly evaluate higher frequencies according to Eq. (6.5). An approximation of modes $\boldsymbol{A}_i$, $i = 1, 2, \cdots, s$, must be found.

Letting $\boldsymbol{\psi}_j$, $j = 1, 2, \cdots, s$, be a set of assumed modes that satisfy the displacement boundary condition of the system and are independent of each other, the ith natural mode $\boldsymbol{A}_i$ of the system can be expressed approximately as

$$\boldsymbol{A}_i = \sum_{j=1}^{s} a_{ji}\boldsymbol{\psi}_j = \boldsymbol{\psi}\boldsymbol{a}_i, i = 1, 2, \cdots, s \tag{6.11}$$

where $a_{ji}, j = 1, 2, \cdots, s$ are coefficients yet undetermined,

$$\boldsymbol{\psi} = \begin{bmatrix} \boldsymbol{\psi}_1 & \boldsymbol{\psi}_2 & \ldots & \boldsymbol{\psi}_s \end{bmatrix} \tag{6.12}$$

$$\boldsymbol{a}_i = \begin{Bmatrix} a_{1i} & a_{2i} & \cdots & a_{si} \end{Bmatrix}^T \tag{6.13}$$

Substituting Eq. (6.11) into Eq. (6.5) yields

$$R_{\mathrm{I}}(\boldsymbol{A}_i) = \frac{\boldsymbol{A}_i^T\boldsymbol{K}\boldsymbol{A}_i}{\boldsymbol{A}_i^T\boldsymbol{M}\boldsymbol{A}_i} = \frac{\boldsymbol{a}_i^T\boldsymbol{\psi}^T\boldsymbol{K}\boldsymbol{\psi}\boldsymbol{a}_i}{\boldsymbol{a}_i^T\boldsymbol{\psi}^T\boldsymbol{M}\boldsymbol{\psi}\boldsymbol{a}_i} \equiv \frac{V_{\mathrm{I}}(\boldsymbol{a}_i)}{T(\boldsymbol{a}_i)} \tag{6.14}$$

where $V_{\mathrm{I}}(\boldsymbol{a}_i) = \boldsymbol{a}_i^T\boldsymbol{\psi}^T\boldsymbol{K}\boldsymbol{\psi}\boldsymbol{a}_i$ and $T(\boldsymbol{a}_i) = \boldsymbol{a}_i^T\boldsymbol{\psi}^T\boldsymbol{M}\boldsymbol{\psi}\boldsymbol{a}_i$.

Thus $R_{\mathrm{I}}(\boldsymbol{A}_i)$ can be regarded as a function of a_{ji}, $j = 1, 2, \cdots, s$, also denoted as $R_{\mathrm{I}}(\boldsymbol{a}_i)$. The mode represented by Eq. (6.11) is only an approximation of the real mode. In order to make the obtained ω_i close to the exact value, $R_{\mathrm{I}}(\boldsymbol{a}_i)$ needs to reach the stationary value, which can only be achieved by varying a_{ji}, $j = 1, 2, \cdots, s$. The necessary condition for $R_{\mathrm{I}}(\boldsymbol{a}_i)$ to be stationary is

$$\frac{\partial R_{\mathrm{I}}(\boldsymbol{a}_i)}{\partial a_{ji}} = 0, j = 1, 2, \cdots, s$$

that is,

$$\frac{\partial R_{\mathrm{I}}(\boldsymbol{a}_i)}{\partial a_{ji}} = \frac{\partial}{\partial a_{ji}}\left(\frac{V_{\mathrm{I}}(\boldsymbol{a}_i)}{T(\boldsymbol{a}_i)}\right) = \frac{1}{T^2(\boldsymbol{a}_i)}\left[T(\boldsymbol{a}_i)\frac{\partial V_{\mathrm{I}}(\boldsymbol{a}_i)}{\partial a_{ji}} - V_{\mathrm{I}}(\boldsymbol{a}_i)\frac{\partial T(\boldsymbol{a}_i)}{\partial a_{ji}}\right] = 0$$

then

$$\frac{\partial V_{\mathrm{I}}(\boldsymbol{a}_i)}{\partial a_{ji}} - \omega_i^2\frac{\partial T(\boldsymbol{a}_i)}{\partial a_{ji}} = 0, j = 1, 2, \cdots, s \tag{6.15}$$

Considering

$$\begin{aligned}\frac{\partial V_{\mathrm{I}}(\boldsymbol{a}_i)}{\partial a_{ji}} &= \frac{\partial}{\partial a_{ji}}\left(\boldsymbol{a}_i^T\boldsymbol{\psi}^T\boldsymbol{K}\boldsymbol{\psi}\boldsymbol{a}_i\right) \\ &= \left[\frac{\partial \boldsymbol{a}_i^T}{\partial a_{ji}}\right]\boldsymbol{\psi}^T\boldsymbol{K}\boldsymbol{\psi}\boldsymbol{a}_i + \boldsymbol{a}_i^T\boldsymbol{\psi}^T\boldsymbol{K}\boldsymbol{\psi}\left[\frac{\partial \boldsymbol{a}_i}{\partial a_{ji}}\right] \\ &= 2\left[\frac{\partial \boldsymbol{a}_i^T}{\partial a_{ji}}\right]\boldsymbol{\psi}^T\boldsymbol{K}\boldsymbol{\psi}\boldsymbol{a}_i = 2\boldsymbol{\psi}_j^T\boldsymbol{K}\boldsymbol{\psi}\boldsymbol{a}_i\end{aligned} \tag{6.16}$$

where $\boldsymbol{\psi}_j^T = \left[\frac{\partial \boldsymbol{a}_i^T}{\partial a_{ji}}\right]\boldsymbol{\psi}^T$, and similarly

$$\frac{\partial T(\boldsymbol{a}_i)}{\partial a_{ji}} = 2\boldsymbol{\psi}_j^T\boldsymbol{M}\boldsymbol{\psi}\boldsymbol{a}_i \tag{6.17}$$

then Eq. (6.15) can be written as

$$\boldsymbol{\psi}_j^T\boldsymbol{K}\boldsymbol{\psi}\boldsymbol{a}_i - \omega_i^2\boldsymbol{\psi}_j^T\boldsymbol{M}\boldsymbol{\psi}\boldsymbol{a}_i = 0, j = 1, 2, \cdots, s$$

Combining these s equations above into a matrix equation leads to

$$\boldsymbol{\psi}^T\boldsymbol{K}\boldsymbol{\psi}\boldsymbol{a}_i - \omega_i^2\boldsymbol{\psi}^T\boldsymbol{M}\boldsymbol{\psi}\boldsymbol{a}_i = \boldsymbol{0} \tag{6.18}$$

which may be written in the abbreviated form

$$(\boldsymbol{K}^* - \omega_i^2\boldsymbol{M}^*)\boldsymbol{a}_i = \boldsymbol{0} \tag{6.19}$$

where $\boldsymbol{K}^* = \boldsymbol{\psi}^T\boldsymbol{K}\boldsymbol{\psi}$ and $\boldsymbol{M}^* = \boldsymbol{\psi}^T\boldsymbol{M}\boldsymbol{\psi}$ are called the generalized stiffness matrix and generalized mass matrix, respectively, which are symmetric matrices of $s \times s$ order.

Thus the following task is to solve the eigenvalue problem represented by Eq. (6.19). From $\left|\boldsymbol{K}^* - \omega_i^2\boldsymbol{M}^*\right| = 0$, the approximate values ω_1^2, $\omega_2^2, \ldots, \omega_s^2$ of the square of the first s frequencies are solved. By substituting them into Eq. (6.19), respectively, s eigenvectors $\boldsymbol{a}_1$, $\boldsymbol{a}_2$, …, $\boldsymbol{a}_s$ can be obtained. Then, $\boldsymbol{a}_1$, $\boldsymbol{a}_2$, ..., $\boldsymbol{a}_s$ are respectively substituted into Eq. (6.11) to obtain the first s approximate modes.

The approximate modes $\boldsymbol{A}_i$, $i = 1, 2, \cdots, s$, determined from the Rayleigh–Ritz method satisfy the orthogonality conditions, which is proven as follows:

The eigenvector $\boldsymbol{a}_i$ determined from Eq. (6.19) must satisfy the following orthogonality conditions

$$\boldsymbol{a}_i^T \boldsymbol{K}^* \boldsymbol{a}_j = 0, \boldsymbol{a}_i^T \boldsymbol{M}^* \boldsymbol{a}_j = 0, i \neq j$$

Substituting $\boldsymbol{K}^* = \boldsymbol{\psi}^T \boldsymbol{K} \boldsymbol{\psi}$ and $\boldsymbol{M}^* = \boldsymbol{\psi}^T \boldsymbol{M} \boldsymbol{\psi}$ into the above two equations gives

$$\boldsymbol{a}_i^T \boldsymbol{\psi}^T \boldsymbol{K} \boldsymbol{\psi} \boldsymbol{a}_j = 0, \boldsymbol{a}_i^T \boldsymbol{\psi}^T \boldsymbol{M} \boldsymbol{\psi} \boldsymbol{a}_j = 0, i \neq j$$

Substituting Eq. (6.11) and its transpose into the above two equations yields

$$\boldsymbol{A}_i^T \boldsymbol{K} \boldsymbol{A}_j = 0, \boldsymbol{A}_i^T \boldsymbol{M} \boldsymbol{A}_j = 0, i \neq j$$

It can be seen that the approximate modes $\boldsymbol{A}_i$, $i = 1, 2, \cdots, s$, also satisfy the orthogonality conditions.

The second Rayleigh quotient $R_{\text{II}}(\boldsymbol{A}_i)$ can also be treated as follows:

Substituting Eq. (6.11) into Eq. (6.7) gives

$$R_{\text{II}}(\boldsymbol{a}_i) = \frac{\boldsymbol{a}_i^T \boldsymbol{\psi}^T \boldsymbol{M} \boldsymbol{\psi} \boldsymbol{a}_i}{\boldsymbol{a}_i^T \boldsymbol{\psi} \boldsymbol{M} \boldsymbol{R} \boldsymbol{M} \boldsymbol{\psi} \boldsymbol{a}_i} \equiv \frac{T(\boldsymbol{a}_i)}{V_{\text{II}}(\boldsymbol{a}_i)} \tag{6.20}$$

where $V_{\text{II}}(\boldsymbol{a}_i) = \boldsymbol{a}_i^T \boldsymbol{\psi}^T \boldsymbol{M} \boldsymbol{R} \boldsymbol{M} \boldsymbol{\psi} \boldsymbol{a}_i$ and $T(\boldsymbol{a}_i) = \boldsymbol{a}_i^T \boldsymbol{\psi}^T \boldsymbol{M} \boldsymbol{\psi} \boldsymbol{a}_i$. The necessary condition for $R_{\text{II}}(\boldsymbol{a}_i)$ to be stationary is

$$\frac{\partial R_{\text{II}}(\boldsymbol{a}_i)}{\partial a_{ji}} = 0, j = 1, 2, \cdots, s$$

that is,

$$\frac{\partial R_{\text{II}}(\boldsymbol{a}_i)}{\partial a_{ji}} = \frac{1}{V_{\text{II}}^2(\boldsymbol{a}_i)} \left[V_{\text{II}}(\boldsymbol{a}_i) \frac{\partial T(\boldsymbol{a}_i)}{\partial a_{ji}} - T(\boldsymbol{a}_i) \frac{\partial V_{\text{II}}(\boldsymbol{a}_i)}{\partial a_{ji}} \right] = 0 \tag{6.21}$$

Considering

$$\frac{\partial V_{\text{II}}(\boldsymbol{a}_i)}{\partial a_{ji}} = 2\boldsymbol{\psi}_j^T \boldsymbol{M} \boldsymbol{R} \boldsymbol{M} \boldsymbol{\psi} \boldsymbol{a}_i, \frac{\partial T(\boldsymbol{a}_i)}{\partial a_{ji}} = 2\boldsymbol{\psi}_j^T \boldsymbol{M} \boldsymbol{\psi} \boldsymbol{a}_i$$

Eq. (6.21) becomes

$$\psi_j^T M\psi a_i - \omega_i^2 \psi_j^T MRM\psi a_i = 0, j = 1, 2, \cdots, s$$

Combining these s equations above into a matrix equation leads to

$$\psi^T M\psi a_i - \omega_i^2 \psi^T MRM\psi a_i = \mathbf{0}$$

which is abbreviated as

$$(M^* - \omega_i^2 R^*)a_i = \mathbf{0} \tag{6.22}$$

where M^* is the same as above, and $R^* = \psi^T MRM\psi$ is called the generalized flexibility matrix. From $|M^* - \omega_i^2 R^*| = 0$, the approximate values of the square of the first s frequencies, ω_1^2, ω_2^2, ..., ω_s^2, are obtained. By substituting them into Eq. (6.22) respectively, s eigenvectors a_1, a_2, ..., a_s can be obtained. Then, a_1, a_2, ..., a_s are respectively substituted into Eq. (6.11) to obtain the first s approximate modes.

It has been pointed out that $R_{\text{I}}(A_i) \geq R_{\text{II}}(A_i)$. Therefore based on the same assumed mode, the frequencies evaluated by Eq. (6.22) are more accurate than those obtained from Eq. (6.19), which will be confirmed in Example 6.2.

Although the Rayleigh–Ritz method still comes down to solve the eigenvalue problem, its order is much lower than that of the original eigenvalue problem described in Chapter 4, Analysis of dynamic response of MDOF systems: mode superposition method, that is, $s \ll n$. Thus it is easy to solve for the first few natural frequencies and modes using this method. However, the accuracy of the estimated eigen-pairs still depends on the degree of approximation of the assumed modes. And the requirement for mode approximation in Rayleigh–Ritz method is more relaxed than that for the Rayleigh energy method. In general, the accuracy of higher-order eigenvalues evaluated by the Rayleigh–Ritz method is lower than that of lower-order eigenvalues. Therefore in order to obtain k modes and frequencies with required accuracy, $2k$ assumed modes are generally adopted.

Example 6.2: Evaluate the first two natural frequencies and modes of the system shown in Fig. 6.1.

Solution:

According to Example 6.1, the mass and stiffness matrices of this system, respectively, are

$$M = J\begin{bmatrix} 1 & 0 & 0 \\ 0 & 1 & 0 \\ 0 & 0 & 1 \end{bmatrix}, K = k\begin{bmatrix} 2 & -1 & 0 \\ -1 & 2 & -1 \\ 0 & -1 & 1 \end{bmatrix}$$

Taking $\boldsymbol{\psi}_1 = \{1 \quad 2 \quad 3\}^T$ and $\boldsymbol{\psi}_2 = \{1 \quad 0 \quad -1\}^T$ leads to

$$\boldsymbol{A}_i = \sum_{j=1}^{2} a_{ji}\boldsymbol{\psi}_j = [\boldsymbol{\psi}_1 \quad \boldsymbol{\psi}_2]\begin{Bmatrix} a_{1i} \\ a_{2i} \end{Bmatrix} = \boldsymbol{\psi}\boldsymbol{a}_i,$$

$$\boldsymbol{M}^* = \boldsymbol{\psi}^T\boldsymbol{M}\boldsymbol{\psi} = J\begin{bmatrix} 14 & -2 \\ -2 & 2 \end{bmatrix} \text{ and } \boldsymbol{K}^* = \boldsymbol{\psi}^T\boldsymbol{K}\boldsymbol{\psi} = k\begin{bmatrix} 3 & -1 \\ -1 & 3 \end{bmatrix}$$

From $(\boldsymbol{K}^* - \omega_i^2\boldsymbol{M}^*)\boldsymbol{a}_i = \boldsymbol{0}$, one obtains

$$\begin{bmatrix} 3k - 14\omega_i^2 J & -k + 2\omega_i^2 J \\ -k + 2\omega_i^2 J & 3k - 2\omega_i^2 J \end{bmatrix}\begin{Bmatrix} a_{1i} \\ a_{2i} \end{Bmatrix} = \begin{Bmatrix} 0 \\ 0 \end{Bmatrix} \tag{6.23}$$

Letting the determinant of the coefficient matrix vanish, that is,

$$\begin{vmatrix} 3k - 14\omega_i^2 J & -k + 2\omega_i^2 J \\ -k + 2\omega_i^2 J & 3k - 2\omega_i^2 J \end{vmatrix} = 0$$

The first two natural frequencies can be solved as follows:

$$\omega_1 = \sqrt{0.2047k/J}, \ \omega_2 = \sqrt{1.6287k/J}$$

Substituting ω_1 and ω_2 into Eq. (6.23), respectively, one obtains

$$\boldsymbol{a}_1 = \begin{Bmatrix} 4.386 \\ 1 \end{Bmatrix}, \boldsymbol{a}_2 = \begin{Bmatrix} 0.114 \\ 1 \end{Bmatrix}$$

Substituting $\boldsymbol{a}_1$ and $\boldsymbol{a}_2$ into Eq. (6.11), respectively, one obtains

$$\boldsymbol{A}_1 = \{1.000 \quad 1.629 \quad 2.257\}^T, \boldsymbol{A}_2 = \{1.000 \quad 0.205 \quad -0.591\}^T$$

The exact natural frequencies of the system are $\omega_1 = \sqrt{0.1981k/J}$ and $\omega_2 = \sqrt{1.555k/J}$, and the corresponding exact modes are $\boldsymbol{A}_1 = \{1.000 \quad 1.802 \quad 2.247\}^T$ and $\boldsymbol{A}_2 = \{1.000 \quad 0.445 \quad -0.802\}^T$, respectively. The relative errors of the first two evaluated frequencies are 1.65% and 2.34%, respectively. Generally, the higher frequencies and modes have lower precision than the lower ones.

6.3 Matrix iteration method

The accuracy of the Rayleigh–Ritz method depends on the degree of approximation of the assumed modes. The accuracy requirement of the assumed modes is relatively high, and the difficulty lies in selecting the assumed modes with sufficient accuracy when using the Rayleigh–Ritz method. The matrix iteration method can overcome this difficulty, and the modes can be roughly assumed. Simple problems can be solved by hand with an electronic calculator, and complex problems can be dealt with using computer programs.

6.3.1 Iteration procedure for fundamental frequency and mode

According to Chapter 4, Analysis of dynamic response of MDOF systems: mode superposition method, the eigenvalue equation for solving the *i*th frequency and mode of the system can be written as

$$(\boldsymbol{K} - \lambda_i \boldsymbol{M})\boldsymbol{A}_i = \boldsymbol{0}$$

which can be transformed into

$$\boldsymbol{K}^{-1}\boldsymbol{M}\boldsymbol{A}_i = \frac{1}{\lambda_i}\boldsymbol{A}_i$$

Letting $\boldsymbol{D} = \boldsymbol{K}^{-1}\boldsymbol{M}$, which is called dynamic matrix, and $\overline{\lambda}_i = 1/\lambda_i = 1/\omega_i^2$, the above eigenvalue equation can also be expressed as

$$\boldsymbol{D}\boldsymbol{A}_i = \overline{\lambda}_i \boldsymbol{A}_i \tag{6.24}$$

To provide an overview of the matrix iteration method, the iteration steps are first introduced, then its basic concept is explained, and finally an example is used to illustrate the procedure. The iteration steps are as follows:

1. Assume a normalized fundamental mode $\boldsymbol{u}_0$. In this section, a certain entry in mode vector is taken to be unity. Generally, it can be roughly taken to be $\{1 \quad 1 \quad 1 \quad \cdots \quad 1\}^T$.
2. Premultiplying $\boldsymbol{u}_0$ by matrix $\boldsymbol{D}$, and normalizing the resulting vector $\boldsymbol{D}\boldsymbol{u}_0$ leads to

$$\boldsymbol{D}\boldsymbol{u}_0 = a_1 \boldsymbol{u}_1$$

   where $\boldsymbol{u}_1$ is the first approximation of the normalized mode, and a_1 is the first normalized factor of the mode.
3. Premultiplying $\boldsymbol{u}_1$ by matrix $\boldsymbol{D}$ and normalizing $\boldsymbol{D}\boldsymbol{u}_1$, one obtains

$$\boldsymbol{D}\boldsymbol{u}_1 = a_2 \boldsymbol{u}_2$$

   where $\boldsymbol{u}_2$ is the second approximation of the normalized mode and a_2 is the second normalized factor of the mode.

4. Continue the above iteration when $|\boldsymbol{u}_2 - \boldsymbol{u}_1|/|\boldsymbol{u}_1| \geq \varepsilon$, where ε represents a specified tolerance. The iteration ends until $|\boldsymbol{u}_k - \boldsymbol{u}_{k-1}|/|\boldsymbol{u}_{k-1}| < \varepsilon$. Note that $|\cdot|$ denotes the Euclidean norm of the vector. In this situation, a_k generally converges to the first eigenvalue $\overline{\lambda}_1 = 1/\omega_1^2$, and the corresponding eigenvector $\boldsymbol{u}_k$ converges to the fundamental mode $\boldsymbol{A}_1$, as demonstrated below.

The assumed mode $\boldsymbol{u}_0$ of n-DOF system can be expressed as a linear combination of the exact modes of the system, that is,

$$\boldsymbol{u}_0 = \sum_{i=1}^{n} C_i \boldsymbol{A}_i \tag{6.25}$$

where $\boldsymbol{A}_i$, $i = 1, 2, \cdots, n$, are the exact mode vectors of the system and C_i, $i = 1, 2, \cdots, n$, are the mode combining coefficients.

Assume that all the eigenvalues $\overline{\lambda}_i$, $i = 1, 2, \cdots, n$, of the system are unequal and they are arranged as $\overline{\lambda}_1 > \overline{\lambda}_2 > \cdots > \overline{\lambda}_n$. After the first cycle of iteration, $\boldsymbol{D}\boldsymbol{u}_0 = \sum_{i=1}^{n} C_i \overline{\lambda}_i \boldsymbol{A}_i = a_1 \boldsymbol{u}_1$, hence $\boldsymbol{u}_1 = 1/a_1 \sum_{i=1}^{n} C_i \overline{\lambda}_i \boldsymbol{A}_i$. After the second cycle of iteration, $\boldsymbol{D}\boldsymbol{u}_1 = 1/a_1 \sum_{i=1}^{n} C_i \overline{\lambda}_i^2 \boldsymbol{A}_i = a_2 \boldsymbol{u}_2$, hence $\boldsymbol{u}_2 = 1/(a_1 a_2) \sum_{i=1}^{n} C_i \overline{\lambda}_i^2 \boldsymbol{A}_i$.

Continuing the above iterations, the result of the kth iteration is obtained as

$$\boldsymbol{u}_k = \frac{1}{a_1 a_2 \cdots a_k} \sum_{i=1}^{n} C_i \overline{\lambda}_i^k \boldsymbol{A}_i \tag{6.26}$$

From Eq. (6.26), it should be noted that the term associated with $\overline{\lambda}_1^k$ becomes dominant (since $\overline{\lambda}_1$ is the largest eigenvalue), and the other terms can be ignored, as the number of iteration increases, that is,

$$\boldsymbol{u}_k \approx \frac{1}{a_1 a_2 \cdots a_k} C_1 \overline{\lambda}_1^k \boldsymbol{A}_1$$

When k is large enough, $\boldsymbol{u}_{k-1} = \boldsymbol{u}_k$ can be obtained with acceptable accuracy, that is,

$$\frac{1}{a_1 a_2 \cdots a_{k-1}} C_1 \overline{\lambda}_1^{k-1} \boldsymbol{A}_1 = \frac{1}{a_1 a_2 \cdots a_k} C_1 \overline{\lambda}_1^k \boldsymbol{A}_1 \tag{6.27}$$

From Eq. (6.27), $\overline{\lambda}_1 = a_k$ is obtained.

Eq. (6.26) can also be written as

$$\boldsymbol{u}_k = \frac{1}{a_1 a_2 \cdots a_k} C_1 \overline{\lambda}_1^k \left(\boldsymbol{A}_1 + \sum_{i=2}^{n} \frac{C_i \overline{\lambda}_i^k}{C_1 \overline{\lambda}_1^k} \boldsymbol{A}_i \right) \tag{6.28}$$

where $\boldsymbol{u}_k$ converges to the fundamental mode $\boldsymbol{A}_1$, and the convergence rate depends on the rate of $\left(\overline{\lambda}_i/\overline{\lambda}_1\right)^k \to 0$. The methods of accelerating the convergence rate of the iteration can be found in Ref. [3].

Example 6.3: Evaluate the fundamental frequency and mode of the system given by Example 6.1 using the matrix iteration method.

Solution:

According to Example 6.1, the flexibility matrix of the system is given by $\boldsymbol{R} = \boldsymbol{K}^{-1} = \frac{1}{k}\begin{bmatrix} 1 & 1 & 1 \\ 1 & 2 & 2 \\ 1 & 2 & 3 \end{bmatrix}$,

and the mass matrix is

$$\boldsymbol{M} = J\begin{bmatrix} 1 & 0 & 0 \\ 0 & 1 & 0 \\ 0 & 0 & 1 \end{bmatrix},$$

thus

$$\boldsymbol{D} = \boldsymbol{K}^{-1}\boldsymbol{M} = \frac{J}{k}\begin{bmatrix} 1 & 1 & 1 \\ 1 & 2 & 2 \\ 1 & 2 & 3 \end{bmatrix}.$$

Assuming the fundamental mode vector $\boldsymbol{u}_0 = \left\{ 1 \quad 1 \quad 1 \right\}^T$, the iteration is carried out as follows:

1. $\boldsymbol{D}\boldsymbol{u}_0 = \frac{J}{k}\begin{bmatrix} 1 & 1 & 1 \\ 1 & 2 & 2 \\ 1 & 2 & 3 \end{bmatrix}\begin{Bmatrix} 1 \\ 1 \\ 1 \end{Bmatrix} = \frac{J}{k}\begin{Bmatrix} 3 \\ 5 \\ 6 \end{Bmatrix} = \frac{3J}{k}\begin{Bmatrix} 1.0000 \\ 1.6667 \\ 2.0000 \end{Bmatrix}$, then $a_1 = 3J/k$ and $\boldsymbol{u}_1 = \begin{Bmatrix} 1.0000 \\ 1.6667 \\ 2.0000 \end{Bmatrix}$
2. $\boldsymbol{D}\boldsymbol{u}_1 = 4.6667\frac{J}{k}\begin{Bmatrix} 1.0000 \\ 1.7857 \\ 2.2143 \end{Bmatrix}$, then $a_2 = 4.6667\,J/k$ and $\boldsymbol{u}_2 = \begin{Bmatrix} 1.0000 \\ 1.7857 \\ 2.2143 \end{Bmatrix}$
3. $\boldsymbol{D}\boldsymbol{u}_2 = 5.0000\frac{J}{k}\begin{Bmatrix} 1.0000 \\ 1.8000 \\ 2.2429 \end{Bmatrix}$, then $a_3 = 5.0000\,J/k$ and $\boldsymbol{u}_3 = \begin{Bmatrix} 1.0000 \\ 1.8000 \\ 2.2429 \end{Bmatrix}$

4. $\boldsymbol{D}\boldsymbol{u}_3 = 5.0429\frac{J}{k}\begin{Bmatrix} 1.0000 \\ 1.8017 \\ 2.2465 \end{Bmatrix}$, then $a_4 = 5.0429J/k$ and

$$\boldsymbol{u}_4 = \begin{Bmatrix} 1.000\ 0 \\ 1.801\ 7 \\ 2.246\ 5 \end{Bmatrix}$$

5. $\boldsymbol{D}\boldsymbol{u}_4 = 5.0667\frac{J}{k}\begin{Bmatrix} 1.0000 \\ 1.8026 \\ 2.2387 \end{Bmatrix}$, then $a_5 = 5.0667J/k$ and

$$\boldsymbol{u}_5 = \begin{Bmatrix} 1.0000 \\ 1.8026 \\ 2.2387 \end{Bmatrix}$$

6. $\boldsymbol{D}\boldsymbol{u}_5 = 5.0413\frac{J}{k}\begin{Bmatrix} 1.0000 \\ 1.8010 \\ 2.2457 \end{Bmatrix}$, then $a_6 = 5.0413J/k$ and

$$\boldsymbol{u}_6 = \begin{Bmatrix} 1.0000 \\ 1.8010 \\ 2.2457 \end{Bmatrix}$$

7. $\boldsymbol{D}\boldsymbol{u}_6 = 5.0467\frac{J}{k}\begin{Bmatrix} 1.0000 \\ 1.8019 \\ 2.2468 \end{Bmatrix}$, then $a_7 = 5.0467J/k$ and

$$\boldsymbol{u}_7 = \begin{Bmatrix} 1.0000 \\ 1.8019 \\ 2.2468 \end{Bmatrix}$$

Note that $\boldsymbol{u}_6 \approx \boldsymbol{u}_7$, hence $\overline{\lambda}_1 = 5.0467J/k$, and $\omega_1^2 = k/(5.0467\,J) = 0.1981\ k/J$. The corresponding fundamental mode vector is

$$\boldsymbol{A}_1 = \begin{Bmatrix} 1.0000 & 1.8019 & 2.2468 \end{Bmatrix}^T$$

Note that the tolerance ε has not been specified in this example, and the convergence is judged by experience. If the iteration is conducted by means of computer programs, a tolerance is required. The exact value of the fundamental frequency of the system is $\omega_1 = \sqrt{0.1981k/J}$, hence, the result has also confirmed the conclusions drawn in this section.

6.3.2 Iteration procedure for higher frequencies and modes

Eq. (6.26) shows that when the linear combination of the assumed mode vector $\boldsymbol{u}_0$ represented by Eq. (6.25) contains the first, second,…, and nth mode components, $\boldsymbol{u}_k$ converges to the first mode. If $\boldsymbol{u}_0$ does not contain the component of the first mode $\boldsymbol{A}_1$, that is, $C_1 = 0$, $\boldsymbol{u}_k$ will converge to the second mode. Similarly, if $\boldsymbol{u}_0$ does not contain the first r mode components, $\boldsymbol{u}_k$ will converge to the $(r+1)$th mode. Therefore when the iteration procedure is applied to evaluate the $(r+1)$th mode and frequency of the system, it is necessary to remove the first r mode components from $\boldsymbol{u}_0$. The details are as follows:

Premultiplying Eq. (6.25) by $\boldsymbol{A}_i^T\boldsymbol{M}$ and considering the orthogonality conditions leads to

$$\boldsymbol{A}_i^T\boldsymbol{M}\boldsymbol{u}_0 = C_i\boldsymbol{A}_i^T\boldsymbol{M}\boldsymbol{A}_i = C_iM_i, i = 1, 2, \cdots, n$$

where $M_i = \boldsymbol{A}_i^T\boldsymbol{M}\boldsymbol{A}_i$ is the ith generalized mass of the system. Thus

$$C_i = \frac{\boldsymbol{A}_i^T\boldsymbol{M}\boldsymbol{u}_0}{M_i} \tag{6.29}$$

When $\sum_{i=1}^{r}\boldsymbol{A}_iC_i = \sum_{i=1}^{r}\boldsymbol{A}_i\boldsymbol{A}_i^T\boldsymbol{M}\boldsymbol{u}_0/M_i$ is subtracted from Eq. (6.25), that is, the first r mode components are removed from Eq. (6.25), the $(r+1)$th mode and frequency can be evaluated using the matrix iteration method. In this case, the assumed mode would be

$$\boldsymbol{u}_0 - \sum_{i=1}^{r}\boldsymbol{A}_iC_i = \boldsymbol{u}_0 - \sum_{i=1}^{r}\boldsymbol{A}_i\frac{\boldsymbol{A}_i^T\boldsymbol{M}\boldsymbol{u}_0}{M_i} = \left(\boldsymbol{I} - \sum_{i=1}^{r}\frac{\boldsymbol{A}_i\boldsymbol{A}_i^T\boldsymbol{M}}{M_i}\right)\boldsymbol{u}_0 = \mathbf{Q}_r\boldsymbol{u}_0 \tag{6.30}$$

where

$$\mathbf{Q}_r = \boldsymbol{I} - \sum_{i=1}^{r}\frac{\boldsymbol{A}_i\boldsymbol{A}_i^T\boldsymbol{M}}{M_i} \tag{6.31}$$

which is called the sweeping matrix. Eq. (6.30) shows that any assumed mode $\boldsymbol{u}_0$ premultiplied by the sweeping matrix $\mathbf{Q}_r$ indicates that the first r mode components are removed from $\boldsymbol{u}_0$. Therefore $\mathbf{Q}_r\boldsymbol{u}_0$ is used as the assumed mode, and the result will converge to the $(r+1)$th mode $\boldsymbol{A}_{r+1}$.

Round-off error is inevitable in the iterative process. When the $(r+1)$ th mode and frequency are evaluated, the iteration starts from the assumed mode $\mathbf{Q}_r\boldsymbol{u}_0$. However, the resulting $\boldsymbol{u}_1$ may still contain the first r mode components. Therefore the effect of the first r mode components should be treated after each cycle of iteration, that is, the approximate mode $\boldsymbol{u}_k$ should be premultiplied by the $\boldsymbol{D}_r = \boldsymbol{D}\mathbf{Q}_r$ in each iteration. When the first frequency and mode are evaluated, $\mathbf{Q}_0 = \boldsymbol{I}$ and $\boldsymbol{D}_0 = \boldsymbol{D}$. Premultiplying Eq. (6.31) by the matrix $\boldsymbol{D}$ yields

$$\boldsymbol{D}_r = \boldsymbol{D}\mathbf{Q}_r = \boldsymbol{D} - \boldsymbol{D}\sum_{i=1}^{r}\frac{\boldsymbol{A}_i\boldsymbol{A}_i^T\boldsymbol{M}}{M_i} = \boldsymbol{D} - \sum_{i=1}^{r}\frac{\lambda_i\boldsymbol{A}_i\boldsymbol{A}_i^T\boldsymbol{M}}{M_i} \tag{6.32}$$

From Eq. (6.32), the recursion formula for $\boldsymbol{D}_r$ can be obtained as follows:

$$\boldsymbol{D}_r = \boldsymbol{D}_{r-1} - \frac{\overline{\lambda}_r \boldsymbol{A}_r \boldsymbol{A}_r^T \boldsymbol{M}}{M_r} \tag{6.33}$$

which will provide the convenience for programming. To facilitate programming, the process of evaluating the $(r+1)$th frequency and mode by the matrix iteration method is summarized as follows:

1. Select the assumed mode $\boldsymbol{u}_0$.
2. Determine $\boldsymbol{D}_r$ according to Eq. (6.33). Note that $\boldsymbol{D}_0 = \boldsymbol{D}$, when $r = 0$.
3. Premultiply $\boldsymbol{u}_0$ by $\boldsymbol{D}_r$, and normalize the resulting mode vector $\boldsymbol{D}_r \boldsymbol{u}_0$ as $\boldsymbol{D}_r \boldsymbol{u}_0 = a_1 \boldsymbol{u}_1$ to get a_1 and $\boldsymbol{u}_1$.
4. Premultiply $\boldsymbol{u}_1$ by $\boldsymbol{D}_r$ continuously, and normalize $\boldsymbol{D}_r \boldsymbol{u}_1$ as $\boldsymbol{D}_r \boldsymbol{u}_1 = a_2 \boldsymbol{u}_2$ to get a_2 and $\boldsymbol{u}_2$.
5. Continue the above iteration when $|\boldsymbol{u}_2 - \boldsymbol{u}_1|/|\boldsymbol{u}_1| \geq \varepsilon$. The iteration ends until $|\boldsymbol{u}_k - \boldsymbol{u}_{k-1}|/|\boldsymbol{u}_{k-1}| < \varepsilon$. In this situation, a_k converges to the $(r+1)$th eigenvalue $\overline{\lambda}_{r+1} = 1/\omega_{r+1}^2$, and $\boldsymbol{u}_k$ converges to the $(r+1)$th mode $\boldsymbol{A}_{r+1}$. Associated symbols have been illustrated in Section 6.3.1.

Example 6.4: Evaluate the second frequency and mode of the system shown in Example 6.1.

Solution:

The first mode $\boldsymbol{A}_1 = \{1.0000 \quad 1.8019 \quad 2.2468\}^T$ has been evaluated in Example 6.3. The corresponding eigenvalue is $\overline{\lambda}_1 = 1/\omega_1^2 = 5.0467J/k$, and the first generalized mass is $M_1 = \boldsymbol{A}_1^T \boldsymbol{M} \boldsymbol{A}_1 = 9.2949J$. From Eqs. (6.31) and (6.33), one obtains

$$\boldsymbol{Q}_1 = \begin{bmatrix} 0.892 & -0.194 & -0.242 \\ -0.194 & 0.651 & -0.436 \\ -0.242 & -0.436 & 0.457 \end{bmatrix}, \boldsymbol{D}_1 = \frac{J}{k} \begin{bmatrix} 0.456 & 0.021 & 0.221 \\ 0.021 & 0.236 & -0.200 \\ 0.221 & -0.200 & 0.257 \end{bmatrix}$$

The assumed mode is selected to be $\boldsymbol{u}_0 = \{1 \quad 1 \quad -1\}^T$. After 12 cycles of iteration, $\overline{\lambda}_2 = 0.6430\,J/k$, $\omega_2^2 = 1.5552\,k/J$, $\boldsymbol{A}_2 = \{1.000 \quad 0.4452 \quad -0.8020\}^T$, and $M_2 = \boldsymbol{A}_2^T \boldsymbol{M} \boldsymbol{A}_2 = 1.8414J$ are obtained.

The two examples in this section demonstrate that the convergence rate of the iteration for higher frequencies is much slower than that for the fundamental frequency. The lower modes and frequencies must be obtained before the mode and frequency of the specified order can be evaluated. Since the first few modes and frequencies are approximate, many iterations may lead to a relatively large accumulative error, which will increase the error of the higher modes and frequencies. Generally, this method is efficient for evaluating the first (five to eight) modes and frequencies. When higher modes and frequencies are further evaluated, the convergence rate will be very slow. In order to evaluate the higher modes and frequencies more efficiently, the subspace iteration method to be introduced in the next section can be adopted.

The prominent advantage of the matrix iteration method is that the assumed modes only affect the convergence rate, rather than the accuracy of the final results. Therefore even if an error occurs during the calculation, it does not affect the final results, which is equivalent to the case that the iteration starts from a new assumed mode.

6.4 Subspace iteration method

When the matrix iteration method is used to evaluate the natural frequencies and modes of multi-DOF systems, the calculation should be conducted order by order from the lowest frequency and mode. When the Rayleigh–Ritz method is applied to solve for the first few frequencies and modes, the original n-order eigenvalue problem is transformed into a lower order one (s-order, $s \ll n$). Since the assumed modes are required to be provided with high accuracy in the Rayleigh–Ritz method, the accuracy of the results mainly depends on the degree of approximation of the assumed modes. Based on the Rayleigh–Ritz and matrix iteration methods, the subspace iteration method was developed, which has the advantages of the two methods above [4]. It is an efficient method for evaluating the first few frequencies and modes of complex structural systems.

Supposing that the system has n DOFs, its stiffness and mass matrices are $\boldsymbol{K}$ and $\boldsymbol{M}$, respectively. The eigenvector (mode vector) of the system is $\boldsymbol{A}_i$, and the corresponding eigenvalue is $\overline{\lambda}_i = 1/\omega_i^2$. The eigenvalue equation is $\boldsymbol{D}\boldsymbol{A}_i = \overline{\lambda}_i\boldsymbol{A}_i$, where $\boldsymbol{D} = \boldsymbol{K}^{-1}\boldsymbol{M}$. Assume $s(s<n)$ n-dimensional mode vectors $\boldsymbol{\psi}_{j0}$, $j = 1, 2, \cdots, s$, which are arranged to form an $n \times s$ matrix $\boldsymbol{\psi}_0 = \begin{bmatrix} \boldsymbol{\psi}_{10} & \boldsymbol{\psi}_{20} & \cdots & \boldsymbol{\psi}_{s0} \end{bmatrix}$. $\boldsymbol{\psi}_0$ is taken to be the initial approximation of the mode matrix $\boldsymbol{A}_{n\times s}$ (subscripts 0, I, II,. . ., denote the serial number of iteration cycle), letting

$$\boldsymbol{A}_0 = \boldsymbol{\psi}_0 \tag{6.34}$$

Similar to the matrix iteration method, premultiplying $\boldsymbol{\psi}_0$ by $\boldsymbol{D}$, which is equivalent to premultiplying each column vector of $\boldsymbol{\psi}_0$ by $\boldsymbol{D}$, one obtains

$$\boldsymbol{\psi}_{\mathrm{I}} = \boldsymbol{D}\boldsymbol{\psi}_{\mathbf{0}} \tag{6.35}$$

Continuously premultiplying $\boldsymbol{\psi}_{\mathrm{I}}$ by $\boldsymbol{D}$ leads to $\boldsymbol{\psi}_{\mathrm{II}}$, and the iterative process continues as such. According to the matrix iteration method, the first column vector in $\boldsymbol{\psi}_i$, $i = 0, \mathrm{I}, \mathrm{II}, \cdots$, converges to the exact fundamental mode by means of enough iterations, as long as $\boldsymbol{\psi}_{10}$ contains the component of the first mode. The second column vector in $\boldsymbol{\psi}_i$ converges to the second exact mode if $\boldsymbol{\psi}_{20}$ does not contain the component of the first mode, but contains the second one. Similarly, the jth column vector in $\boldsymbol{\psi}_i$ converges to the jth exact mode, if $\boldsymbol{\psi}_{j0}$ does not contain the components of the first $(j-1)$ modes, but contains the jth one. However, it is almost impossible to satisfy the above requirements for each column vector of $\boldsymbol{\psi}_0$ from the beginning. This goal can be achieved step by step through the iterative process. Namely, after $\boldsymbol{\psi}_{\mathrm{I}}$ is obtained, it is not used directly for the next iteration. Instead, it is treated using the Rayleigh–Ritz method before the next iteration to find the substituting matrix $\boldsymbol{A}_{\mathrm{I}}$ of $\boldsymbol{\psi}_{\mathrm{I}}$. In doing so, each column vector of the matrix $\boldsymbol{A}_{\mathrm{I}}$ becomes close to the corresponding mode. According to the proof in Section 6.2, the substituting matrix $\boldsymbol{A}_{\mathrm{I}}$ is orthogonal with respect to $\boldsymbol{K}$ and $\boldsymbol{M}$, hence, the treatment above is called the orthogonalizing process, given as follows:

According to Eq. (6.11), the first approximate mode $\boldsymbol{A}_{i\mathrm{I}}$ of the first s modes can be expressed as

$$\boldsymbol{A}_{i\mathrm{I}} = \sum_{j=1}^{s} a_{ji\mathrm{I}}\boldsymbol{\psi}_{j\mathrm{I}} = \boldsymbol{\psi}_{\mathrm{I}}\boldsymbol{a}_{i\mathrm{I}}, i = 1, 2, \cdots, s \tag{6.36}$$

where $\boldsymbol{\psi}_{\mathrm{I}}$ is calculated from Eq. (6.35), and $\boldsymbol{a}_{i\mathrm{I}} = \left\{ a_{1i\mathrm{I}} \quad a_{2i\mathrm{I}} \quad \cdots \quad a_{si\mathrm{I}} \right\}^T$ is the vector of undetermined coefficients. The numeric subscript of each element in the vector $\boldsymbol{a}_{i\mathrm{I}}$ corresponds to the column number, $1, 2, \cdots, s$, in $\boldsymbol{\psi}_{\mathrm{I}}$, and the subscript i represents the mode serial number.

Substituting Eq. (6.36) into Eq. (6.5) (expression of the first Rayleigh quotient) yields

$$R_{\mathrm{I}}(\boldsymbol{A}_{i\mathrm{I}}) = \lambda_{i\mathrm{I}} = \frac{\boldsymbol{a}_{i\mathrm{I}}^T\boldsymbol{\psi}_{\mathrm{I}}^T\boldsymbol{K}\boldsymbol{\psi}_{\mathrm{I}}\boldsymbol{a}_{i\mathrm{I}}}{\boldsymbol{a}_{i\mathrm{I}}^T\boldsymbol{\psi}_{\mathrm{I}}^T\boldsymbol{M}\boldsymbol{\psi}_{\mathrm{I}}\boldsymbol{a}_{i\mathrm{I}}}, i = 1, 2, \cdots, s \tag{6.37}$$

where $R_{\mathrm{I}}(\boldsymbol{A}_{i\mathrm{I}})$ is a function of $\boldsymbol{a}_{i\mathrm{I}}$ that can be denoted as $R_{\mathrm{I}}(\boldsymbol{a}_{i\mathrm{I}})$. To make the approximate modes given by Eq. (6.36) be closer to real modes, $\boldsymbol{a}_{i\mathrm{I}}$ should be taken properly so that $R_{\mathrm{I}}(\boldsymbol{a}_{i\mathrm{I}})$ will be stationary. Using the same operation as in the Rayleigh–Ritz method, one obtains

$$\boldsymbol{\psi}_{\mathrm{I}}^{T}\boldsymbol{K}\boldsymbol{\psi}_{\mathrm{I}}\boldsymbol{a}_{i\mathrm{I}} - \lambda_{i\mathrm{I}}\boldsymbol{\psi}_{\mathrm{I}}^{T}\boldsymbol{M}\boldsymbol{\psi}_{\mathrm{I}}\boldsymbol{a}_{i\mathrm{I}} = \boldsymbol{0}, i = 1, 2, \cdots, s \tag{6.38}$$

The corresponding approximate generalized stiffness and mass matrices are

$$\boldsymbol{K}_{\mathrm{I}}^{*} = \boldsymbol{\psi}_{\mathrm{I}}^{T}\boldsymbol{K}\boldsymbol{\psi}_{\mathrm{I}}, \boldsymbol{M}_{\mathrm{I}}^{*} = \boldsymbol{\psi}_{\mathrm{I}}^{T}\boldsymbol{M}\boldsymbol{\psi}_{\mathrm{I}} \tag{6.39}$$

where $\boldsymbol{K}_{\mathrm{I}}^{*}$ and $\boldsymbol{M}_{\mathrm{I}}^{*}$ are both $s \times s$ order symmetric matrix.

Thus Eq. (6.38) becomes

$$\left(\boldsymbol{K}_{\mathrm{I}}^{*} - \lambda_{i\mathrm{I}}\boldsymbol{M}_{\mathrm{I}}^{*}\right)\boldsymbol{a}_{i\mathrm{I}} = \boldsymbol{0}, i = 1, 2, \cdots, s \tag{6.40}$$

Solving the frequency equation $\left|\boldsymbol{K}_{\mathrm{I}}^{*} - \lambda_{i\mathrm{I}}\boldsymbol{M}_{\mathrm{I}}^{*}\right| = 0$ leads to the first approximation of $\lambda_{i\mathrm{I}}$, $i = 1, 2, \cdots, s$. Substituting $\lambda_{i\mathrm{I}}$ into Eq. (6.40) successively, the corresponding s eigenvectors $\boldsymbol{a}_{i\mathrm{I}}$, $i = 1, 2, \cdots, s$, can be solved. To simplify the calculation, $\boldsymbol{a}_{i\mathrm{I}}$ is generally normalized.

Substituting the solved $\boldsymbol{a}_{i\mathrm{I}}$, $i = 1, 2, \cdots, s$, into Eq. (6.36), the first approximation of $\boldsymbol{A}_{i\mathrm{I}}$, $i = 1, 2, \cdots, s$, can be obtained, and the substituting matrix $\boldsymbol{A}_{\mathrm{I}}$ of $\boldsymbol{\psi}_{\mathrm{I}}$ is as follows:

$$\boldsymbol{A}_{\mathrm{I}} = \begin{bmatrix} \boldsymbol{\psi}_{\mathrm{I}}\boldsymbol{a}_{1\mathrm{I}} & \boldsymbol{\psi}_{\mathrm{I}}\boldsymbol{a}_{2\mathrm{I}} & \cdots & \boldsymbol{\psi}_{\mathrm{I}}\boldsymbol{a}_{s\mathrm{I}} \end{bmatrix}_{n \times s} \tag{6.41}$$

Up to now, the first iteration has been completed. By means of the orthogonalizing process, $\boldsymbol{A}_{\mathrm{I}}$ satisfies the orthogonality conditions and is much closer to the real mode matrix than $\boldsymbol{\psi}_{\mathrm{I}}$. Therefore $\boldsymbol{A}_{\mathrm{I}}$, rather than $\boldsymbol{\psi}_{\mathrm{I}}$, is used for next iteration. To avoid large values in the calculation, it is often necessary to normalize each column vector of $\boldsymbol{A}_{\mathrm{I}}$ before replacing $\boldsymbol{\psi}_{\mathrm{I}}$ by $\boldsymbol{A}_{\mathrm{I}}$.

The second iteration can be conducted as above. Premultiplying $\boldsymbol{A}_{\mathrm{I}}$ by $\boldsymbol{D}$ gives $\boldsymbol{\psi}_{\mathrm{II}}$. The second approximations of the generalized mass matrix $\boldsymbol{M}_{\mathrm{II}}^{*} = \boldsymbol{\psi}_{\mathrm{II}}^{T}\boldsymbol{M}\boldsymbol{\psi}_{\mathrm{II}}$ and stiffness matrix $\boldsymbol{K}_{\mathrm{II}}^{*} = \boldsymbol{\psi}_{\mathrm{II}}^{T}\boldsymbol{K}\boldsymbol{\psi}_{\mathrm{II}}$ are obtained, and the second eigenvalue equations $\left(\boldsymbol{K}_{\mathrm{II}}^{*} - \lambda_{i\mathrm{II}}\boldsymbol{M}_{\mathrm{II}}^{*}\right)\boldsymbol{a}_{i\mathrm{II}} = \boldsymbol{0}$, $i = 1, 2, \cdots, s$, are also obtained. The second approximations of $\lambda_{i\mathrm{II}}$, $i = 1, 2, \cdots, s$, are solved from $\left|\boldsymbol{K}_{\mathrm{II}}^{*} - \lambda_{i\mathrm{II}}\boldsymbol{M}_{\mathrm{II}}^{*}\right| = 0$. Substituting $\lambda_{i\mathrm{II}}$ into $\left(\boldsymbol{K}_{\mathrm{II}}^{*} - \lambda_{i\mathrm{II}}\boldsymbol{M}_{\mathrm{II}}^{*}\right)\boldsymbol{a}_{i\mathrm{II}} = \boldsymbol{0}$, $\boldsymbol{a}_{iII}$, $i = 1, 2, \cdots, s$, are evaluated, with which $\boldsymbol{A}_{\mathrm{II}}$ ($\boldsymbol{A}_{i\mathrm{II}} = \boldsymbol{\psi}_{\mathrm{II}}\boldsymbol{a}_{i\mathrm{II}}$, $i = 1, 2, \cdots, s$) can be obtained. This is the end of the second iteration.

The above iterative process is repeated continuously. Theoretically, when the number of iteration cycle tends to be infinite ($N \to \infty$), $\boldsymbol{A}_{N}$

will converge to real mode matrix $\boldsymbol{A}_{n\times s}$, and λ_{iN}, $i=1,2,\cdots,s$, will converge to the first s eigenvalues corresponding to natural frequencies.

In practice, the first few modes of the system generally converge to their exact values very quickly. When s frequencies and modes are required to be evaluated, r ($r>s$) assumed modes are taken to conduct the iterative process. The iteration ends when the desired s frequencies and modes are obtained with the required accuracy. It will accelerate the iterative process for the first s modes by taking $(r-s)$ more assumed modes for iteration. However, the computational effort in each iteration step will increase due to the $(r-s)$ more modes. Therefore the advantages and disadvantages should be weighed to select a reasonable number of assumed modes. By experience, r can be taken as the smaller of $2s$ and $s+8$.

Note that the degree of approximation of the assumed mode vectors directly affects the convergence rate of the iteration. The selected s mode vectors are only required to be linearly independent, and there is no other special requirement for the selection of the assumed mode vectors. In general, the procedure used in the following example can be adopted to determine the assumed mode vectors.

Suppose that

$$\boldsymbol{K}=\begin{bmatrix} 2 & \times & \times & \times & \times \\ \times & 3 & \times & \times & \times \\ \times & \times & 2 & \times & \times \\ \times & \times & \times & 1 & \times \\ \times & \times & \times & \times & 1 \end{bmatrix},\boldsymbol{M}=\begin{bmatrix} 8 & \times & \times & \times & \times \\ \times & 9 & \times & \times & \times \\ \times & \times & 10 & \times & \times \\ \times & \times & \times & 6 & \times \\ \times & \times & \times & \times & 4 \end{bmatrix}$$

where "$\times$" denotes arbitrary value.

Dividing the diagonal elements of $\boldsymbol{M}$ by the diagonal elements of $\boldsymbol{K}$ in the same position, respectively, the following diagonal matrix can be obtained

$$\boldsymbol{B}=\text{diag}\,(4,3,5,6,4)$$

Based on matrix $\boldsymbol{B}$, the assumed mode matrix can be determined as

$$\boldsymbol{\psi}_0=\begin{bmatrix} 1 & 0 & 0 & 1 & 0 \\ 1 & 0 & 0 & 0 & 0 \\ 1 & 0 & 1 & 0 & 0 \\ 1 & 1 & 0 & 0 & 0 \\ 1 & 0 & 0 & 0 & 1 \end{bmatrix}$$

The detailed process is described below. Each element of the first column of ψ_0 is taken as unity. Only one nonzero element, taken as unity, appears in each of the other columns. The position of the nonzero element in the second column corresponds to the row serial number (4th row) of the maximum element (6) of $\boldsymbol{B}$. Similarly, the position of the nonzero element in the third column corresponds to the row serial number (3rd row) of the second maximum element (5) of $\boldsymbol{B}$. The position of the nonzero element in the fourth column corresponds to the row serial number (1st row) of the third maximum element (4) of $\boldsymbol{B}$. The position of the nonzero element in the fifth column corresponds to the row serial number (5th row) of the fourth maximum element (4) of $\boldsymbol{B}$. Note that the fourth and fifth columns of ψ_0 can be exchanged with each other due to the same element (4). By experience, the convergence rate can be increased with the assumed mode vectors obtained by the above procedure. The effect will be better if the diagonal elements are dominant in $\boldsymbol{K}$ and $\boldsymbol{M}$.

In summary, the subspace iteration method makes use of the advantage of the Rayleigh–Ritz method to reduce the order of the eigenvalue problem, and avoid the latter's disadvantage that the computational accuracy is greatly affected by the assumed mode vectors. It also applies the iterative technique to make the evaluated frequencies and modes gradually approach their exact solutions. Different from the matrix iteration method, a set of mode vectors are iterated simultaneously in the subspace iteration method. When higher frequencies and modes are determined from the matrix iteration method, the results are affected by the accumulative errors of the obtained lower frequencies and modes. Since the subspace iteration method is a combination of the Rayleigh–Ritz method and the matrix iteration method, the first few frequencies and modes can be evaluated simultaneously with required accuracy by means of repeated iterations. When the repeated natural frequencies or close natural frequencies are present for a system, the convergence rate will be very slow using the matrix iteration method. This difficulty can be overcome effectively by using the subspace iteration method [4].

In the vibration analysis of large and complex structures, the number of DOFs of the systems may be as many as tens of thousands, and the natural frequencies and modes of practical interest are often the first few frequencies and modes only. Therefore the subspace iteration method has become one of the most effective methods for the vibration analysis of large and complex structures.

Example 6.5: The first two natural frequencies and modes of the system shown in Fig. 6.2 are evaluated using the subspace iteration method. It is known that $m_1 = m_2 = m_3 = m_4 = m$, $k_1 = k_2 = k_3 = k_4 = k$, and the masses can only move in the horizontal direction.

Solution:

The matrices of the dynamic system can be determined with the method used in Example 6.1.

Mass matrix is $\boldsymbol{M} = m\boldsymbol{I}$.

Stiffness matrix is

$$\boldsymbol{K} = k\begin{bmatrix} 2 & -1 & 0 & 0 \\ -1 & 2 & -1 & 0 \\ 0 & -1 & 2 & -1 \\ 0 & 0 & -1 & 1 \end{bmatrix}, \text{and then } \boldsymbol{K}^{-1} = \frac{1}{k}\begin{bmatrix} 1 & 1 & 1 & 1 \\ 1 & 2 & 2 & 2 \\ 1 & 2 & 3 & 3 \\ 1 & 2 & 3 & 4 \end{bmatrix}.$$

Hence

$$\boldsymbol{D} = \boldsymbol{K}^{-1}\boldsymbol{M} = \frac{m}{k}\begin{bmatrix} 1 & 1 & 1 & 1 \\ 1 & 2 & 2 & 2 \\ 1 & 2 & 3 & 3 \\ 1 & 2 & 3 & 4 \end{bmatrix}$$

The initial mode matrix is assumed to be

$$\boldsymbol{A}_0 = \boldsymbol{\psi}_0 = \begin{bmatrix} 1 & 0 \\ 1 & 0 \\ 1 & 0 \\ 1 & 1 \end{bmatrix}$$

Premultiplying $\boldsymbol{\psi}_0$ by $\boldsymbol{D}$ leads to

$$\boldsymbol{D}\boldsymbol{\psi}_0 = \frac{m}{k}\begin{bmatrix} 4 & 1 \\ 7 & 2 \\ 9 & 3 \\ 10 & 4 \end{bmatrix}$$

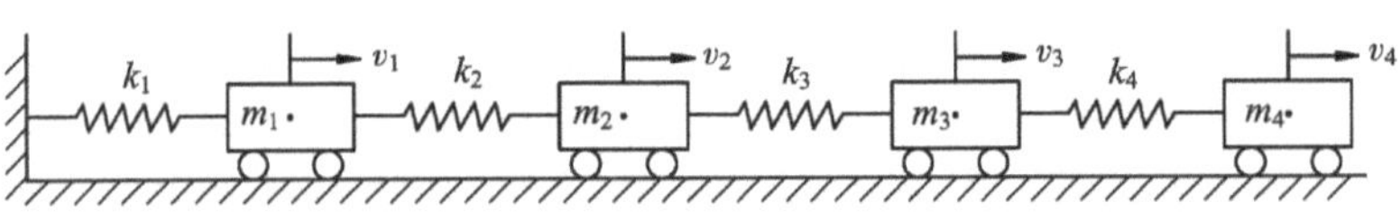

Figure 6.2 Multimass–spring system.

Normalizing $\boldsymbol{D}\boldsymbol{\psi}_0$ yields (in order to illustrate the iterative effect, the more significant figures are retained in the calculated results)

$$\boldsymbol{\psi}_{\mathrm{I}} = \begin{bmatrix} 0.4 & 0.25 \\ 0.7 & 0.5 \\ 0.9 & 0.75 \\ 1.0 & 1.0 \end{bmatrix}$$

According to Eq. (6.39), the first approximations of generalized mass matrix $\boldsymbol{M}_{\mathrm{I}}^*$ and stiffness matrix $\boldsymbol{K}_{\mathrm{I}}^*$ are calculated as

$$\boldsymbol{M}_{\mathrm{I}}^* = \boldsymbol{\psi}_{\mathrm{I}}^T \boldsymbol{M} \boldsymbol{\psi}_{\mathrm{I}} = m \begin{bmatrix} 2.460 & 2.125 \\ 2.125 & 1.875 \end{bmatrix}$$

$$\boldsymbol{K}_{\mathrm{I}}^* = \boldsymbol{\psi}_{\mathrm{I}}^T \boldsymbol{K} \boldsymbol{\psi}_{\mathrm{I}} = k \begin{bmatrix} 0.30 & 0.25 \\ 0.25 & 0.25 \end{bmatrix}$$

Substituting them into the eigenvalue equation $\left(\boldsymbol{K}_{\mathrm{I}}^* - \lambda_{i\mathrm{I}} \boldsymbol{M}_{\mathrm{I}}^*\right) \boldsymbol{a}_{i\mathrm{I}} = \boldsymbol{0}$, one obtains

$$\begin{bmatrix} 0.30 - 2.460\alpha_{i\mathrm{I}} & 0.25 - 2.125\alpha_{i\mathrm{I}} \\ 0.25 - 2.125\alpha_{i\mathrm{I}} & 0.25 - 1.875\alpha_{i\mathrm{I}} \end{bmatrix} \begin{Bmatrix} a_{1i\mathrm{I}} \\ a_{2i\mathrm{I}} \end{Bmatrix} = \begin{Bmatrix} 0 \\ 0 \end{Bmatrix}$$

where $\alpha_{i\mathrm{I}} = \lambda_{i\mathrm{I}} m/k$.

Considering that the above equations have a nontrivial solution, the determinant of the coefficients in this set of equations should vanish, and then a frequency equation can be obtained as

$$0.096875\alpha_{i\mathrm{I}}^2 - 0.115\alpha_{i\mathrm{I}} + 0.0125 = 0$$

Solving the frequency equation yields

$$\alpha_{1\mathrm{I}} = 0.121037, \alpha_{2\mathrm{I}} = 1.066060$$

Substituting $\alpha_{1\mathrm{I}}$ and $\alpha_{2\mathrm{I}}$ into the above eigenvalue equation, respectively, and solving the corresponding equations leads to

$$\boldsymbol{a}_{1\mathrm{I}} = \begin{Bmatrix} 1.000000 \\ 0.312393 \end{Bmatrix}, \boldsymbol{a}_{2\mathrm{I}} = \begin{Bmatrix} -0.867759 \\ 1.000000 \end{Bmatrix}$$

Substituting $\boldsymbol{a}_{1\mathrm{I}}$ and $\boldsymbol{a}_{2\mathrm{I}}$ into Eq. (6.36), respectively, and normalizing the resulting vectors, one obtains

$$\boldsymbol{A}_{1\mathrm{I}}=\begin{Bmatrix}0.364295\\0.652393\\0.864295\\1.000000\end{Bmatrix},\boldsymbol{A}_{2\mathrm{I}}=\begin{Bmatrix}-0.734295\\-0.812393\\-0.234295\\1.000000\end{Bmatrix}$$

Therefore the first approximation of the mode matrix is

$$\boldsymbol{A}_{\mathrm{I}}=\begin{bmatrix}0.364295 & -0.734295\\0.652393 & -0.812393\\0.864295 & -0.234295\\1.000000 & 1.000000\end{bmatrix}$$

Repeating the iteration as above, the fourth approximations of eigenvalues and the mode matrix are as follows:

$$\alpha_{1\mathrm{IV}}=0.120615,\alpha_{2\mathrm{IV}}=1.000278$$

$$\boldsymbol{A}_{\mathrm{IV}}=\begin{bmatrix}0.347298 & -0.974687\\0.652702 & -0.991800\\0.879382 & -0.016083\\1.000000 & 1.000000\end{bmatrix}$$

The fifth approximations of eigenvalues and the mode matrix are as follows:

$$\alpha_{1\mathrm{V}}=0.120615,\alpha_{2\mathrm{V}}=1.000049$$

$$\boldsymbol{A}_{\mathrm{V}}=\begin{bmatrix}0.347296 & -0.989035\\0.652704 & -0.996970\\0.879385 & -0.006579\\1.000000 & 1.000000\end{bmatrix}$$

The difference of the frequencies and the modes between the fourth and fifth iteration cycles is negligibly small, and the iterative process ends. Note that the convergence is judged by experience in this example. When computer programs are used to conduct the iterations above, a tolerance similar to that adopted in the matrix iteration method is required.

Finally, the approximations of the first two natural frequencies are obtained as

$$\omega_1=\sqrt{0.120615k/m},\omega_2=\sqrt{1.000049k/m}$$

The first two mode vectors are obtained as

$$\boldsymbol{A}_1 = \begin{Bmatrix} 0.347296 \\ 0.652704 \\ 0.879385 \\ 1.000000 \end{Bmatrix}, \boldsymbol{A}_2 = \begin{Bmatrix} -0.989035 \\ -0.996970 \\ -0.006579 \\ 1.000000 \end{Bmatrix}$$

The first two exact natural frequencies are $\omega_1 = \sqrt{0.120615k/m}$ and $\omega_2 = \sqrt{k/m}$, respectively, and the corresponding exact mode vectors are $\boldsymbol{A}_1 = \{0.347296 \quad 0.652704 \quad 0.879385 \quad 1\}^T$ and $\boldsymbol{A}_2 = \{-1 \quad -1 \quad 0 \quad 1\}^T$. Note that the calculated results converge to the corresponding exact solutions by means of five iteration cycles.

6.5 Reduction of degrees of freedom in dynamic analysis

6.5.1 Preliminary comments

An important objective of structural dynamic analysis is to evaluate the structural dynamic response to dynamic loads. In practice, a structural dynamic analysis is usually preceded by static analysis. The idealized model of the structure for static analysis is dictated by the complexity of the structure. To accurately evaluate the internal element forces and stresses in a complex structure, several hundred to a few thousand DOFs may be necessary.

The same refined idealization may be used for structural dynamic analysis, but this may be unnecessarily refined and drastically fewer DOFs could suffice. Such is the case because the displacement dynamic responses of many structures can be well represented by the first few natural modes, and these modes can be accurately evaluated from an idealized model with drastically fewer DOFs than required for static analysis. Therefore the number of DOFs can be reduced as much as reasonably possible before proceeding the evaluation of natural frequencies and modes, which is perhaps the most demanding phase of dynamic analysis. The internal element forces and stresses required for structural design can be determined from the structural static analysis at each instant of time, and additional dynamic analysis is not required.

The methods for reducing the number of DOFs of the system include the kinematic constraints method, the static condensation method (mass lumping method), and the Rayleigh–Ritz method, etc [1,2]. These three methods are introduced below.

6.5.2 Kinematic constraints method

The configuration and properties of a structure may suggest kinematic constraints which express the displacements of many DOFs in terms of a much smaller set of primary displacement variables. For example, the floor diaphragms (or slabs) of a multistory building, although flexible in the vertical direction, are usually very stiff in their own plane and can be assumed to be rigid without introducing significant error. This assumption is the kinematic constraint imposed on the structure. With these constraints, the horizontal displacements of all positions at one floor level are related to the three rigid-body DOFs of the diaphragm in its own plane (two horizontal displacement components and one rotation component about the vertical axis).

The 20-story building shown in Fig. 6.3 consists of eight frames in the x direction and four frames in the y direction. With 640 joints and six DOFs (three translations and three rotations) per joint, the system has 3840 DOFs. Assume the floor diaphragms to be rigid in their own planes, the system has only 1980 DOFs, including the vertical displacement and two rotations (in xz and yz planes) of each joint, and three rigid-body DOFs per floor.

Another kinematic constraint sometimes assumed in the building analysis is that the columns are axially rigid. This constraint should be used with discretion, because it may be reasonable only in certain special

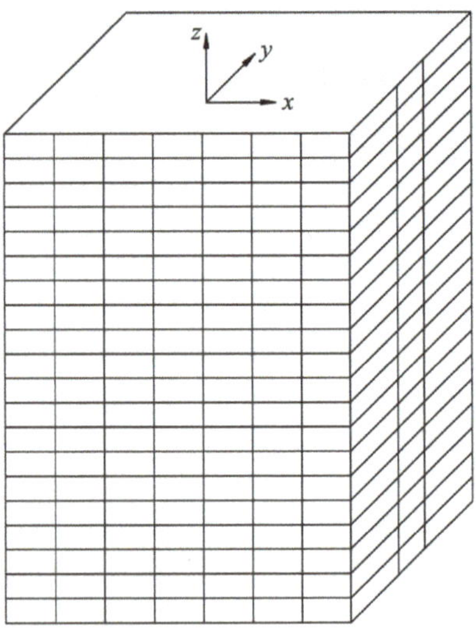

Figure 6.3 Twenty-story building.

situations, such as nonslender buildings. If justifiable, the assumption will lead to further reduction in the number of DOFs. For the static analysis of the multistory building in Fig. 6.3, the number of DOFs reduces to 1340.

Additionally, when the stiffness property of the space beam element is analyzed, some displacement interpolation functions are used to represent element displacements with node-displacement variables. It indicates that kinematic constraints are imposed to each element. Thus 12 primary displacement variables of two nodes are used to express the displacements of beam element, but the actual number of DOFs of each element is infinite.

6.5.3 Static condensation method

Some inertial forces can often be ignored in dynamic analysis. For example, the inertial forces related to the rotation and vertical displacement at joints are often ignored in antiseismic analysis for multistory buildings. Thus the primary displacement variables $\boldsymbol{q}$ representing the configuration of the structural system are divided into two categories: the displacement variables $\boldsymbol{q}_0$ in which no mass participates so that inertial forces are not developed, and the displacement variables $\boldsymbol{q}_t$ having mass that induces inertial forces. The displacement variables $\boldsymbol{q}_0$ are removed in dynamic analysis using the static condensation method. However, all displacement variables are considered in static analysis. The detailed implementation process is as follows:

The equation of motion of the undamped system can be written in partitioned matrix form

$$\begin{bmatrix} \boldsymbol{M}_{tt} & \boldsymbol{0} \\ \boldsymbol{0} & \boldsymbol{0} \end{bmatrix} \begin{Bmatrix} \ddot{\boldsymbol{q}}_t \\ \ddot{\boldsymbol{q}}_0 \end{Bmatrix} + \begin{bmatrix} \boldsymbol{K}_{tt} & \boldsymbol{K}_{t0} \\ \boldsymbol{K}_{0t} & \boldsymbol{K}_{00} \end{bmatrix} \begin{Bmatrix} \boldsymbol{q}_t \\ \boldsymbol{q}_0 \end{Bmatrix} = \begin{Bmatrix} \boldsymbol{Q}_t \\ \boldsymbol{0} \end{Bmatrix} \tag{6.42}$$

where $\boldsymbol{q}_0$ represents the displacement variables with zero mass (or condensed displacement variables), $\boldsymbol{q}_t$ represents displacement variables with mass (or dynamic displacement variables), and $\boldsymbol{K}_{t0} = \boldsymbol{K}_{0t}^T$. $\boldsymbol{q}_t$ are the independent displacement variables required to express all inertial forces (or positions of moving masses) completely, and the number of these variables is called dynamic DOF.

Eq. (6.42) can be written as two separate equations

$$\boldsymbol{M}_{tt}\ddot{\boldsymbol{q}}_t + \boldsymbol{K}_{tt}\boldsymbol{q}_t + \boldsymbol{K}_{t0}\boldsymbol{q}_0 = \boldsymbol{Q}_t \tag{6.43}$$

$$\boldsymbol{K}_{0t}\boldsymbol{q}_t + \boldsymbol{K}_{00}\boldsymbol{q}_0 = \boldsymbol{0} \tag{6.44}$$

Since no inertial terms and external forces are associated with $\boldsymbol{q}_0$, Eq. (6.44) gives the static relationship between $\boldsymbol{q}_0$ and $\boldsymbol{q}_t$

$$\boldsymbol{q}_0 = -\boldsymbol{K}_{00}^{-1}\boldsymbol{K}_{0t}\boldsymbol{q}_t \tag{6.45}$$

Substituting Eq. (6.45) into Eq. (6.43) leads to

$$\boldsymbol{M}_{tt}\ddot{\boldsymbol{q}}_t + \hat{\boldsymbol{K}}_{tt}\boldsymbol{q}_t = \boldsymbol{Q}_t \tag{6.46}$$

where $\hat{\boldsymbol{K}}_{tt}$ is the condensed stiffness matrix, given by

$$\hat{\boldsymbol{K}}_{tt} = \boldsymbol{K}_{tt} - \boldsymbol{K}_{0t}^T\boldsymbol{K}_{00}^{-1}\boldsymbol{K}_{0t} \tag{6.47}$$

$\boldsymbol{q}_0$ and $\boldsymbol{q}_t$ can be solved directly from Eq. (6.42). The dynamic displacement variables $\boldsymbol{q}_t$ can also be solved from Eq. (6.46), and the order of Eq. (6.46) is less than that of Eq. (6.42). If necessary, $\boldsymbol{q}_0$ can be obtained from Eq. (6.45). Based on the resulting $\boldsymbol{q}_0$ and $\boldsymbol{q}_t$, the internal element forces and stresses can be evaluated by static analysis at each instant of time (see Section 9.10 in Ref. [1]).

Letting the load vector be a zero vector in Eq. (6.42), the natural frequencies and modes can be evaluated. Since the diagonal elements in mass matrix corresponding to $\boldsymbol{q}_0$ are all zero, that is, the mass matrix is singular, the rank is equal to the number of dynamic DOF. Thus only the natural frequencies and modes related to $\boldsymbol{q}_t$ can be evaluated. Additionally, According to the mass and stiffness matrices in Eq. (6.46), the natural frequencies and modes mentioned above can also be obtained by solving a reduced eigenvalue problem.

However, the reduction in the actual computational effort may be much less significant than the reduction of the number of DOFs. This is because the computational efficiency permitted by the narrow banding of the stiffness matrix $\boldsymbol{K}$ in Eq. (6.48) is in part lost in using the fully populated condensed stiffness matrix $\hat{\boldsymbol{K}}_{tt}$ in Eq. (6.46).

$$\boldsymbol{M}\ddot{\boldsymbol{q}} + \boldsymbol{C}\dot{\boldsymbol{q}} + \boldsymbol{K}\boldsymbol{q} = \boldsymbol{Q} \tag{6.48}$$

The static condensation method is particularly effective for the seismic analysis of multistory buildings to horizontal ground motion because of three special features of this type of structures and excitations. First, floor diaphragms (or floor slabs) are usually assumed to be rigid in their own plane. Second, the effective seismic forces associated with rotations and vertical displacements of the joints are zero. Third, the inertial forces associated with these same displacement variables are usually not significant in lower modes that contribute dominantly to structural responses. Assigning

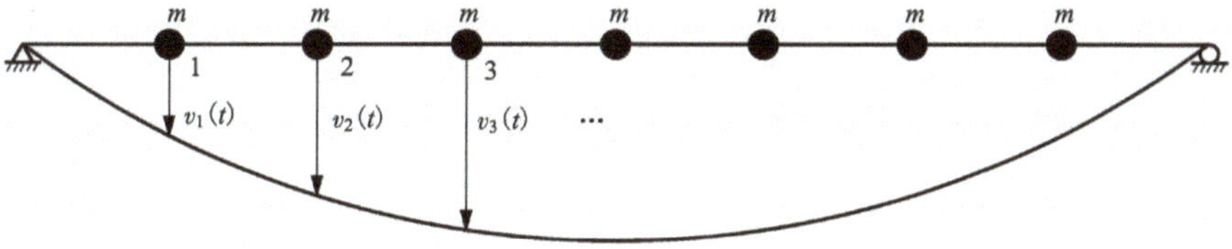

Figure 6.4 Lumped mass beam.

zero mass to these displacement variables leaves only three rigid-body DOFs of each floor diaphragm for dynamic analysis. For the 20-story building in Fig. 6.3, this method reduces the number of DOFs from 1980 to 60.

As shown in Fig. 6.4, infinite DOFs are theoretically required to represent the vibration configuration of a simply supported beam in the plane with uniformly distributed mass. The reduction of DOFs can be achieved by means of mass lumping for the dynamic analysis. The mass of the whole beam is concentrated on a number of positions, and the mass at these positions is called lumped mass. The influence of the moment of inertia of each lumped mass is ignored, that is, the rotation of each mass is not considered. Only the vertical displacements $v_i(t)$ of each mass are used as the dynamic displacement variables of the system, which are the dynamic DOFs. The condensed equation of motion can be formulated using the influence coefficient method, and the number of DOFs of plane beam for the vibration analysis is considerably reduced.

6.5.4 Rayleigh–Ritz method

Eq. (6.48) is the equations of motion of an n-DOF system. In the Rayleigh method, the structural displacements are expressed as $\boldsymbol{q} = a(t)\boldsymbol{\psi}_0$, where $\boldsymbol{\psi}_0$ is the assumed mode vector, and $a(t)$ is the generalized coordinate. With this expression, the system is simplified to a single-DOF system and the fundamental natural frequency can be approximately evaluated. In the Rayleigh–Ritz method, the displacements are expressed as a linear combination of several assumed mode vectors $\boldsymbol{\psi}_j$, $j = 1, 2, \cdots, s$, as follows:

$$\boldsymbol{q} = \sum_{j=1}^{s} a_j(t)\boldsymbol{\psi}_j = \boldsymbol{\psi}\boldsymbol{a}(t) \tag{6.49}$$

where $a_j(t)$, $j = 1, 2, \cdots, s$, are the generalized coordinates, and $\boldsymbol{\psi}_j$, $j = 1, 2, \cdots, s$, are the assumed mode vectors which must satisfy geometric boundary conditions and be linearly independent of each other. For the system to be analyzed, appropriate assumed mode vectors should be selected. All the vectors ψ_j, $j = 1, 2, \cdots, s$, are arranged to form an $n \times s$ matrix ψ, that is, $\psi = \begin{bmatrix} \psi_1 & \psi_2 & \cdots & \psi_s \end{bmatrix}$. $a(t)$ is the vector of s generalized coordinates, that is, $a(t) = \begin{Bmatrix} a_1 & a_2 & \cdots & a_s \end{Bmatrix}^T$.

Substituting Eq. (6.49) into Eq. (6.48) yields

$$M\psi\ddot{a} + C\psi\dot{a} + K\psi a = Q \tag{6.50}$$

Premultiplying each term of the above equation by ψ^T gives

$$M^*\ddot{a} + C^*\dot{a} + K^* a = Q^* \tag{6.51}$$

where $M^* = \psi^T M\psi$, $C^* = \psi^T C\psi$, $K^* = \psi^T K\psi$ and $Q^* = \psi^T Q$.

Eq. (6.51) is a set of s differential equations in s generalized coordinates a. Since the matrix ψ is generally different from the exact mode matrix, M^* and K^* are generally not diagonal matrices. When the matrix ψ is an exact mode matrix and Rayleigh damping is assumed, M^*, C^*, and K^* are all diagonal matrices, and Eq. (6.51) is a set of s independent differential equations. This treatment is essentially the mode superposition method.

In summary, the Ritz transformation of Eq. (6.49) has made it possible to reduce the original set of n equations in primary displacement variables q to a smaller set of s equations in the generalized coordinates a.

References

[1] Chopra AK. Dynamics of structures theory and applications to earthquake engineering. 4th ed. NJ: Prentice Hall; 2012.
[2] Clough RW, Penzien J. Dynamics of structures. 3rd ed. Berkeley, CA: Computers & Structures, Inc; 2003.
[3] Zhang Z, Zhou X, Jiang D. Dynamics of structures. Beijing: China Electric Power Press; 2009.
[4] Zhou Z, Wen Y, Zeng Q. Lectures on dynamics of structures. 2nd ed. Beijing: China Communications Press Co., Ltd; 2017.

Problems

6.1. Describe briefly the principle of solving the frequencies of a system by the Rayleigh energy method.

6.2. What conditions should be satisfied for the assumed modes in the Rayleigh and the Rayleigh–Ritz methods?

6.3. Are the values of frequencies obtained by using the Rayleigh or Rayleigh–Ritz method always an upper bound of the real natural frequencies?

6.4. Why are the first frequency and mode usually obtained when using the matrix iteration method to evaluate the dynamic properties of a system? What measures should be taken to evaluate higher frequencies and modes?

6.5. Describe briefly the relationship of the four methods for evaluating the dynamic properties of a system introduced in this chapter.

6.6. For the structural system given by Problem 4.6, calculate the deflection curve produced by self-weight, evaluate the fundamental frequency using the resulting deflection curve as the assumed mode, and compare the result with that obtained from Problem 4.6.

6.7. The assumed mode is $\boldsymbol{A}_1 = \left\{ \sqrt{3} \quad \sqrt{2} \quad 1 \right\}^T$. Evaluate the fundamental frequency of the 3-DOF system shown in Figure P6.1 using the Rayleigh method.

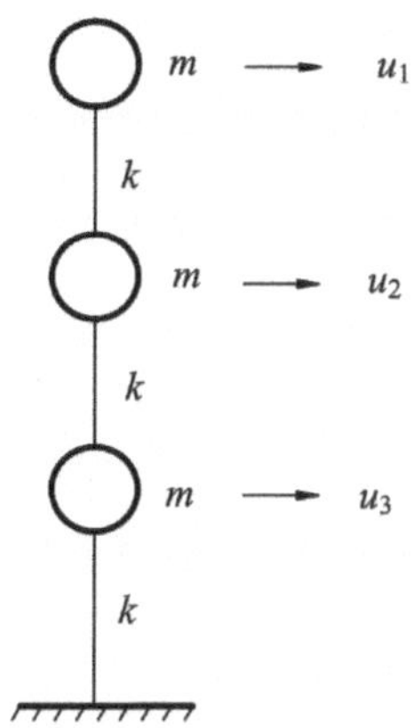

Figure P6.1 Figure of problem 6.7.

6.8. A 7-DOF mass-spring system is shown in Figure P6.2. The assumed mode vectors are $\psi_1 = \left\{ 1 \quad 2 \quad 3 \quad 4 \quad 5 \quad 6 \quad 7 \right\}^T$, and $\psi_2 = \left\{ 1 \quad 4 \quad 9 \quad 16 \quad 25 \quad 36 \quad 49 \right\}^T$. Evaluate the first two frequencies and modes of the system using the Rayleigh–Ritz method and the matrix iteration method, respectively.

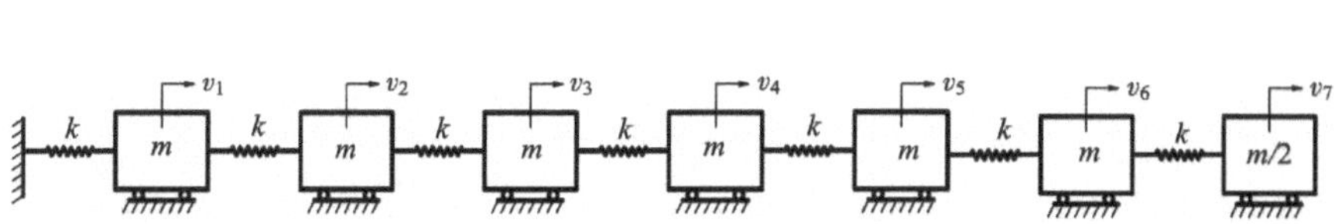

Figure P6.2 Figure of problem 6.8.

6.9. As shown in Problem 2.7, the equation of motion of the simply supported beam has been formulated in previous chapter. Evaluate the first few frequencies and modes using the subspace iteration method, compare the results with the analytical solution given by Chapter 5, Analysis of dynamic response of continuous systems: straight beam, and analyze the effect of the number of the selected elements on the calculated frequencies.

CHAPTER 7

Step-by-step integration method

The principle and application of the mode superposition method have been introduced in Chapter 4, Analysis of dynamic response of MDOF systems: mode superposition method, and Chapter 5, Analysis of dynamic response of continuous systems: straight beam. This method is easy to implement, and has a clear physical meaning. The contribution of each mode to the total response is distinct. Higher natural modes can be incorporated according to the accuracy requirement. However, it is based on the principle of superposition. Hence, it only applies to the vibration analysis of linear systems. To remove this restriction, this chapter introduces the commonly used step-by-step integration method, which is applicable to the dynamic response analysis of linear and nonlinear systems. Finally, the Wilson-θ method is used as the example to illustrate the procedure for analyzing the stability and accuracy, and how to select relevant parameters is discussed for the commonly used integration methods.

7.1 Basic idea of step-by-step integration method

Due to the presence of physical, geometric, or damping nonlinearities, many systems are often subjected to nonlinear vibrations. For example, during strong earthquakes, buildings usually exhibit elastic–plastic behavior and hence undergo nonlinear vibrations since they may be severely damaged. The motion of a nonlinear system is governed by nonlinear differential equations, which cannot produce an analytical solution. The successive approximation method and the Ritz averaging method are generally used to obtain approximate solutions. When using the former method, a good approximation can be obtained only if a small time interval is adopted. Nonetheless, after a certain period of time, the solution is often divergent. Although the latter method can produce a better solution than the former, it is necessary to presume an approximate series. For complex nonlinear systems, this approximate series is often difficult to obtain. Therefore it is urgent to develop numerical approaches which are applicable to the dynamic response analysis of nonlinear systems.

Fundamentals of Structural Dynamics
DOI: https://doi.org/10.1016/B978-0-12-823704-5.00007-0

For nonlinear dynamic analysis, the most effective method of solving the nonlinear equations of motion is the step-by-step integration method. The basic principle involves dividing the complete time history into many small time periods Δt, as shown in Fig. 7.1, which is called the time interval or time step. For the convenience of calculation, Δt is usually kept fixed throughout the entire time history. When the system properties vary drastically during a certain time period, for example, a plastic hinge appears in a specific cross section of a frame, the selected time interval Δt may be subdivided into much smaller time intervals $\Delta t'$ for required accuracy. The dynamic properties of the system (mass, stiffness, and damping) are assumed to be invariant during each time interval. The system properties, however, may be changeable in various time intervals. In this case, the system properties at the midpoint of the time interval should be selected as the representative properties during this time interval. In this regard, iterative calculations must be performed during each time interval. To simplify the calculation, the system properties at the starting point of each time interval (Fig. 7.1) are generally taken as the representative properties of the system during this time interval. Thus the equations of motion of the system are piecewise linear differential equations with constant coefficients. The system response at the starting point, say t_i, represents the initial conditions of the system during this time interval, which is used to solve for the response at the end point t_{i+1} (Fig. 7.1). Therefore the system responses from time t_i to t_{i+1} can be evaluated. From the beginning of loading, the system responses are successively evaluated during each time interval to obtain the entire time-history responses. The nonlinear vibration of the system is analyzed by solving a series of linear vibration problems. The step-by-step integration method is obviously applicable to the vibration analysis of linear systems. In this case, the system properties in each time interval keep invariant, and the calculation process is considerably simplified.

The integration methods of step-by-step dynamic response analysis make use of the integration to step forward from the initial conditions to

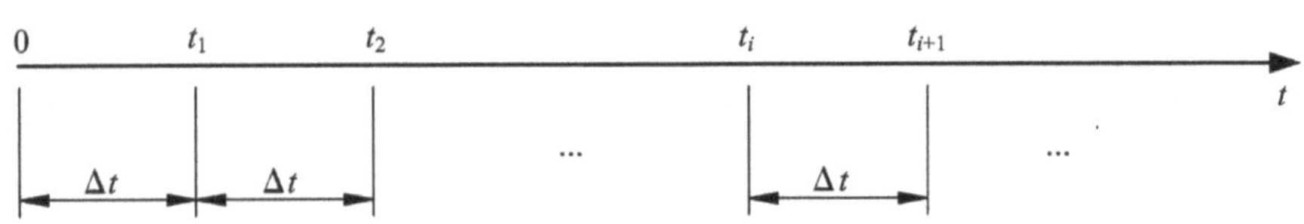

Figure 7.1 Time interval of the step-by-step integration method.

the final responses for each time interval. The essential concept is represented by the following equations [1]

$$\dot{q}_{t+\Delta t} = \dot{q}_t + \int_t^{t+\Delta t} \ddot{q}(\tau)d\tau \tag{7.1}$$

$$q_{t+\Delta t} = q_t + \int_t^{t+\Delta t} \dot{q}(\tau)d\tau \tag{7.2}$$

where the subscripts t and $t+\Delta t$ denote the instants of time of the corresponding responses. Eqs. (7.1) and (7.2) express the final velocity and displacement in terms of the initial values of these quantities plus an integral expression, respectively. The change of velocity depends on the integral of the acceleration history, and the change of displacement depends on the corresponding velocity integral. In order to carry out this type of analysis, it is necessary to first assume how the acceleration varies during a time interval, and this assumption controls the variation of the velocity as well as the displacement. Thus the final responses at time $t+\Delta t$ can be obtained by considering the initial conditions at time t, as well as the acceleration assumption and the dynamic equilibrium relationship at time $t+\Delta t$.

A variety of step-by-step integration methods has been developed. In this chapter, three methods will be introduced: the linear acceleration method, the Wilson-θ method, and the Newmark method.

7.2 Linear acceleration method

The linear acceleration method adopts the following two assumptions: (1) each entry of the acceleration vector $\ddot{\mathbf{q}}$, which is denoted as $\ddot{q}$, varies linearly during each time interval Δt, as shown in Fig. 7.2A; (2) the system properties do not vary within each time interval.

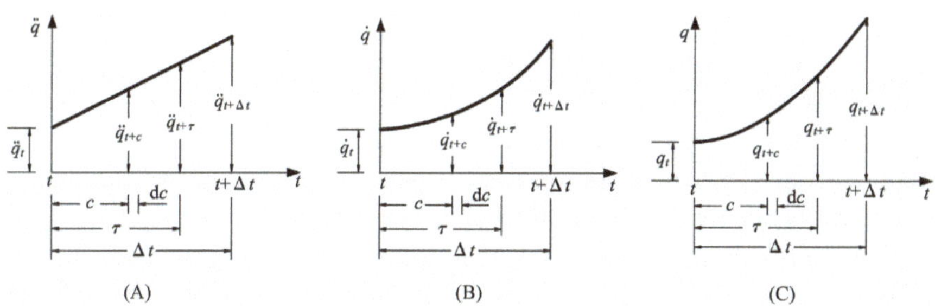

Figure 7.2 Motion based on linearly varying acceleration: (A) acceleration; (B) velocity; (C) displacement.

According to the above assumption, the acceleration, velocity, and displacement at time $t+\tau$, $0 \le \tau \le \Delta t$, are obtained, respectively, as

$$\ddot{q}_{t+\tau} = \ddot{q}_t + \frac{\ddot{q}_{t+\Delta t} - \ddot{q}_t}{\Delta t}\tau \tag{7.3}$$

$$\dot{q}_{t+\tau} = \dot{q}_t + \int_0^\tau \ddot{q}_{t+c} dc = \dot{q}_t + \int_0^\tau \left(\ddot{q}_t + \frac{\ddot{q}_{t+\Delta t} - \ddot{q}_t}{\Delta t} c\right) dc = \dot{q}_t + \tau\ddot{q}_t + \frac{\tau^2}{2\Delta t}(\ddot{q}_{t+\Delta t} - \ddot{q}_t) \tag{7.4}$$

$$\begin{aligned} q_{t+\tau} &= q_t + \int_0^\tau \dot{q}_{t+c} dc = q_t + \int_0^\tau \left(\dot{q}_t + \ddot{q}_t c + \frac{\ddot{q}_{t+\Delta t} - \ddot{q}_t}{\Delta t}\frac{c^2}{2}\right) dc \\ &= q_t + \tau\dot{q}_t + \frac{\tau^2}{2}\ddot{q}_t + \frac{\tau^3}{6\Delta t}(\ddot{q}_{t+\Delta t} - \ddot{q}_t) \end{aligned} \tag{7.5}$$

which are shown in Fig. 7.2A, B, and C, respectively.

Evaluating Eqs. (7.4) and (7.5) at $\tau = \Delta t$ gives the velocity and displacement at the time $t+\Delta t$

$$\dot{q}_{t+\Delta t} = \dot{q}_t + \Delta t\ddot{q}_t + \frac{\Delta t}{2}(\ddot{q}_{t+\Delta t} - \ddot{q}_t) = \dot{q}_t + \frac{\Delta t}{2}(\ddot{q}_{t+\Delta t} + \ddot{q}_t) \tag{7.6}$$

$$q_{t+\Delta t} = q_t + \Delta t\dot{q}_t + \frac{\Delta t^2}{2}\ddot{q}_t + \frac{\Delta t^2}{6}(\ddot{q}_{t+\Delta t} - \ddot{q}_t) = q_t + \Delta t\dot{q}_t + \frac{\Delta t^2}{3}\ddot{q}_t + \frac{\Delta t^2}{6}\ddot{q}_{t+\Delta t} \tag{7.7}$$

From Eq. (7.7), one obtains

$$\ddot{q}_{t+\Delta t} = b_0 q_{t+\Delta t} - b_0 q_t - b_2\dot{q}_t - 2\ddot{q}_t \tag{7.8}$$

where $b_0 = 6/\Delta t^2$, $b_2 = 6/\Delta t$.

Substituting Eq. (7.8) into Eq. (7.6) leads to

$$\dot{q}_{t+\Delta t} = b_1 q_{t+\Delta t} - b_1 q_t - 2\dot{q}_t - b_3\ddot{q}_t \tag{7.9}$$

where $b_1 = 3/\Delta t$, $b_3 = \Delta t/2$.

Note that the quantities associated with acceleration, velocity, and displacement in the above expressions should be rewritten in the vector form for multidegree-of-freedom (MDOF) systems.

For the instant of time $t+\Delta t$, the equation of motion of MDOF systems is

$$\boldsymbol{M}\ddot{\boldsymbol{q}}_{t+\Delta t} + \boldsymbol{C}\dot{\boldsymbol{q}}_{t+\Delta t} + \boldsymbol{K}\boldsymbol{q}_{t+\Delta t} = \boldsymbol{Q}_{t+\Delta t} \tag{7.10}$$

Substituting Eqs. (7.8) and (7.9) into Eq. (7.10) in the vector form leads to

$$\boldsymbol{M}\left(b_0\boldsymbol{q}_{t+\Delta t} - b_0\boldsymbol{q}_t - b_2\dot{\boldsymbol{q}}_t - 2\ddot{\boldsymbol{q}}_t\right) + \boldsymbol{C}\left(b_1\boldsymbol{q}_{t+\Delta t} - b_1\boldsymbol{q}_t - 2\dot{\boldsymbol{q}}_t - b_3\ddot{\boldsymbol{q}}_t\right) + \boldsymbol{K}\boldsymbol{q}_{t+\Delta t} = \boldsymbol{Q}_{t+\Delta t}$$

After some rearrangement, one obtains

$$\overline{\boldsymbol{K}}\boldsymbol{q}_{t+\Delta t} = \overline{\boldsymbol{Q}}_{t+\Delta t} \tag{7.11}$$

where

$$\overline{\boldsymbol{K}} = \boldsymbol{K} + b_0\boldsymbol{M} + b_1\boldsymbol{C} \tag{7.12}$$

$$\overline{\boldsymbol{Q}}_{t+\Delta t} = \boldsymbol{Q}_{t+\Delta t} + \boldsymbol{M}\left(b_0\boldsymbol{q}_t + b_2\dot{\boldsymbol{q}}_t + 2\ddot{\boldsymbol{q}}_t\right) + \boldsymbol{C}\left(b_1\boldsymbol{q}_t + 2\dot{\boldsymbol{q}}_t + b_3\ddot{\boldsymbol{q}}_t\right) \tag{7.13}$$

where $\overline{\boldsymbol{K}}$ is called the effective stiffness matrix of the system at time $t + \Delta t$ and $\overline{\boldsymbol{Q}}_{t+\Delta t}$ is called the effective load vector of the system at time $t + \Delta t$.

Finally, Eq. (7.11) is solved for $\boldsymbol{q}_{t+\Delta t}$. $\ddot{\boldsymbol{q}}_{t+\Delta t}$ and $\dot{\boldsymbol{q}}_{t+\Delta t}$ can subsequently be obtained by substituting $\boldsymbol{q}_{t+\Delta t}$ into Eqs. (7.8) and (7.9), respectively.

The above calculation process indicates that $\boldsymbol{q}_{t+\Delta t}$, $\dot{\boldsymbol{q}}_{t+\Delta t}$, and $\ddot{\boldsymbol{q}}_{t+\Delta t}$ can be calculated on the basis of $\boldsymbol{q}_t$, $\dot{\boldsymbol{q}}_t$, and $\ddot{\boldsymbol{q}}_t$. Generally, the initial displacements $\boldsymbol{q}_0$ and velocities $\dot{\boldsymbol{q}}_0$ are known, and the initial accelerations $\ddot{\boldsymbol{q}}_0$ can be determined from the equation of motion $\boldsymbol{M}\ddot{\boldsymbol{q}}_0 + \boldsymbol{C}\dot{\boldsymbol{q}}_0 + \boldsymbol{K}\boldsymbol{q}_0 = \boldsymbol{Q}_0$ at initial time. Therefore system responses can be evaluated in a step-by-step manner.

For the convenience of programming, the time-stepping solution using the linear acceleration method is summarized as follows:

1. Initial calculations.
   - **a.** Input the matrices $\boldsymbol{M}$, $\boldsymbol{C}$, and $\boldsymbol{K}$.
   - **b.** Evaluate initial accelerations $\ddot{\boldsymbol{q}}_0$ from $\boldsymbol{M}\ddot{\boldsymbol{q}}_0 + \boldsymbol{C}\dot{\boldsymbol{q}}_0 + \boldsymbol{K}\boldsymbol{q}_0 = \boldsymbol{Q}_0$.
   - **c.** Select appropriate time step Δt.
   - **d.** Calculate associated coefficients $b_0 = 6/\Delta t^2$, $b_1 = 3/\Delta t$, $b_2 = 6/\Delta t$, $b_3 = \Delta t/2$.
   - **e.** Calculate the effective stiffness matrix $\overline{\boldsymbol{K}} = \boldsymbol{K} + b_0\boldsymbol{M} + b_1\boldsymbol{C}$.

2. Calculations for each time step, $t = \Delta t, 2\Delta t, \cdots$
   - **a.** Calculate the effective load vector

$$\overline{\boldsymbol{Q}}_{t+\Delta t} = \boldsymbol{Q}_{t+\Delta t} + \boldsymbol{M}\left(b_0\boldsymbol{q}_t + b_2\dot{\boldsymbol{q}}_t + 2\ddot{\boldsymbol{q}}_t\right) + \boldsymbol{C}\left(b_1\boldsymbol{q}_t + 2\dot{\boldsymbol{q}}_t + b_3\ddot{\boldsymbol{q}}_t\right)$$

   - **b.** Solve Eq. (7.11) for $\boldsymbol{q}_{t+\Delta t}$ at time $t + \Delta t$.

c. Calculate the accelerations and velocities at time $t + \Delta t$:

$$\ddot{\boldsymbol{q}}_{t+\Delta t} = b_0 \boldsymbol{q}_{t+\Delta t} - b_0 \boldsymbol{q}_t - b_2 \dot{\boldsymbol{q}}_t - 2\ddot{\boldsymbol{q}}_t$$

$$\dot{\boldsymbol{q}}_{t+\Delta t} = b_1 \boldsymbol{q}_{t+\Delta t} - b_1 \boldsymbol{q}_t - 2\dot{\boldsymbol{q}}_t - b_3 \ddot{\boldsymbol{q}}_t$$

3. Repetition for the next time step. Replace t by $t + \Delta t$ and implement step 2 for the next time step.

It should be noted that the matrices $\boldsymbol{M}$, $\boldsymbol{C}$, and $\boldsymbol{K}$ are not time-varying for the linear systems, and they are time-varying and need to be modified at each time step for the nonlinear systems.

The selection of the time step is associated with the computational efficiency, accuracy and stability of the solution. If Δt is relatively large, the computational efficiency will be improved, but the requirements of accuracy and stability of the solution may not be satisfied. The stability of the solution can be classified as conditionally stable and unconditionally stable. For any selected ratio of time step to period, $\Delta t/T$, where T is a natural period of the system, if the obtained solution is bounded, the adopted algorithm is unconditionally stable. If the solution is bounded only when $\Delta t/T$ is less than a certain limit, the algorithm is conditionally stable, as shown in Fig. 7.3. The linear acceleration method is stable if $\Delta t \leq T_{\min}/1.8$, where $T_{\min}$ is the shortest natural period corresponding to the highest mode of the system.

The accuracy of the step-by-step integration method also depends on the value of the time step. Three factors need to be considered when selecting Δt: (1) the rate of change of the dynamic load; (2) the

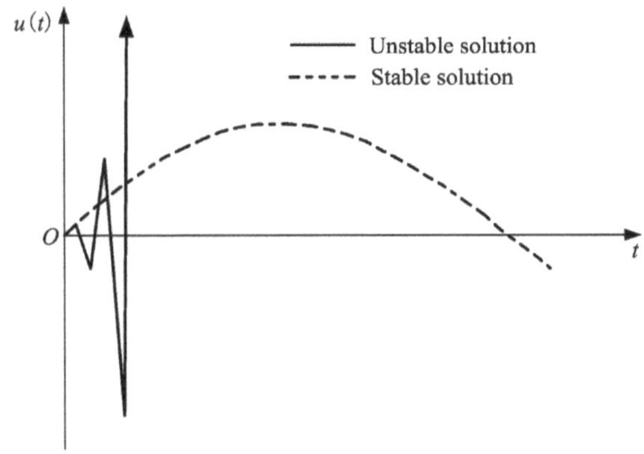

Figure 7.3 Stability of the solution.

complexity of the nonlinear damping and stiffness properties, and (3) the natural periods T_i, $i = 1, 2, \cdots, n$. For a better consideration of these factors, the time step Δt must be short enough. In general, the changes in damping and stiffness properties are not critical. If a major change occurs, for example, a plastic hinge appears in a frame, a much smaller time interval $\Delta t'$ may be used for required accuracy. Additionally, it is not difficult to estimate the time step to represent dynamic loads properly. Therefore the natural periods T_i, $i = 1, 2, \cdots, n$, are the main factor to be considered when selecting Δt. The accuracy of the linear acceleration method is related to the value of Δt. The larger the Δt, the larger the error will be, which is expressed in terms of the period elongation (*PE*) and the amplitude decay (*AD*) shown in Fig. 7.4.

In many cases, the system responses are primarily contributed by the lower mode components corresponding to longer natural periods, hence a very short time step is not necessary for required accuracy. However, the linear acceleration method is only conditionally stable. If it is employed to analyze the dynamic response of the system of which a certain natural period is less than $1.8\Delta t$, the calculated response will be divergent. In this situation, regardless of the contribution of the higher modes to the dynamic response, the response associated with higher mode components will increase continuously, which will make the obtained response meaningless. Therefore the time step Δt of the linear acceleration method must be much smaller than the shortest period of the system.

For certain types of MDOF structures, such as multistory buildings which can be idealized to have only one DOF per layer, this restriction on the size of time step is insignificant. In the seismic analysis of such structures, the time step must be quite small to properly represent the

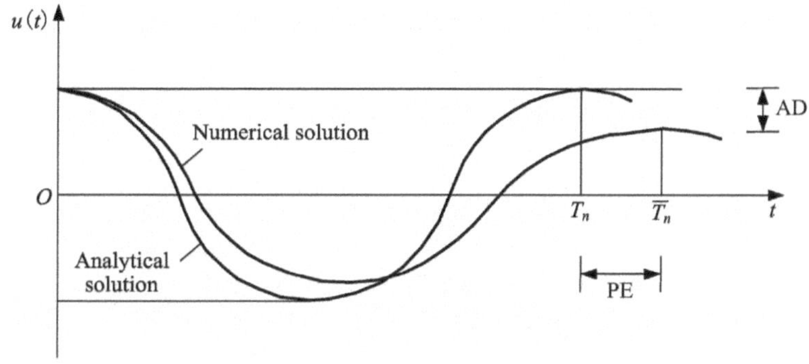

Figure 7.4 Period elongation and amplitude decay.

ground motion. However, the shortest natural period of the idealized model is usually much larger than the time step. Therefore the linear acceleration method was proven to be effective for the linear and nonlinear seismic analysis of frame buildings. For the finite element models of the structures with complex geometry, the shortest natural period may be extremely small compared with the periods that play a major role in structural responses. In this case, in order to ensure the solution stable, the time step must be very short, which often makes the linear acceleration method inapplicable. Therefore a variety of unconditionally-stable methods have been developed for the structural dynamic analysis. The Wilson-θ and Newmark methods, which are used widely, are introduced in the following two sections.

7.3 Wilson-θ method

The Wilson-θ method is the extension of the linear acceleration method. It is assumed in this method that each entry of the acceleration vector $\ddot{\mathbf{q}}$, which is denoted as $\ddot{q}$, varies linearly during the extended time interval $\theta\Delta t$, shown in Fig. 7.5 (when $\theta > 1.37$, the Wilson-θ method is unconditionally stable), that is,

$$\ddot{q}_{t+\tau} = \ddot{q}_t + \frac{\tau}{\theta\Delta t}(\ddot{q}_{t+\theta\Delta t} - \ddot{q}_t) \tag{7.14}$$

where $0 \leq \tau \leq \theta\Delta t$. When $\theta = 1$, Eq. (7.14) becomes Eq. (7.3), that is, the Wilson-θ method reduces to the linear acceleration method.

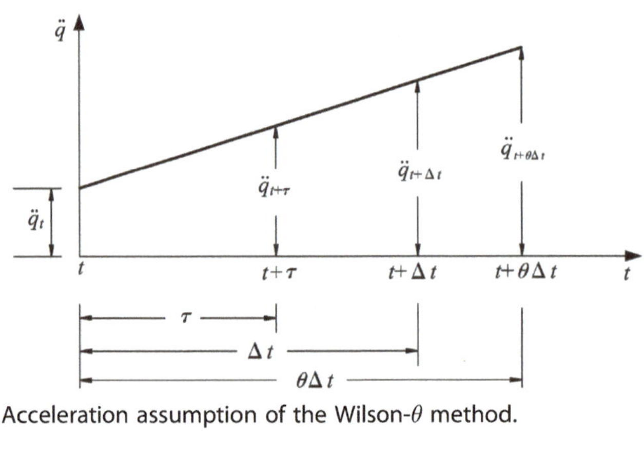

Figure 7.5 Acceleration assumption of the Wilson-θ method.

By means of similar integral in Section 7.2, one obtains

$$\dot{q}_{t+\tau} = \dot{q}_t + \ddot{q}_t\tau + \frac{\tau^2}{2\theta\Delta t}(\ddot{q}_{t+\theta\Delta t} - \ddot{q}_t) \tag{7.15}$$

$$q_{t+\tau} = q_t + \dot{q}_t\tau + \frac{\tau^2}{2}\ddot{q}_t + \frac{\tau^3}{6\theta\Delta t}(\ddot{q}_{t+\theta\Delta t} - \ddot{q}_t) \tag{7.16}$$

Evaluating Eqs. (7.15) and (7.16) at $\tau = \theta\Delta t$ gives the velocity and displacement at the time $t + \theta\Delta t$

$$\dot{q}_{t+\theta\Delta t} = \dot{q}_t + \frac{\theta\Delta t}{2}(\ddot{q}_{t+\theta\Delta t} + \ddot{q}_t) \tag{7.17}$$

$$q_{t+\theta\Delta t} = q_t + \theta\Delta t\dot{q}_t + \frac{(\theta\Delta t)^2}{6}(\ddot{q}_{t+\theta\Delta t} + 2\ddot{q}_t) \tag{7.18}$$

From Eq. (7.18), one obtains

$$\ddot{q}_{t+\theta\Delta t} = b_0(q_{t+\theta\Delta t} - q_t) - b_2\dot{q}_t - 2\ddot{q}_t \tag{7.19}$$

where $b_0 = 6/(\theta\Delta t)^2$, $b_2 = 6/(\theta\Delta t)$.

Substituting Eq. (7.19) into Eq. (7.17) leads to

$$\dot{q}_{t+\theta\Delta t} = b_1(q_{t+\theta\Delta t} - q_t) - 2\dot{q}_t - b_3\ddot{q}_t \tag{7.20}$$

where $b_1 = 3/(\theta\Delta t)$, $b_3 = \theta\Delta t/2$.

At the instant of time $t + \theta\Delta t$, the equation of motion of MDOF systems is

$$\boldsymbol{M}\ddot{\boldsymbol{q}}_{t+\theta\Delta t} + \boldsymbol{C}\dot{\boldsymbol{q}}_{t+\theta\Delta t} + \boldsymbol{K}\boldsymbol{q}_{t+\theta\Delta t} = \mathbf{Q}_{t+\theta\Delta t} \tag{7.21}$$

Substituting Eqs. (7.19) and (7.20) into Eq. (7.21) in the vector form, and rearranging the equation leads to

$$\overline{\boldsymbol{K}}\boldsymbol{q}_{t+\theta\Delta t} = \overline{\mathbf{Q}}_{t+\theta\Delta t} \tag{7.22}$$

where

$$\overline{\boldsymbol{K}} = \boldsymbol{K} + b_0\boldsymbol{M} + b_1\boldsymbol{C} \tag{7.23}$$

$$\overline{\mathbf{Q}}_{t+\theta\Delta t} = \mathbf{Q}_{t+\theta\Delta t} + \boldsymbol{M}\left(b_0\boldsymbol{q}_t + b_2\dot{\boldsymbol{q}}_t + 2\ddot{\boldsymbol{q}}_t\right) + \boldsymbol{C}\left(b_1\boldsymbol{q}_t + 2\dot{\boldsymbol{q}}_t + b_3\ddot{\boldsymbol{q}}_t\right) \tag{7.24}$$

$\boldsymbol{q}_{t+\theta\Delta t}$ can be obtained by solving Eq. (7.22). Letting $\tau = \Delta t$ in Eq. (7.14) leads to $\ddot{q}_{t+\theta\Delta t} = \theta\ddot{q}_{t+\Delta t} + (1-\theta)\ddot{q}_t$. Substituting this equation into Eq. (7.19), one obtains the acceleration at $t + \Delta t$ given by Eq. (7.25).

Substituting this equation into Eqs. (7.15) and (7.16) and letting $\tau = \Delta t$, one obtains the velocity and displacement at $t + \Delta t$, respectively, which are given by Eqs. (7.26) and (7.27). Note that the response quantities in the following three equations are expressed in the vector form for the dynamic analysis of MDOF systems.

$$\ddot{\boldsymbol{q}}_{t+\Delta t} = b_4\left(\boldsymbol{q}_{t+\theta\Delta t} - \boldsymbol{q}_t\right) + b_5\dot{\boldsymbol{q}}_t + b_6\ddot{\boldsymbol{q}}_t \tag{7.25}$$

$$\dot{\boldsymbol{q}}_{t+\Delta t} = \dot{\boldsymbol{q}}_t + b_7\left(\ddot{\boldsymbol{q}}_{t+\Delta t} + \ddot{\boldsymbol{q}}_t\right) \tag{7.26}$$

$$\boldsymbol{q}_{t+\Delta t} = \boldsymbol{q}_t + 2b_7\dot{\boldsymbol{q}}_t + b_8\left(\ddot{\boldsymbol{q}}_{t+\Delta t} + 2\ddot{\boldsymbol{q}}_t\right) \tag{7.27}$$

where $b_4 = 6/(\theta^3\Delta t^2)$, $b_5 = -6/(\theta^2\Delta t)$, $b_6 = 1 - 3/\theta$, $b_7 = \Delta t/2$, and $b_8 = \Delta t^2/6$.

Substituting the obtained $\boldsymbol{q}_{t+\theta\Delta t}$ into Eqs. (7.25)–(7.27) leads to $\ddot{\boldsymbol{q}}_{t+\Delta t}$, $\dot{\boldsymbol{q}}_{t+\Delta t}$, and $\boldsymbol{q}_{t+\Delta t}$, respectively, which are also the initial conditions for the next time step. By repeating the calculation above, the response history of the system can be obtained in a step-by-step manner. The time-stepping solution using the Wilson-θ method is summarized as follows:

1. Initial calculations.
   - **a.** Input the matrices $\boldsymbol{M}$, $\boldsymbol{C}$, and $\boldsymbol{K}$.
   - **b.** Evaluate initial accelerations $\ddot{\boldsymbol{q}}_0$ from $\boldsymbol{M}\ddot{\boldsymbol{q}}_0 + \boldsymbol{C}\dot{\boldsymbol{q}}_0 + \boldsymbol{K}\boldsymbol{q}_0 = \boldsymbol{Q}_0$.
   - **c.** Select appropriate time step Δt and parameter θ (usually $\theta = 1.4$).
   - **d.** Calculate associated coefficients $b_0 = 6/(\theta\Delta t)^2$, $b_1 = 3/(\theta\Delta t)$, $b_2 = 6/(\theta\Delta t)$, $b_3 = \theta\Delta t/2$, $b_4 = 6/(\theta^3\Delta t^2)$, $b_5 = -6/(\theta^2\Delta t)$, $b_6 = 1 - 3/\theta$, $b_7 = \Delta t/2$, $b_8 = \Delta t^2/6$.
   - **e.** Calculate the effective stiffness matrix $\overline{\boldsymbol{K}} = \boldsymbol{K} + b_0\boldsymbol{M} + b_1\boldsymbol{C}$.

2. Calculations for each time step, $t = \Delta t, 2\Delta t, \cdots$
   - **a.** Calculate the effective load vector

$$\overline{\boldsymbol{Q}}_{t+\theta\Delta t} = \boldsymbol{Q}_{t+\theta\Delta t} + \boldsymbol{M}\left(b_0\boldsymbol{q}_t + b_2\dot{\boldsymbol{q}}_t + 2\ddot{\boldsymbol{q}}_t\right) + \boldsymbol{C}\left(b_1\boldsymbol{q}_t + 2\dot{\boldsymbol{q}}_t + b_3\ddot{\boldsymbol{q}}_t\right)$$

   - **b.** Solve Eq. (7.22) for $\boldsymbol{q}_{t+\theta\Delta t}$ at time $t + \theta\Delta t$.
   - **c.** Calculate the accelerations, velocities, and displacements at time $t + \Delta t$

$$\ddot{\boldsymbol{q}}_{t+\Delta t} = b_4\left(\boldsymbol{q}_{t+\theta\Delta t} - \boldsymbol{q}_t\right) + b_5\dot{\boldsymbol{q}}_t + b_6\ddot{\boldsymbol{q}}_t$$

$$\dot{\boldsymbol{q}}_{t+\Delta t} = \dot{\boldsymbol{q}}_t + b_7\left(\ddot{\boldsymbol{q}}_{t+\Delta t} + \ddot{\boldsymbol{q}}_t\right)$$

$$\boldsymbol{q}_{t+\Delta t} = \boldsymbol{q}_t + 2b_7\dot{\boldsymbol{q}}_t + b_8\left(\ddot{\boldsymbol{q}}_{t+\Delta t} + 2\ddot{\boldsymbol{q}}_t\right)$$

3. Repetition for the next time step. Replace t by $t + \Delta t$ and implement step 2 for the next time step.

In many cases, $\mathbf{Q}(t)$ is known at discrete instants of time, $t = 0, \Delta t, 2\Delta t, \cdots$, and $\mathbf{Q}_{t+\theta\Delta t}$ can be calculated approximately as follows:

$$\mathbf{Q}_{t+\theta\Delta t} = \mathbf{Q}_t + \theta(\mathbf{Q}_{t+\Delta t} - \mathbf{Q}_t) \tag{7.28}$$

The Wilson-θ method is unconditionally stable if $\theta > 1.37$ (usually $\theta = 1.4$). In order to consider the contribution of the higher mode components of interest to the total response, Δt should be smaller than $1/10$ of the natural period of interest.

7.4 Newmark method

Based on the linear acceleration method, two parameters are introduced in the Newmark method to express the displacements and velocities at time $t + \Delta t$ in terms of the known $\boldsymbol{q}_t$, $\dot{\boldsymbol{q}}_t$, and $\ddot{\boldsymbol{q}}_t$ at time t, that is,

$$\dot{\boldsymbol{q}}_{t+\Delta t} = \dot{\boldsymbol{q}}_t + (1 - \delta)\Delta t \ddot{\boldsymbol{q}}_t + \delta \Delta t \ddot{\boldsymbol{q}}_{t+\Delta t} \tag{7.29}$$

$$\boldsymbol{q}_{t+\Delta t} = \boldsymbol{q}_t + \dot{\boldsymbol{q}}_t \Delta t + \left(\frac{1}{2} - \alpha\right)\Delta t^2 \ddot{\boldsymbol{q}}_t + \alpha \Delta t^2 \ddot{\boldsymbol{q}}_{t+\Delta t} \tag{7.30}$$

The parameters δ and α define the variation of acceleration over a time step and determine the stability and accuracy characteristics of the method. Typical selection of $\delta = 1/2$ and $1/6 \leq \alpha \leq 1/4$ is satisfactory from all points of view, including that of accuracy. Two special cases of the Newmark method that are commonly used are (1) $\delta = 1/2$ and $\alpha = 1/6$, which gives the linear acceleration method; and (2) $\delta = 1/2$ and $\alpha = 1/4$, corresponding to the constant average acceleration method.

From Eq. (7.30), one obtains

$$\ddot{\boldsymbol{q}}_{t+\Delta t} = b_0(\boldsymbol{q}_{t+\Delta t} - \boldsymbol{q}_t) - b_2\dot{\boldsymbol{q}}_t - b_3\ddot{\boldsymbol{q}}_t \tag{7.31}$$

where, $b_0 = 1/\left(\alpha \Delta t^2\right)$, $b_2 = 1/(\alpha \Delta t)$, $b_3 = 1/(2\alpha) - 1$.

Substituting Eq. (7.31) into Eq. (7.29) leads to

$$\dot{\boldsymbol{q}}_{t+\Delta t} = b_1(\boldsymbol{q}_{t+\Delta t} - \boldsymbol{q}_t) - b_4\dot{\boldsymbol{q}}_t - b_5\ddot{\boldsymbol{q}}_t \tag{7.32}$$

where $b_1 = \delta/(\alpha \Delta t)$, $b_4 = \delta/\alpha - 1$, $b_5 = \left(\delta/\alpha - 2\right)\Delta t/2$.

The equation of the motion of MDOF systems at time $t + \Delta t$ is

$$\boldsymbol{M}\ddot{\boldsymbol{q}}_{t+\Delta t} + \boldsymbol{C}\dot{\boldsymbol{q}}_{t+\Delta t} + \boldsymbol{K}\boldsymbol{q}_{t+\Delta t} = \mathbf{Q}_{t+\Delta t}$$

Substituting Eqs. (7.31) and (7.32) into the equation of the motion, and rearranging the equation leads to

$$\overline{\boldsymbol{K}}\boldsymbol{q}_{t+\Delta t} = \overline{\boldsymbol{Q}}_{t+\Delta t} \tag{7.33}$$

where

$$\overline{\boldsymbol{K}} = \boldsymbol{K} + b_0\boldsymbol{M} + b_1\boldsymbol{C} \tag{7.34}$$

$$\overline{\boldsymbol{Q}}_{t+\Delta t} = \boldsymbol{Q}_{t+\Delta t} + \boldsymbol{M}\left(b_0\boldsymbol{q}_t + b_2\dot{\boldsymbol{q}}_t + b_3\ddot{\boldsymbol{q}}_t\right) + \boldsymbol{C}\left(b_1\boldsymbol{q}_t + b_4\dot{\boldsymbol{q}}_t + b_5\ddot{\boldsymbol{q}}_t\right) \tag{7.35}$$

$\boldsymbol{q}_{t+\Delta t}$ can be solved from Eq. (7.33). Substituting the obtained $\boldsymbol{q}_{t+\Delta t}$ into Eqs. (7.31) and (7.32) yields $\ddot{\boldsymbol{q}}_{t+\Delta t}$ and $\dot{\boldsymbol{q}}_{t+\Delta t}$, respectively.

The Newmark method is stable if [2]

$$(2\alpha - \delta)\left(\frac{2\pi}{T_{\min}}\right)^2 \Delta t^2 + 2 \geq 0 \tag{7.36}$$

If $2\alpha - \delta \geq 0$ is satisfied, Eq. (7.36) holds definitely, that is, the Newmark method is unconditionally stable. In practical applications, $2\alpha - \delta = 0$ is usually satisfied to determine δ and α, say $\delta = 1/2$ and $\alpha = 1/4$. When $2\alpha - \delta \geq 0$ is not satisfied, Eq. (7.36) can be used to determine the time step. For example, when $\delta = 1/2$ and $\alpha = 1/6$, the Newmark method reduces to the linear acceleration method, and the condition for stable solution is $\Delta t \leq \sqrt{3}T_{\min}/\pi \approx T_{\min}/1.8$. Additionally, in order to consider the contribution of the higher mode components of interest to the total responses, Δt should be smaller than $1/7$ of the natural period of interest.

The time-stepping solution using the Newmark method is summarized as follows:

1. Initial calculations.
   - **a.** Input the matrices $\boldsymbol{M}$, $\boldsymbol{C}$, and $\boldsymbol{K}$.
   - **b.** Evaluate initial accelerations $\ddot{\boldsymbol{q}}_0$ from $\boldsymbol{M}\ddot{\boldsymbol{q}}_0 + \boldsymbol{C}\dot{\boldsymbol{q}}_0 + \boldsymbol{K}\boldsymbol{q}_0 = \boldsymbol{Q}_0$.
   - **c.** Select appropriate time step Δt and parameters δ and α (usually $\delta = 1/2$, $\alpha = 1/4$).
   - **d.** Calculate associated coefficients $b_0 = 1/\left(\alpha\Delta t^2\right)$, $b_1 = \delta/(\alpha\Delta t)$, $b_2 = 1/(\alpha\Delta t)$, $b_3 = 1/(2\alpha) - 1$, $b_4 = \delta/\alpha - 1$, $b_5 = \left(\delta/\alpha - 2\right)\Delta t/2$.
   - **e.** Calculate the effective stiffness matrix $\overline{\boldsymbol{K}} = \boldsymbol{K} + b_0\boldsymbol{M} + b_1\boldsymbol{C}$.
2. Calculations for each time step, $t = \Delta t, 2\Delta t, \cdots$
   - **a.** Calculate the effective load vector

$$\overline{\boldsymbol{Q}}_{t+\Delta t} = \boldsymbol{Q}_{t+\Delta t} + \boldsymbol{M}\left(b_0\boldsymbol{q}_t + b_2\dot{\boldsymbol{q}}_t + b_3\ddot{\boldsymbol{q}}_t\right) + \boldsymbol{C}\left(b_1\boldsymbol{q}_t + b_4\dot{\boldsymbol{q}}_t + b_5\ddot{\boldsymbol{q}}_t\right)$$

b. Solve Eq. (7.33) for $q_{t+\Delta t}$ at time $t+\Delta t$.
c. Calculate the accelerations and velocities at time $t+\Delta t$

$$\ddot{q}_{t+\Delta t} = b_0(q_{t+\Delta t} - q_t) - b_2\dot{q}_t - b_3\ddot{q}_t$$

$$\dot{q}_{t+\Delta t} = b_1(q_{t+\Delta t} - q_t) - b_4\dot{q}_t - b_5\ddot{q}_t$$

3. Repetition for the next time step. Replace t by $t+\Delta t$ and implement step 2 for the next time step.

7.5 Stability and accuracy of step-by-step integration method

For various step-by-step integration methods, appropriate time steps Δt (or $\Delta t/T$, T is a natural period of a system) have been presented for the requirement of stability and accuracy in previous sections. However, the explanations have not been provided. The Wilson-θ method is used as the example to illustrate the procedure for analyzing the stability and accuracy, and to show how to select associated parameters [3,4].

The equation of motion of n-DOF systems is

$$\boldsymbol{M}\ddot{\boldsymbol{q}} + \boldsymbol{C}\dot{\boldsymbol{q}} + \boldsymbol{K}\boldsymbol{q} = \boldsymbol{Q} \tag{7.37}$$

As discussed in Chapter 4, Analysis of dynamic response of MDOF systems: mode superposition method, considering the assumption of $\boldsymbol{C} = a_0\boldsymbol{M} + a_1\boldsymbol{K}$, and using normal coordinate transformation $\boldsymbol{q} = \overline{\boldsymbol{A}}\,\overline{\boldsymbol{T}}$, Eq. (7.37) becomes

$$\ddot{\overline{\boldsymbol{T}}} + (a_0\boldsymbol{I} + a_1\boldsymbol{\lambda})\dot{\overline{\boldsymbol{T}}} + \boldsymbol{\lambda}\overline{\boldsymbol{T}} = \overline{\boldsymbol{P}} \tag{7.38}$$

Letting $a_0 + a_1\omega_i^2 = 2\xi_i\omega_i$, the ith equation in Eq. (7.38) is

$$\ddot{\overline{T}}_i + 2\xi_i\omega_i\dot{\overline{T}}_i + \omega_i^2\overline{T}_i = \overline{P}_i, \quad i = 1, 2, \cdots, n \tag{7.39}$$

The solutions of Eq. (7.37) are equivalent to those obtained from the n independent equations given by Eq. (7.39) by using the step-by-step integration method with the same time step. Since each equation in Eq. (7.39) has similar form, one of them is selected to analyze the stability and accuracy. For simplicity, the subscript i is dropped, and Eq. (7.39) can be rewritten as

$$\ddot{\overline{T}} + 2\xi\omega\dot{\overline{T}} + \omega^2\overline{T} = \overline{P} \tag{7.40}$$

According to Eq. (7.19), one obtains

$$\ddot{\overline{T}}_{t+\theta\Delta t} = \frac{6}{(\theta\Delta t)^2}(\overline{T}_{t+\theta\Delta t} - \overline{T}_t) - \frac{6}{\theta\Delta t}\dot{\overline{T}}_t - 2\ddot{\overline{T}}_t \tag{7.41}$$

Considering Eq. (7.20) leads to

$$\dot{\overline{T}}_{t+\theta\Delta t} = \frac{3}{\theta\Delta t}(\overline{T}_{t+\theta\Delta t} - \overline{T}_t) - 2\dot{\overline{T}}_t - \frac{\theta\Delta t}{2}\ddot{\overline{T}}_t \tag{7.42}$$

For the instant of time $t + \theta\Delta t$, Eq. (7.40) can be written as

$$\ddot{\overline{T}}_{t+\theta\Delta t} + 2\xi\omega\dot{\overline{T}}_{t+\theta\Delta t} + \omega^2\overline{T}_{t+\theta\Delta t} = \overline{P}_{t+\theta\Delta t} \tag{7.43}$$

Substituting Eqs. (7.41) and (7.42) into Eq. (7.43), one can solve for $\overline{T}_{t+\theta\Delta t}$. Then $\ddot{\overline{T}}_{t+\Delta t}$, $\dot{\overline{T}}_{t+\Delta t}$, and $\overline{T}_{t+\Delta t}$ can be obtained respectively from Eq. (7.25) to Eq. (7.27), that is,

$$\begin{Bmatrix} \ddot{\overline{T}}_{t+\Delta t} \\ \dot{\overline{T}}_{t+\Delta t} \\ \overline{T}_{t+\Delta t} \end{Bmatrix} = \boldsymbol{A}\begin{Bmatrix} \ddot{\overline{T}}_t \\ \dot{\overline{T}}_t \\ \overline{T}_t \end{Bmatrix} + \boldsymbol{L}\overline{P}_{t+\theta\Delta t} \tag{7.44}$$

where $\boldsymbol{A}$ is called the amplification matrix, and $\boldsymbol{L}$ is called the load operator, which are respectively given by:

$$\boldsymbol{A} = \begin{bmatrix} 1 - \dfrac{\beta\theta^2}{3} - \dfrac{1}{\theta} - K\theta & \dfrac{-\beta\theta - 2K}{\Delta t} & -\dfrac{\beta}{\Delta t^2} \\ \Delta t\left(1 - \dfrac{1}{2\theta} - \dfrac{\beta\theta^2}{6} - \dfrac{K\theta}{2}\right) & 1 - \dfrac{\beta\theta}{2} - K & -\dfrac{\beta}{2\Delta t} \\ \Delta t^2\left(\dfrac{1}{2} - \dfrac{1}{6\theta} - \dfrac{\beta\theta^2}{18} - \dfrac{K\theta}{6}\right) & \Delta t\left(1 - \dfrac{\beta\theta}{6} - \dfrac{K}{3}\right) & 1 - \dfrac{\beta}{6} \end{bmatrix} \tag{7.45}$$

$$\boldsymbol{L} = \begin{Bmatrix} \dfrac{\beta}{\omega^2\Delta t^2} \\ \dfrac{\beta}{2\omega^2\Delta t} \\ \dfrac{\beta}{6\omega^2} \end{Bmatrix} \tag{7.46}$$

where $\beta = \left(\frac{\theta}{\omega^2\Delta t^2} + \frac{\xi\theta^2}{\omega\Delta t} + \frac{\theta^3}{6}\right)^{-1}$ and $K = \xi\beta/(\omega\Delta t)$.

Eq. (7.44) is the recursion relation between the responses at time t and $t+\Delta t$. The response at time $t+2\Delta t$ can be obtained from the response at time $t+\Delta t$ by replacing t in Eq. (7.44) by $t+\Delta t$, that is,

$$\begin{Bmatrix} \ddot{\overline{T}}_{t+2\Delta t} \\ \dot{\overline{T}}_{t+2\Delta t} \\ \overline{T}_{t+2\Delta t} \end{Bmatrix} = \boldsymbol{A} \begin{Bmatrix} \ddot{\overline{T}}_{t+\Delta t} \\ \dot{\overline{T}}_{t+\Delta t} \\ \overline{T}_{t+\Delta t} \end{Bmatrix} + \boldsymbol{L}\overline{\boldsymbol{P}}_{t+2\Delta t+(\theta-1)\Delta t}$$

$$= \boldsymbol{A}\left(\boldsymbol{A}\begin{Bmatrix} \ddot{\overline{T}}_{t} \\ \dot{\overline{T}}_{t} \\ \overline{T}_{t} \end{Bmatrix} + \boldsymbol{L}\overline{\boldsymbol{P}}_{t+\theta\Delta t}\right) + \boldsymbol{L}\overline{\boldsymbol{P}}_{t+2\Delta t+(\theta-1)\Delta t}$$

$$= \boldsymbol{A}^2\begin{Bmatrix} \ddot{\overline{T}}_{t} \\ \dot{\overline{T}}_{t} \\ \overline{T}_{t} \end{Bmatrix} + \boldsymbol{A}\boldsymbol{L}\overline{\boldsymbol{P}}_{t+\Delta t+(\theta-1)\Delta t} + \boldsymbol{L}\overline{\boldsymbol{P}}_{t+2\Delta t+(\theta-1)\Delta t}$$

Applying the above recursion relation successively leads to

$$\begin{Bmatrix} \ddot{\overline{T}}_{t+n\Delta t} \\ \dot{\overline{T}}_{t+n\Delta t} \\ \overline{T}_{t+n\Delta t} \end{Bmatrix} = \boldsymbol{A}^n\begin{Bmatrix} \ddot{\overline{T}}_{t} \\ \dot{\overline{T}}_{t} \\ \overline{T}_{t} \end{Bmatrix} + \boldsymbol{A}^{n-1}\boldsymbol{L}\overline{\boldsymbol{P}}_{t+\Delta t+(\theta-1)\Delta t} + \cdots + \boldsymbol{L}\overline{\boldsymbol{P}}_{t+n\Delta t+(\theta-1)\Delta t} \tag{7.47}$$

which is the basic relation for analyzing the stability of the Wilson-θ method.

Since the stability of the algorithm is independent of the external load, only the free vibration response of the system is considered with arbitrary initial conditions, that is, $\overline{P}(t)=0$. In this case, Eq. (7.47) becomes

$$\begin{Bmatrix} \ddot{\overline{T}}_{t+n\Delta t} \\ \dot{\overline{T}}_{t+n\Delta t} \\ \overline{T}_{t+n\Delta t} \end{Bmatrix} = \boldsymbol{A}^n\begin{Bmatrix} \ddot{\overline{T}}_{t} \\ \dot{\overline{T}}_{t} \\ \overline{T}_{t} \end{Bmatrix} \tag{7.48}$$

1. Stability analysis

It can be seen from the definition of stability and Eq. (7.48) that if $\boldsymbol{A}^n$ is bounded when $n\to\infty$, the algorithm is unconditionally stable.

According to the principle of linear algebra, any matrix $\boldsymbol{A}$ can be expressed as

$$\boldsymbol{A}=\boldsymbol{P}^{-1}\boldsymbol{J}\boldsymbol{P} \tag{7.49}$$

where $\boldsymbol{P}$ is a nonsingular matrix, $\boldsymbol{J}$ is the Jordan form of $\boldsymbol{A}$ with eigenvalues λ_i, $i = 1, 2, 3$, of $\boldsymbol{A}$ on its diagonal.

Since $\boldsymbol{A}^2 = \boldsymbol{P}^{-1}\boldsymbol{J}\boldsymbol{P}\boldsymbol{P}^{-1}\boldsymbol{J}\boldsymbol{P} = \boldsymbol{P}^{-1}\boldsymbol{J}^2\boldsymbol{P}$, one can easily obtain

$$\boldsymbol{A}^n = \boldsymbol{P}^{-1}\boldsymbol{J}^n\boldsymbol{P} \tag{7.50}$$

The diagonal elements of $\boldsymbol{J}^n$ are λ_i^n, $i = 1, 2, 3$. Eq. (7.50) shows that $\boldsymbol{J}^n$ must be bounded in response to the bounded $\boldsymbol{A}^n$ when $n \to \infty$. Let $\rho(\boldsymbol{A})$ be the spectral radius of $\boldsymbol{A}$ defined as

$$\rho(\boldsymbol{A}) = \max|\lambda_i|, i = 1, 2, 3 \tag{7.51}$$

Note that $|\cdot|$ denotes the module of λ_i. Then $\boldsymbol{J}^n$ is bounded for $n \to \infty$ if and only if $\rho(\boldsymbol{A}) \le 1$. This is the stability criteria. Furthermore, $\boldsymbol{J}^n \to 0$ if $\rho(\boldsymbol{A}) < 1$ and the smaller $\rho(\boldsymbol{A})$, the more rapid is the convergence rate.

For convenience, a similarity transformation of $\boldsymbol{A}$ is conducted first

$$\overline{\boldsymbol{A}} = \boldsymbol{D}^{-1}\boldsymbol{A}\boldsymbol{D} \tag{7.52}$$

where

$$\boldsymbol{D} = \begin{bmatrix} \Delta t & 0 & 0 \\ 0 & \Delta t^2 & 0 \\ 0 & 0 & \Delta t^3 \end{bmatrix} \tag{7.53}$$

Since $\overline{\boldsymbol{A}} \sim \boldsymbol{A}$, and similar matrices have the same eigenvalues, the eigenvalues of $\overline{\boldsymbol{A}}$ are identical to those of $\boldsymbol{A}$. From Eq. (7.52), one gets

$$\overline{\boldsymbol{A}} = \begin{bmatrix} 1 - \dfrac{\beta\theta^2}{3} - \dfrac{1}{\theta} - K\theta & -\beta\theta - 2K & -\beta \\ 1 - \dfrac{1}{2\theta} - \dfrac{\beta\theta^2}{6} - \dfrac{K\theta}{2} & 1 - \dfrac{\beta\theta}{2} - K & -\dfrac{\beta}{2} \\ \dfrac{1}{2} - \dfrac{1}{6\theta} - \dfrac{\beta\theta^2}{18} - \dfrac{K\theta}{6} & 1 - \dfrac{\beta\theta}{6} - \dfrac{K}{3} & 1 - \dfrac{\beta}{6} \end{bmatrix} \tag{7.54}$$

According to Eq. (7.54), the eigenvalues of $\overline{\boldsymbol{A}}$ only depend on $\Delta t/T$, ξ, and θ. Note that β and K are the functions of these quantities and the known quantity ω which is the ith natural frequency. Once all quantities in Eq. (7.54) are determined, the eigenvalues of $\overline{\boldsymbol{A}}$ are easily calculated, and $\rho(\boldsymbol{A})$ can be determined. The plots of $\rho(\boldsymbol{A})$ versus θ are shown in Fig. 7.6 for discrete values of the ratio of $\Delta t/T$ and damping ratio ξ. This figure shows that the Wilson-θ method is unconditionally stable when

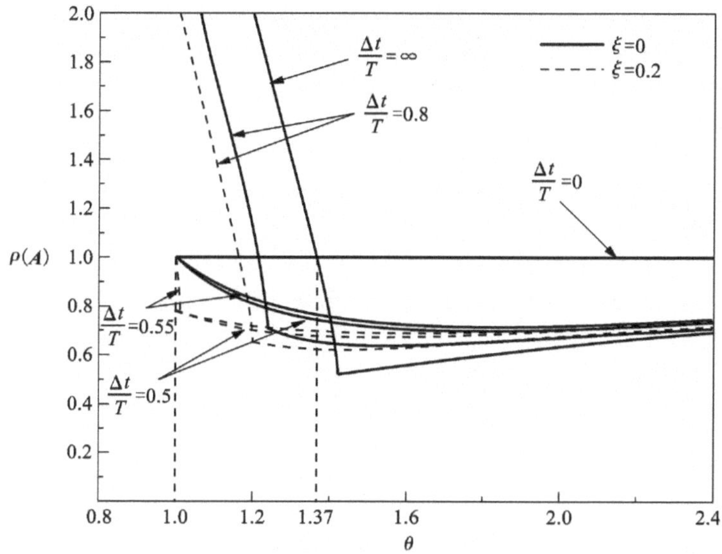

Figure 7.6 Plots of $\rho(\boldsymbol{A})$ versus θ (Wilson-θ method).

$\theta \geq 1.37$, since $\rho(\boldsymbol{A}) \leq 1$ regardless of the magnitudes of $\Delta t/T$ and ξ. When $\theta = 1$, the Wilson-θ method reduces to the linear acceleration method, which is stable only if $\Delta t/T \leq 0.55 \approx \sqrt{3}/\pi$ with which $\rho(\boldsymbol{A}) \leq 1$.

In Fig. 7.6, when $\Delta t/T = 0$ or ∞, $\rho(\boldsymbol{A})$ is independent of the damping ratio ξ (the plots corresponding to different damping ratios coincide in the figure). This is because when $\Delta t/T \to 0$, $\beta \to 0$, $K \to 0$, and thus

$$\overline{\boldsymbol{A}} \to \begin{bmatrix} 1 - \dfrac{1}{\theta} & 0 & 0 \\ 1 - \dfrac{1}{2\theta} & 1 & 0 \\ \dfrac{1}{2} - \dfrac{1}{6\theta} & 1 & 1 \end{bmatrix}$$

Obviously, $\rho(\boldsymbol{A}) = 1$.

When $\Delta t/T \to \infty$, $\beta \to \left(\theta^3/6\right)^{-1}$, and $K \to 0$ (because $\Delta t/T \to \infty$, $\Delta t \to \infty$, and T cannot approach zero). Thus $\overline{\boldsymbol{A}}$ and $\rho(\boldsymbol{A})$ are independent of ξ, but only the function of θ.

The amplification matrix $\boldsymbol{A}$ of the Newmark method can also be obtained using the procedure mentioned above similarly, and the stability criteria is still $\rho(\boldsymbol{A}) \leq 1$.

The plots of $\rho(\boldsymbol{A})$ versus $\Delta t/T$ are shown in Fig. 7.7 for discrete values of θ (Wilson-θ method), δ and α (Newmark method). Note that appropriate parameters will make the algorithms unconditionally stable. For example, the Newmark method is unconditionally stable, when $\delta = 1/2$ is combined with $\alpha = 1/4$; or $\delta = 11/20$ is combined with $\alpha = 3/10$. The Wilson-θ method is unconditionally stable, when $\theta = 1.4$ or $\theta = 2.0$.

2. Accuracy analysis

In addition to stability, the accuracy is another important factor to be considered in the selection of appropriate parameters, say θ, δ, and α. Therefore the accuracy analysis of the algorithm is described below.

To understand the essence of the accuracy of the Wilson-θ and the Newmark methods, two algorithms are used to obtain the response of the undamped free vibration governed by

$$\ddot{\overline{T}} + \omega^2 \overline{T} = 0 \tag{7.55}$$

Two kinds of initial conditions are considered

1. $\overline{T}_0 = 1.0$, $\dot{\overline{T}}_0 = 0$. In this case, $\ddot{\overline{T}}_0 = -\omega^2$, and the exact solution of Eq. (7.55) is $\overline{T} = \cos \omega t$.

2. $\overline{T}_0 = 0$, $\dot{\overline{T}}_0 = \omega$. In this case, $\ddot{\overline{T}}_0 = 0$, and the corresponding exact solution is $\overline{T} = \sin \omega t$.

The free vibration governed by Eq. (7.55) has no *AD* and its period remains unchanged. In this sense, the accuracy of the algorithms can be

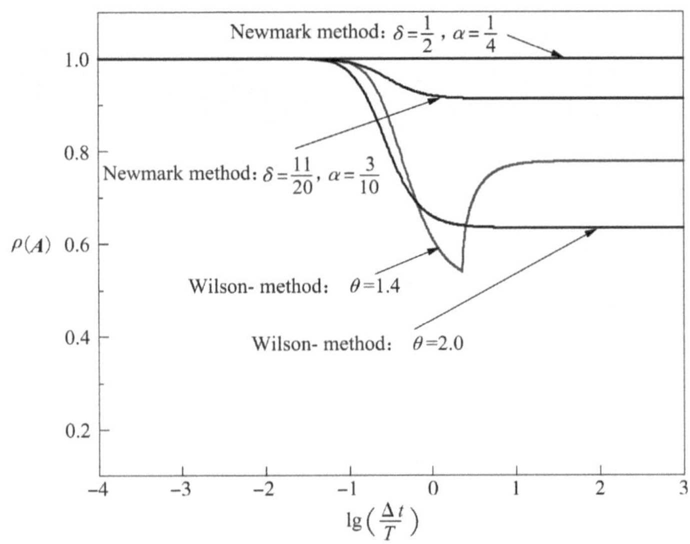

Figure 7.7 Plots of $\rho(\boldsymbol{A})$ versus $\lg\left(\frac{\Delta t}{T}\right)$ ($\xi = 0$).

evaluated by comparing the integrated response curves with the analytical solutions. The comparison shows that the step-by-step integration method produces *PE* and *AD*. The plots of *PE* and *AD* versus $\Delta t/T$ are shown in Figs. 7.8 and 7.9, respectively. Some conclusions can be drawn from these plots as follows:

1. When $\Delta t/T \leq 0.01$, the *PE* and *AD* are very small, so the Wilson-θ and the Newmark methods can predict accurate solutions.
2. When $\delta = 1/2$ and $\alpha = 1/4$, the Newmark method introduces only *PE*, and no *AD*.
3. For the Wilson-θ method, the accuracy of the algorithm in the case of $\theta = 1.4$ is higher than that of $\theta = 2.0$.

When the step-by-step integration method is used to solve Eq. (7.37), a smaller ratio of time step to period, say $\Delta t/T_n \leq 0.01$, where T_n is the natural period of the highest mode, should be adopted to ensure the accuracy of response of each mode component. However, for most dynamic systems, the higher modes have little contribution to the total response of the system. Therefore to ensure the accuracy of the response of higher mode components is unnecessary and will reduce the computational

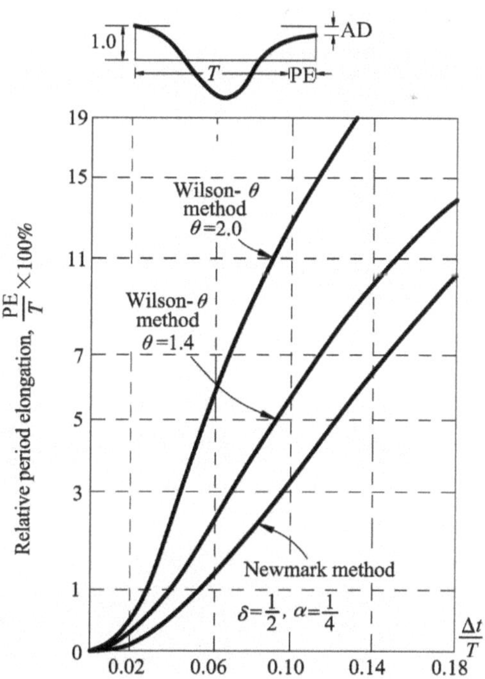

Figure 7.8 Period elongation versus $\Delta t/T$.

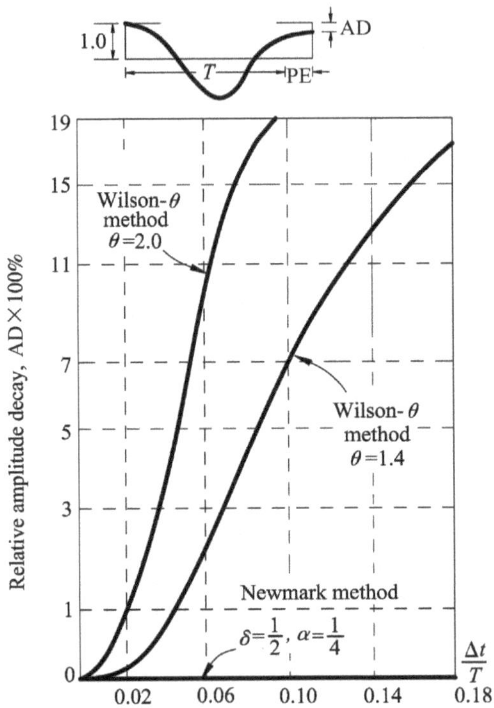

Figure 7.9 Amplitude decay versus $\Delta t/T$.

efficiency. The appropriate time step can be determined according to the actual situation as follows:

Using various ratios of time step to period to solve Eq. (7.55) is equivalent to predicting responses of the systems with various natural frequencies $\omega's$, while Δt remains unchanged. In a sense, various natural frequencies $\omega's$ represent distinct principal vibrations of the given system. Therefore the *PE* and *AD* corresponding to various values of $\Delta t/T$ in Figs. 7.8 and 7.9 can be interpreted as the *PE* and *AD* of various mode responses of a system. Fig. 7.10 shows the solutions of Eq. (7.55) starting from the initial condition 1 by using the Wilson-θ method, which indicates that the amplitudes of the higher mode responses, say the case of $\Delta t/T \geq 1.0$, decay rapidly.

Based on the above discussion, the following concepts can be obtained [4]:

1. The *AD* caused by the Wilson-θ method is equivalent to the effect of extra damping of the system, which is called numerical damping. Fig. 7.9 shows that when $\Delta t/T < 0.1$, the *AD* is below 7%, 1% of the critical damping (i.e., $\xi = 1\%$) will produce the *AD* at the rate of 6%

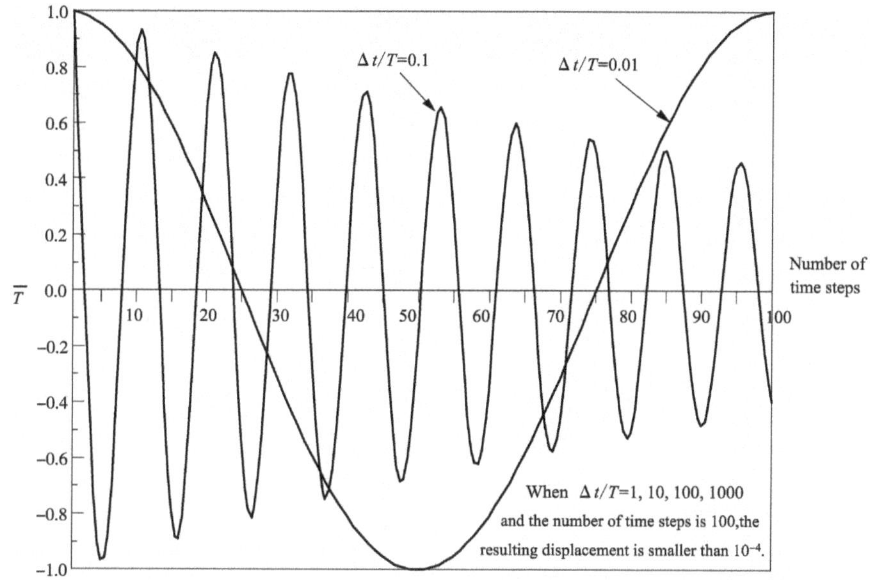

Figure 7.10 Displacement response within 100 time steps for discrete values of $\Delta t/T$(Wilson-θ method, $\theta = 1.4$).

per cycle approximately. Thus for the structures with damping ratio $\xi \geq 5\%$, if $\Delta t/T < 0.1$, the numerical damping is equivalent to an increment of 1% of practical damping. Such error is completely negligible. Therefore in the dynamic analysis of the MDOF systems, the time step Δt is recommended as $1/10$ of the natural period of interest.

2. The effect of the numerical damping should be of concern in the dynamic analysis. On one hand, a sufficiently short time step Δt must be adopted to obtain all the mode responses of interest with least *AD*. On the other hand, it should be noted that the mathematical idealization of any complex structure tends to be less reliable in predicting the higher modes, and the applied loads primarily produce lower mode responses in many cases, as discussed in Chapter 4, Analysis of dynamic response of MDOF systems: mode superposition method. Therefore it is desirable to use numerical damping to filter out the responses of higher modes, which is similar to the truncation of the higher modes in the mode superposition method. For this consideration, it is unnecessary to select a very small time step.
3. In the mode superposition method, the equations of motion of the MDOF systems are uncoupled into a series of independent equations in modal coordinates. Only the independent equations corresponding

to lower modes are required to solve for accurate responses, which is the advantage of this method. As mentioned previously, the equation of motion in the modal coordinate can also be solved by the step-by-step integration method. Various time steps can be used to obtain the distinct mode response, and a larger time step can be adopted to solve the lower mode responses. Thus the combined application of the mode superposition method and the step-by-step integration method will achieve optimum trade-off between the accuracy and the efficiency of the vibration analysis.

Problems

7.1. What are the basic ideas and calculation steps of the step-by-step integration methods? Are these methods applicable to the evaluation of the dynamic response of both linear and nonlinear systems?

7.2. What is the stability of the step-by-step integration method? What factors are related to the stability?

7.3. Derive the recursion relation for the stability analysis of the Newmark method, and investigate the condition of stability for this method.

7.4. Solve the linear elastic response of Problem 3.6 using the linear acceleration method.

7.5. For the structure given by Problem 2.7, the equations of motion of the simply supported beam are formulated by using the finite element method and considering the damping ($\xi = 0.01$). Use the mode superposition method and the Wilson-θ method (or Newmark method) to respectively evaluate the dynamic responses to the following two type of loads: (1) harmonic load $P_0 \sin\omega t$ acting at mid-span of the beam; (2) concentrated force P moving from the left end to the right end of the beam with an uniform speed v. Analyze the influence of the speed of the moving load on the response to the load (2).

References

[1] Clough RW, Penzien J. Dynamics of structures. 3rd ed. Berkeley, CA: Computers & Structures, Inc; 2003.

[2] Zhen Z. Mechanical vibration. Beijing: China Machine Press; 1986.

[3] Bathe K-J, Wilson EL. Numerical methods in finite element analysis. 2nd ed. Englewood Cliffs, NJ: Prentice Hall; 1976.

[4] Zhou Z, Wen Y, Zeng Q. Lectures on dynamics of structures. 2nd ed. Beijing: China Communications Press Co., Ltd; 2017.

Index

Note: Page numbers followed by "*f*" and "*t*" refer to figures and tables, respectively.

E

T

U